UNITEXT - La Matematica per il 3+2

Volume 104

Editor-in-chief

A. Quarteroni

Series editors

L. Ambrosio
P. Biscari
C. Ciliberto
M. Ledoux
W.J. Runggaldier

More information about this series at http://www.springer.com/series/5418

Elisabetta Fortuna · Roberto Frigerio
Rita Pardini

Projective Geometry

Solved Problems and Theory Review

 Springer

Elisabetta Fortuna
Dipartimento di Matematica
Università di Pisa
Pisa
Italy

Rita Pardini
Dipartimento di Matematica
Università di Pisa
Pisa
Italy

Roberto Frigerio
Dipartimento di Matematica
Università di Pisa
Pisa
Italy

ISSN 2038-5722 ISSN 2038-5757 (electronic)
UNITEXT - La Matematica per il 3+2
ISBN 978-3-319-42823-9 ISBN 978-3-319-42824-6 (eBook)
DOI 10.1007/978-3-319-42824-6

Library of Congress Control Number: 2016946954

Cover illustration: Tito Fornasiero, Punto di fuga, 34 × 49 cm, watercolor, 2010. http://bluoltremare.blogspot.com. Reproduced with permission.

Printed on acid-free paper

This Springer imprint is published by Springer Nature
The registered company is Springer International Publishing AG Switzerland

Preface

Projective geometry topics are taught in many degree programs in Mathematics, Physics and Engineering, also because of their practical applications in areas such as engineering, computer vision, architecture and cryptography. The literature on this subject includes, besides the classical extensive treatises, by now very dated from the point of view of language and terminology, some modern textbooks. Among these we mention only 'E. Casas-Alvero, *Analytic Projective Geometry*, EMS Textbooks in Mathematics (2014)', whose approach is close to ours and to which we refer the reader for bibliographical references.

This is not one further textbook, in particular it has not been conceived for sequential reading from the first to the last page; rather it aims to complement a standard textbook, accompanying the reader in her/his journey through the subject according to the philosophy of "learning by doing". For this reason we make no claim to be systematic; rather we have the ambition, or at least the hope, not only to ease and reinforce the understanding of the material by presenting completely worked out examples and applications of the theory, but also to awaken the curiosity of the reader, to challenge her/him to find original solutions and develop the ability of looking at a question from different perspectives. In addition, the book presents among the solved problems some classical geometric results, whose proofs are accessible within the relatively elementary techniques presented here. Hopefully these examples will encourage some readers to learn more about the topics treated here and undertake the study of classical algebraic geometry.

Indeed the starting point of the original Italian version of this work has been our experience in teaching the Projective Geometry course in the undergraduate Mathematics degree program in Pisa, which brought us to realize the difficulties the students encounter. However, the book contains also topics that, although usually not treated in undergraduate courses, can be useful and interesting for a reader wishing to learn more on the subject. Another feature of the text is that not only complex hypersurfaces and algebraic curves are studied, as it is traditional, but considerable attention is paid also to the real case.

The first chapter of the book contains a concise but exhaustive review of the basic results of projective geometry; the interested reader can find the proofs in any textbook on the subject. The goal of this first part is to provide the reader with an overview of the subject matter and to fix the notations and the concepts used later. The following three chapters are collections of solved problems concerning, respectively: the linear properties of projective spaces, the study of hypersurfaces and plane algebraic curves and, finally, conics and quadrics. In solving the problems we have given preference neither to the analytic nor to the synthetic approach, but every time we have chosen the solution that seemed to us more interesting, or more elegant or quicker; sometimes we have presented more than one solution. The difficulty level varies, ranging from merely computational exercises to more challenging theoretical problems. The exercises that in our opinion are harder are marked with the symbol 🍵, whose meaning is: "take it easy, get yourself a cup of coffee or tea, arm yourself with patience and determination, and you will succeed in the end". In other cases, besides giving the solution that is objectively the simplest, we have proposed alternative solutions that require longer or more involved arguments, but have the merit of giving a deeper conceptual understanding or highlighting connections with other phenomena that would not be evident otherwise. We have marked these, so to say, "enlightening" solutions with the symbol ϟ.

We have tried to offer in this way a guide to reading the text; indeed, we believe that the best way of using it, once the basic notions given in the theory review have been acquired, is to try to solve the problems presented on one's own, using the solutions that we give for checking and complementing one's personal work.

The necessary prerequisites are limited to some basic notions, mostly in linear algebra, that are usually taught in the first year courses of the Mathematics, Physics or Engineering degree programs.

We are extremely grateful to Ciro Ciliberto for believing in this project, encouraging us and constantly sharing with us his experience as mathematician, book author and editor. We also wish to thank Dr. Francesca Bonadei of Springer Italia for suggesting us to produce an English version of our book and assisting us in its preparation.

Pisa, Italy   Elisabetta Fortuna
April 2016   Roberto Frigerio
                                                                        Rita Pardini

Contents

About the Authors

Elisabetta Fortuna was born in Pisa in 1955. In 1977 she received her Diploma di Licenza in Mathematics from Scuola Normale Superiore in Pisa. Since 2001 she is Associate Professor at the University of Pisa. Her areas of research are real and complex analytic geometry, real algebraic geometry, computational algebraic geometry.

Roberto Frigerio was born in Como in 1977. In 2005 he received his Ph.D. in Mathematics at Scuola Normale Superiore in Pisa. Since 2014 he is Associate Professor at the University of Pisa. His primary scientific interests are focused on low-dimensional topology, hyperbolic geometry and geometric group theory.

Rita Pardini was born in Lucca in 1960. She received her Ph.D. in Mathematics from Scuola Normale Superiore in Pisa in 1990; she is Full Professor at the University of Pisa since 2004. Her area of research is classical algebraic geometry, in particular algebraic surfaces and their moduli, irregular varieties and coverings.

Symbols

Chapter 1
Theory Review

Projective spaces. Projective transformations. Duality. The projective line. Affine and projective hypersurfaces. Affine charts and projective closure. Conics and quadrics. Plane algebraic curves and linear systems.

Abstract Review of basic results of projective geometry: projective spaces and their transformations, duality, affine and projective hypersurfaces, plane curves, conics and quadrics.

1.1 Standard Notation

Besides a number of special conventions introduced whenever necessary, throughout the text we shall use standard notation and symbols, as commonly found in the literature. To avoid misunderstandings, however, we recall below some of the most frequent notations used in the sequel.

The symbols $\mathbb{N}, \mathbb{Z}, \mathbb{Q}, \mathbb{R}, \mathbb{C}$ shall denote the sets of natural, integer, rational, real and complex numbers, respectively. The complex conjugate $a - ib$ of a complex number $z = a + ib$ is denoted by $\bar{z}$.

We will denote by $\mathbb{K}$ a subfield of $\mathbb{C}$ and by $\mathbb{K}[x_1, \ldots, x_n]$ the ring of polynomials in the indeterminates (or variables) $x_1, \ldots, x_n$ with coefficients in $\mathbb{K}$. For any polynomial $f \in \mathbb{K}[x_1, \ldots, x_n]$ we will denote by f_{x_i} the partial derivative of f with respect to x_i; a similar notation will be used for partial derivatives of higher order. For a polynomial in a single variable $p(x)$ the derivative will also be denoted by $p'(x)$.

Let us set $\mathbb{K}^* = \mathbb{K} \setminus \{0\}$.

The identity map of a set A will be denoted by Id_A, or simply by Id.

The symbol $M(p, q, \mathbb{K})$ shall denote the vector space of $p \times q$ matrices with entries in the field $\mathbb{K}$ and $M(n, \mathbb{K})$ the space of square matrices of order n. The invertible matrices of order n form the group $\mathrm{GL}(n, \mathbb{K})$ and the real orthogonal matrices of order n the subgroup $O(n)$ of $\mathrm{GL}(n, \mathbb{R})$. The identity matrix of $\mathrm{GL}(n, \mathbb{K})$ is denoted by I (or by I_n if it is important to specify its order).

© Springer International Publishing Switzerland 2016

E. Fortuna et al., *Projective Geometry*, UNITEXT - La Matematica per il 3+2 104, DOI 10.1007/978-3-319-42824-6_1

Since the map $\mathbb{K}^n \to M(n, 1, \mathbb{K})$ associating to any vector $X = (x_1, \ldots, x_n) \in \mathbb{K}^n$
the column $\begin{pmatrix} x_1 \\ \vdots \\ x_n \end{pmatrix}$ is bijective, we will write vectors of $\mathbb{K}^n$ either as n-tuples or as
columns.

For any $A \in M(p, q, \mathbb{K})$ we will write rk A for the rank of A and, in case $p = q$, det A and tr A for the determinant and the trace of A, respectively. If $A = (a_{i,j}) \in M(n, \mathbb{K})$, we will denote by $c_{i,j}(A)$ the square submatrix of order $n - 1$ obtained by deleting the row and the column of A containing $a_{i,j}$. Moreover we will write $A_{i,j} = (-1)^{i+j} \det(c_{i,j}(A))$.

For any vector space V, we will denote by $\mathrm{GL}(V)$ the group of linear automorphisms $V \to V$.

1.2 Projective Spaces and Subspaces, Projective Transformations

1.2.1 Projective Spaces and Subspaces

If V is a finite-dimensional $\mathbb{K}$-vector space, the *projective space* associated to V is the quotient set $\mathbb{P}(V) = (V \setminus \{0\})/ \sim$, where $\sim$ is the equivalence relation on $V \setminus \{0\}$ defined by

$$v \sim w \quad \Longleftrightarrow \quad \exists k \in \mathbb{K}^* \text{ such that } v = kw.$$

The integer $\dim \mathbb{P}(V) = \dim V - 1$ is called the *dimension* of $\mathbb{P}(V)$. If $V = \{0\}$, we have $\mathbb{P}(V) = \emptyset$ and $\dim(\emptyset) = -1$.

From now on, unless otherwise specified, V will denote a $\mathbb{K}$-vector space of dimension $n + 1$ and $\mathbb{P}(V)$ the n-dimensional projective space associated to it.

Let us denote by $\pi \colon V \setminus \{0\} \to \mathbb{P}(V)$ the quotient map and by $[v]$ the equivalence class of the vector $v \in V \setminus \{0\}$.

The space $\mathbb{P}(\mathbb{K}^{n+1})$ is also denoted by $\mathbb{P}^n(\mathbb{K})$ and called the *standard projective space* of dimension n over $\mathbb{K}$. For any $(x_0, \ldots, x_n) \in \mathbb{K}^{n+1}$ we will denote $[(x_0, \ldots, x_n)]$ simply by $[x_0, \ldots, x_n]$.

A *projective subspace* of $\mathbb{P}(V)$ is a subset of the form $\mathbb{P}(W)$, where W is a linear subspace of V. Thus one has $\mathbb{P}(W) = \pi(W \setminus \{0\})$, and hence $\dim \mathbb{P}(W) = \dim W - 1$. If $W = \{0\}$, the subspace $\mathbb{P}(W)$ is empty and its dimension is -1. A projective subspace is called a *projective line* if its dimension is 1, a *projective plane* if its dimension is 2, a *projective hyperplane* if its dimension is $n - 1$. Following the usual terminology, for any projective subspace S of $\mathbb{P}(V)$ the integer codim $S = \dim \mathbb{P}(V) - \dim S$ is called the *codimension* of S.

1.2.2 Projective Transformations

Let V and W be two $\mathbb{K}$-vector spaces. A map $f\colon \mathbb{P}(V) \to \mathbb{P}(W)$ is called a *projective transformation* if there exists an injective linear map $\varphi\colon V \to W$ such that $f([v]) = [\varphi(v)]$ for every $v \in V \setminus \{0\}$. In this case we write $f = \overline{\varphi}$.

If $\varphi\colon V \to W$ is a linear map inducing f, the set of linear maps from V to W inducing f coincides with the family $\{k\varphi \mid k \in \mathbb{K}^*\}$; in other words, the linear map which induces a projective transformation is determined only up to multiplication by a non-zero scalar.

If f is induced by a linear isomorphism φ (and so, in particular, $\dim \mathbb{P}(V) = \dim \mathbb{P}(W)$), we say that f is a *projective isomorphism*. Two projective spaces over a field $\mathbb{K}$ are said to be *isomorphic* if there exists a projective isomorphism between them; this is the case if and only if they have the same dimension.

A projective isomorphism $f\colon \mathbb{P}(V) \to \mathbb{P}(V)$ is called a *projectivity* of $\mathbb{P}(V)$, and an *involution* if, additionally, $f^2 = \mathrm{Id}$. An involution f is said to be *non-trivial* if $f \neq \mathrm{Id}$.

If $f\colon \mathbb{P}(V) \to \mathbb{P}(W)$ is a projective transformation and H is a subspace of $\mathbb{P}(V)$, then $f(H)$ is a subspace of $\mathbb{P}(W)$ having the same dimension as H and $f|_H\colon H \to f(H)$ is a projective isomorphism.

Two subsets A, B of $\mathbb{P}(V)$ are said to be *projectively equivalent* if there exists a projectivity f of $\mathbb{P}(V)$ such that $f(A) = B$.

1.2.3 Operations on Subspaces

Let $S_1 = \mathbb{P}(W_1)$ and $S_2 = \mathbb{P}(W_2)$ be projective subspaces of $\mathbb{P}(V)$. As $\mathbb{P}(W_1) \cap \mathbb{P}(W_2) = \mathbb{P}(W_1 \cap W_2)$, the intersection of two (as a matter of fact, of any number of) projective subspaces is a projective subspace. The subspaces S_1 and S_2 are called *incident* if $S_1 \cap S_2 \neq \emptyset$, while *skew* if $S_1 \cap S_2 = \emptyset$.

For any non-empty subset $A \subseteq \mathbb{P}(V)$ the *subspace generated by A* is defined as the projective subspace $L(A)$ obtained as intersection of all subspaces of $\mathbb{P}(V)$ containing A. If $A = S_1 \cup S_2$ with $S_1 = \mathbb{P}(W_1)$, $S_2 = \mathbb{P}(W_2)$ projective subspaces, the subspace generated by A will be denoted by $L(S_1, S_2)$ and called the *join* of S_1 and S_2; of course one has $L(S_1, S_2) = \mathbb{P}(W_1 + W_2)$. We will denote by $L(P_1, \ldots, P_m)$ the subspace generated by the points $P_1, \ldots, P_m$.

Proposition 1.2.1 (Grassmann's formula) *Let S_1, S_2 be projective subspaces of $\mathbb{P}(V)$. Then $\dim L(S_1, S_2) = \dim S_1 + \dim S_2 - \dim(S_1 \cap S_2)$.*

Grassmann's formula implies that two subspaces S_1, S_2 cannot be skew if $\dim S_1 + \dim S_2 \geq \dim \mathbb{P}(V)$, so that for instance two lines in a projective plane or a line and a plane in a three-dimensional projective space always meet.

1.2.4 The Projective Linear Group

The projectivities of $\mathbb{P}(V)$ form a group with respect to composition, called the *projective linear group* and denoted by $\mathbb{PGL}(V)$.

Since the automorphism that induces a projectivity $f \in \mathbb{PGL}(V)$ is determined only up to a non-zero scalar, the map $GL(V) \to \mathbb{PGL}(V)$ sending the automorphism φ to the projectivity $f = \overline{\varphi}$ induces a group isomorphism between the quotient group $GL(V)/\sim$ and $\mathbb{PGL}(V)$, where $\sim$ denotes the equivalence relation on $GL(V)$ defined as follows: $\varphi \sim \psi$ if and only if there exists $k \in \mathbb{K}^*$ such that $\varphi = k\psi$.

If $\dim V = n + 1$, the group $GL(V)$ is isomorphic to the multiplicative group $GL(n + 1, \mathbb{K})$ consisting of invertible square matrices of order $n + 1$ with entries in $\mathbb{K}$. As a consequence $\mathbb{PGL}(V)$ is isomorphic to the group $\mathbb{PGL}(n + 1, \mathbb{K}) = GL(n + 1, \mathbb{K})/\sim$, where $\sim$ denotes the equivalence relation which identifies two matrices A, B if and only if there exists $k \in \mathbb{K}^*$ such that $A = kB$.

1.2.5 Fixed Points

If f is a projectivity of $\mathbb{P}(V)$, a point $P \in \mathbb{P}(V)$ is a *fixed point* of f if $f(P) = P$. More generally we say that $A \subset \mathbb{P}(V)$ is *invariant under f* (or f-invariant for short) if $f(A) = A$ (of course a point lying in an invariant subset need not be a fixed point).

If S, S_1, S_2 are subspaces of $\mathbb{P}(V)$ invariant under f, then $f|_S$ is a projectivity of S, and $S_1 \cap S_2$ and $L(S_1, S_2)$ are f-invariant.

If $f \in \mathbb{PGL}(V)$ is induced by the linear isomorphism φ and $P = [v]$, then P is a fixed point of f if and only if v is an eigenvector of φ. As a consequence the fixed-point set of a projectivity is the union of projective subspaces $S_1, \ldots, S_m$ of $\mathbb{P}(V)$ such that, for each $j = 1, \ldots, m$, the subspaces S_j and $L(S_1, \ldots, S_{j-1}, S_{j+1}, \ldots, S_m)$ are skew.

If $\mathbb{K}$ is algebraically closed, any projectivity of $\mathbb{P}^n(\mathbb{K})$ has at least one fixed point. Similarly, any projectivity of a real projective space of even dimension has at least one fixed point.

1.2.6 Degenerate Projective Transformations

It is possible to extend the definition of projective transformation so to include transformations induced by linear maps which fail to be injective. If $\varphi \colon V \to W$ is a non-zero linear map between $\mathbb{K}$-vector spaces and $H = \mathbb{P}(\ker \varphi) \subset \mathbb{P}(V)$, the map $f \colon \mathbb{P}(V) \setminus H \to \mathbb{P}(W)$ defined by $f([v]) = [\varphi(v)]$, for any $[v] \in \mathbb{P}(V) \setminus H$, is called the *degenerate projective transformation* induced by φ. As with projective transformations, the linear map φ is determined by f up to a non-zero multiplicative constant. Observe that, when φ is injective, the projective subspace H is empty and

f is the projective transformation induced by φ. If $S = \mathbb{P}(U) \subseteq \mathbb{P}(V)$ is a subspace not contained in H, the restriction of f to $S \setminus (S \cap H)$ is a degenerate projective transformation to $\mathbb{P}(W)$ and, if $S \cap H = \emptyset$, it is a projective transformation. The image under f of $S \setminus (S \cap H)$ is a projective subspace, more precisely it is the projective quotient of the subspace $\varphi(U) \subseteq W$ and its dimension is $\dim S - \dim(S \cap H) - 1$ (cf. Exercise 28).

1.2.7 Projection Centred at a Subspace

Let $S = \mathbb{P}(U)$ and $H = \mathbb{P}(W)$ be projective subspaces of $\mathbb{P}(V)$ such that $S \cap H = \emptyset$ and $L(S, H) = \mathbb{P}(V)$. If we set $k = \dim S$ and $h = \dim H$, Grassmann's formula (cf. Proposition 1.2.1) implies that $k + h = n - 1$. For any $P \in \mathbb{P}(V) \setminus H$, the space $L(H, P)$ has dimension $h + 1$ and hence, using Grassmann's formula again, $\dim(S \cap L(H, P)) = 0$. Therefore $L(H, P)$ intersects S exactly at one point. The map $\pi_H \colon \mathbb{P}(V) \setminus H \to S$ which associates to P the point $L(H, P) \cap S$ is called the *projection onto S centred at H*. One can easily check that π_H is a degenerate projective transformation (cf. Exercise 29).

1.2.8 Perspectivities

Assume that r and s are two distinct lines in the projective plane $\mathbb{P}^2(\mathbb{K})$ and let $A = r \cap s$. For any point $O \notin r \cup s$, the restriction to r of the projection onto s centred at O is a projective isomorphism $f \colon r \to s$, called *perspectivity centred at O*. Clearly $f(A) = A$ by construction, and the inverse isomorphism $f^{-1} \colon s \to r$ is a perspectivity centred at O, too.

More generally, assume that S_1, S_2 are two subspaces of an n-dimensional projective space $\mathbb{P}(V)$ with $\dim S_1 = \dim S_2 = k$, and let H be a subspace such that $H \cap S_1 = H \cap S_2 = \emptyset$ and $\dim H = n - k - 1$. Then the restriction to S_1 of the projection onto S_2 centred at H is a projective isomorphism $f \colon S_1 \to S_2$, called *perspectivity centred at H*. As with perspectivities between lines in the projective plane, the inverse isomorphism $f^{-1} \colon S_2 \to S_1$ is a perspectivity centred at H and the restriction of f to $S_1 \cap S_2$ is the identity map.

1.3 Projective Frames and Homogeneous Coordinates

1.3.1 General Position and Projective Frames

The points $P_0 = [v_0], \ldots, P_k = [v_k]$ of the projective space $\mathbb{P}(V)$ are said to be *projectively independent* if the vectors $v_0, \ldots, v_k \in V$ are linearly independent. Thus the points $P_0, \ldots, P_k$ are projectively independent if and only if the subspace

$L(P_0, \ldots, P_k)$ has dimension k; moreover the largest number of projectively independent points in $\mathbb{P}(V)$ is $\dim \mathbb{P}(V) + 1$.

More generally, if $\dim \mathbb{P}(V) = n$, we say that the points $P_0, \ldots, P_k$ are *in general
position* if either they are projectively independent (when $k \leq n$) or $k > n$ and no
subset of the P_i's consisting of $n + 1$ points is contained in a hyperplane of $\mathbb{P}(V)$.

Any ordered set $\mathcal{R} = \{P_0, \ldots, P_n, P_{n+1}\}$ consisting of $n + 2$ points in general
position is called a *projective frame* of $\mathbb{P}(V)$; the points $P_0, \ldots, P_n$ are called the
fundamental points of the frame, while P_{n+1} is called the *unit point* of the frame.

1.3.2 Systems of Homogeneous Coordinates

Let $\mathcal{R} = \{P_0, \ldots, P_n, P_{n+1}\}$ be a projective frame of $\mathbb{P}(V)$. For any $u \in V \setminus \{0\}$
such that $[u] = P_{n+1}$ there exists a unique basis $\mathcal{B}_u = \{v_0, \ldots, v_n\}$ of V such that
$[v_i] = P_i$ for each $i = 0, \ldots, n$ and $u = v_0 + \cdots + v_n$. Moreover, for any $\lambda \in \mathbb{K}^*$
we have $\mathcal{B}_{\lambda u} = \{\lambda v_0, \ldots, \lambda v_n\}$.

Any basis $\mathcal{B}_u$ obtained as above is called a *normalized basis* of V and can
be used to define a system of *homogeneous coordinates* in $\mathbb{P}(V)$: if $P = [v]$
and if $(x_0, \ldots, x_n)$ are the coordinates of the vector v with respect to the linear basis $\mathcal{B}_u$, we say that $(x_0, \ldots, x_n)$ is an $(n + 1)$-tuple of homogeneous coordinates of P with respect to the frame $\mathcal{R}$. These coordinates are uniquely determined up to a non-zero scalar, hence by homogeneous coordinates of P we actually
mean the homogeneous $(n + 1)$-tuple $[P]_{\mathcal{R}} = [x_0, \ldots, x_n] \in \mathbb{P}^n(\mathbb{K})$. In particular, it turns out that $[P_0]_{\mathcal{R}} = [1, 0, \ldots, 0], [P_1]_{\mathcal{R}} = [0, 1, \ldots, 0], \ldots, [P_n]_{\mathcal{R}} =
[0, 0, \ldots, 1], [P_{n+1}]_{\mathcal{R}} = [1, 1, \ldots, 1]$.

Once a projective frame (and hence a system of homogeneous coordinates) in
$\mathbb{P}(V)$ has been chosen, instead of $[P]_{\mathcal{R}} = [x_0, \ldots, x_n]$ we can simply write $P =
[x_0, \ldots, x_n]$.

Choosing a projective frame $\mathcal{R}$ is equivalent to choosing a projective isomor-
phism between $\mathbb{P}(V)$ and $\mathbb{P}^n(\mathbb{K})$. Namely the map $\phi_{\mathcal{R}} \colon \mathbb{P}(V) \to \mathbb{P}^n(\mathbb{K})$ defined by
$\phi_{\mathcal{R}}(P) = [P]_{\mathcal{R}}$ turns out to be the projective isomorphism induced by the linear
isomorphism $\phi_{\mathcal{B}} \colon V \to \mathbb{K}^{n+1}$ sending $v \in V$ to its coordinates with respect to the
basis $\mathcal{B}$, where $\mathcal{B}$ is any normalized basis associated to $\mathcal{R}$.

In $\mathbb{P}^n(\mathbb{K})$ the frame $\{[1, 0, \ldots, 0], [0, 1, \ldots, 0], \ldots, [0, 0, \ldots, 1], [1, 1, \ldots, 1]\}$
is called the *standard projective frame*; a corresponding normalized basis is the
canonical basis of $\mathbb{K}^{n+1}$.

Theorem 1.3.1 (Fundamental theorem of projective transformations) *Let $\mathbb{P}(V)$ and
$\mathbb{P}(W)$ be projective spaces over the field $\mathbb{K}$ such that $\dim \mathbb{P}(V) = n \leq \dim \mathbb{P}(W)$.
Assume that $\mathcal{R} = \{P_0, \ldots, P_{n+1}\}$ is a projective frame of $\mathbb{P}(V)$ and that $\mathcal{R}' =
\{Q_0, \ldots, Q_{n+1}\}$ is a projective frame of an n-dimensional subspace S of $\mathbb{P}(W)$.
Then there exists a unique projective transformation $f \colon \mathbb{P}(V) \to \mathbb{P}(W)$ such that
$f(P_i) = Q_i$ for each $i = 0, \ldots, n + 1$. If $S = \mathbb{P}(W)$, then f is a projective
isomorphism.*

In particular, given two projective frames $\mathcal{R}$ and $\mathcal{R}'$ of $\mathbb{P}(V)$, there is exactly one projectivity of $\mathbb{P}(V)$ mapping $\mathcal{R}$ to $\mathcal{R}'$; moreover, the only projectivity of $\mathbb{P}(V)$ fixing $n + 2$ points in general position is the identity map.

1.3.3 Analytic Representation of a Projective Transformation

Having homogeneous coordinates allows us to give an analytic representation of subspaces and projective transformations. For instance, let $f : \mathbb{P}(V_1) \rightarrow \mathbb{P}(V_2)$ be a projective transformation induced by the injective linear map $\varphi : V_1 \rightarrow V_2$, with $\dim \mathbb{P}(V_1) = n$ and $\dim \mathbb{P}(V_2) = m$. Let $\mathcal{R}_1$ and $\mathcal{R}_2$ be projective frames in $\mathbb{P}(V_1)$ and $\mathbb{P}(V_2)$, respectively, with associated normalized bases $\mathcal{B}_1$ and $\mathcal{B}_2$. Denote by A the matrix associated to φ with respect to the bases $\mathcal{B}_1$ and $\mathcal{B}_2$. If $P \in \mathbb{P}(V_1)$, assume that $[P]_{\mathcal{R}_1} = [x_0, \ldots, x_n]$ and $[f(P)]_{\mathcal{R}_2} = [y_0, \ldots, y_m]$. If we let $X = (x_0, \ldots, x_n)$ and $Y = (y_0, \ldots, y_n)$, then there exists $k \in \mathbb{K}^*$ such that $kY = AX$. The matrix A, called the *matrix associated to the transformation f with respect to the frames $\mathcal{R}_1$ and $\mathcal{R}_2$*, is determined up to a non-zero scalar, because both the linear map φ and the normalized bases $\mathcal{B}_1$ and $\mathcal{B}_2$ are so.

1.3.4 Change of Projective Frames

Let $\mathcal{R}_1$ and $\mathcal{R}_2$ be two projective frames with associated normalized bases $\mathcal{B}_1$ and $\mathcal{B}_2$. Let $A \in GL(n + 1, \mathbb{K})$ denote the coordinate change between the bases $\mathcal{B}_1$ and $\mathcal{B}_2$. If $P \in \mathbb{P}(V)$, assume that $[P]_{\mathcal{R}_1} = [x_0, \ldots, x_n]$ and $[P]_{\mathcal{R}_2} = [y_0, \ldots, y_n]$. If we let $X = (x_0, \ldots, x_n)$ and $Y = (y_0, \ldots, y_n)$, then there exists $k \in \mathbb{K}^*$ such that $kY = AX$. The matrix A, determined up to a non-zero scalar, is the one governing the change of frame (or change of homogeneous coordinates) between $\mathcal{R}_1$ and $\mathcal{R}_2$.

If $\phi_{\mathcal{R}_1} : \mathbb{P}(V) \rightarrow \mathbb{P}^n(\mathbb{K})$ and $\phi_{\mathcal{R}_2} : \mathbb{P}(V) \rightarrow \mathbb{P}^n(\mathbb{K})$ are the projective isomorphisms induced by the frames $\mathcal{R}_1$ and $\mathcal{R}_2$, the relation $kY = AX$ represents in coordinates the projective transformation $\phi_{\mathcal{R}_2} \circ \phi_{\mathcal{R}_1}^{-1} : \mathbb{P}^n(\mathbb{K}) \rightarrow \mathbb{P}^n(\mathbb{K})$.

1.3.5 Cartesian Representation of Subspaces

Let $S = \mathbb{P}(W)$ be a k-dimensional projective subspace of $\mathbb{P}(V)$. Let $\mathcal{R}$ be a projective frame in $\mathbb{P}(V)$ and $\mathcal{B}$ a normalized basis of V associated to the frame. Since W is a linear subspace of V of dimension $k + 1$, it admits a Cartesian representation in coordinates. More precisely, there exists a matrix $A \in M(n - k, n + 1, \mathbb{K})$ such that $\operatorname{rk} A = n - k$ and W is the set of all vectors $w \in V$ whose coordinates X with respect to the basis $\mathcal{B}$ satisfy the relation $AX = 0$. Then

$$S = \{P \in \mathbb{P}(V) \mid AX = 0, \text{ where } [P]_{\mathcal{R}} = [x_0, \ldots, x_n] \text{ and } X = (x_0, \ldots, x_n)\}.$$

The equations of the homogeneous linear system $AX = 0$ are called the *Cartesian equations of the subspace S* with respect to the frame $\mathcal{R}$.

In this way S is regarded as the set of points in $\mathbb{P}(V)$ whose homogeneous coordinates with respect to $\mathcal{R}$ satisfy a homogeneous linear system of $n - k$ independent equations, where $n - k$ is the codimension of S in $\mathbb{P}(V)$. In particular, every hyperplane of $\mathbb{P}(V)$ can be represented by means of a homogeneous linear equation in $n + 1$ variables $a_0 x_0 + \cdots + a_n x_n = 0$ where at least one coefficient a_i is nonzero. The hyperplanes H_i of equation $x_i = 0$ are called *coordinate hyperplanes*, or *fundamental hyperplanes*.

1.3.6   Parametric Representation of Subspaces

A subspace $S = \mathbb{P}(W)$ of dimension k can also be represented parametrically. Namely, if we view W as the image of an injective linear map $\varphi \colon \mathbb{K}^{k+1} \to V$, then S is the image of the projective transformation $f \colon \mathbb{P}^k(\mathbb{K}) \to \mathbb{P}(V)$ induced by φ. The components of the analytic representation of f (with respect to the standard projective frame of $\mathbb{P}^k(\mathbb{K})$ and a projective frame $\mathcal{R}$ of $\mathbb{P}(V)$) are called *parametric equations of the subspace S* with respect to $\mathcal{R}$. In this way the homogeneous coordinates of the points of S are homogeneous linear functions of $k + 1$ parameters.

In order to determine these equations explicitly it suffices to choose $k + 2$ points $Q_0, \ldots, Q_{k+1}$ of S in general position (so that $S = L(Q_0, \ldots, Q_k)$) and construct the projective transformation f which maps the standard projective frame of $\mathbb{P}^k(\mathbb{K})$ to the projective frame $\{Q_0, \ldots, Q_{k+1}\}$ of S (the existence of f is guaranteed by the Fundamental theorem of projective transformations, cf. Theorem 1.3.1). If $[Q_i]_{\mathcal{R}} = [q_{i,0}, \ldots, q_{i,n}]$, then P lies in S if and only if there exist $\lambda_0, \ldots, \lambda_k \in \mathbb{K}$ such that

$$\begin{cases} x_0 = \lambda_0 q_{0,0} + \ldots + \lambda_k q_{k,0} \\ \vdots \\ x_n = \lambda_0 q_{0,n} + \ldots + \lambda_k q_{k,n} \end{cases},$$

where $[x_0, \ldots, x_n]$ are homogeneous coordinates of P. This occurs if and only if the matrix

$$M = \begin{pmatrix} q_{0,0} & \cdots & q_{k,0} & x_0 \\ \vdots & \vdots & \vdots & \vdots \\ q_{0,n} & \cdots & q_{k,n} & x_n \end{pmatrix}$$

(whose first $k + 1$ columns are linearly independent) has rank $k + 1$. By a well-known linear algebra result, $\mathrm{rk}(M) = k + 1$ if and only if M contains an invertible $(k + 1) \times (k + 1)$ submatrix M' such that the determinant of every $(k + 2) \times (k + 2)$

submatrix of M containing M' vanishes; since there are $n - k$ such matrices, this provides a Cartesian representation of S with $n - k$ equations.

1.3.7 Extension of Projective Frames

If S is a projective subspace of $\mathbb{P}(V)$ of dimension k, we may choose $k+1$ projectively independent points $P_0, \ldots, P_k$ in S and then extend this set to a projective frame of $\mathbb{P}(V)$. In the system of homogeneous coordinates $[x_0, \ldots, x_n]$ thus induced, the subspace S is given by the equations $x_{k+1} = \cdots = x_n = 0$ and the points of S have homogeneous coordinates of the form $[x_0, \ldots, x_k, 0, \ldots, 0]$. The map

$$S \ni P = [x_0, \ldots, x_k, 0, \ldots, 0] \to [x_0, \ldots, x_k] \in \mathbb{P}^k(\mathbb{K})$$

is a projective isomorphism and defines on S a system of homogeneous coordinates, called *system of homogeneous coordinates induced on S*.

1.3.8 Affine Charts

Consider the fundamental hyperplane $H_0 = \{x_0 = 0\}$ of $\mathbb{P}^n(\mathbb{K})$ and let $U_0 = \mathbb{P}^n(\mathbb{K}) \setminus H_0$. The map $j_0 \colon \mathbb{K}^n \to U_0$ defined by

$$j_0(x_1, \ldots, x_n) = [1, x_1, \ldots, x_n]$$

is bijective and its inverse map $U_0 \to \mathbb{K}^n$ is

$$j_0^{-1}([x_0, \ldots, x_n]) = \left(\frac{x_1}{x_0}, \ldots, \frac{x_n}{x_0} \right).$$

The pair (U_0, j_0^{-1}) is called *standard affine chart* of $\mathbb{P}^n(\mathbb{K})$.

We can generalize the notion of affine chart of a projective space $\mathbb{P}(V)$ of dimension n starting with any hyperplane H.

To do that, let $f \colon \mathbb{P}^n(\mathbb{K}) \to \mathbb{P}(V)$ be a projective isomorphism such that $f(H_0) = H$ (and hence $f(U_0) = U_H$, where $U_H = \mathbb{P}(V) \setminus H$); the map $j_H \colon \mathbb{K}^n \to U_H$ defined by $j_H = f \circ j_0$ is bijective and the pair (U_H, j_H^{-1}) is called an *affine chart* of $\mathbb{P}(V)$. Sometimes, by abuse of terminology, one uses the word "chart" to indicate either the subset U_H or the bijection with $\mathbb{K}^n$ alone, if no confusion arises.

For instance, let H be a hyperplane of $\mathbb{P}(V)$ given by the equation $a_0 x_0 + \cdots + a_n x_n = 0$ in the system of homogeneous coordinates induced by a projective frame $\mathcal{R}$. Since not all a_i are zero, we can assume $a_0 \neq 0$ for instance. Then the map $f \colon \mathbb{P}^n(\mathbb{K}) \to \mathbb{P}(V)$ which associates to the point $[x_0, \ldots, x_n] \in \mathbb{P}^n(\mathbb{K})$

the point of $\mathbb{P}(V)$ of coordinates $\left[\dfrac{x_0 - a_1 x_1 - \cdots - a_n x_n}{a_0}, x_1, \ldots, x_n\right]$ with respect to $\mathcal{R}$ is a projective isomorphism such that $f(H_0) = H$. In this case the map $j_H = f \circ j_0 \colon \mathbb{K}^n \to U_H$ is given by

$$j_H(x_1, \ldots, x_n) = f([1, x_1, \ldots, x_n]) = \left[\dfrac{1 - a_1 x_1 - \cdots - a_n x_n}{a_0}, x_1, \ldots, x_n\right]$$

and its inverse map $U_H \to \mathbb{K}^n$ by

$$j_H^{-1}([x_0, \ldots, x_n]) = \left(\dfrac{x_1}{\sum_{i=0}^n a_i x_i}, \ldots, \dfrac{x_n}{\sum_{i=0}^n a_i x_i}\right).$$

If $P \in U_H$, the components of the vector $j_H^{-1}(P) = \left(\dfrac{x_1}{\sum a_i x_i}, \ldots, \dfrac{x_n}{\sum a_i x_i}\right)$ are called *affine coordinates* of the point P in the chart U_H. Points in H are called *points at infinity* (or sometimes *improper points*) with respect to the chart U_H, while points in U_H are called *proper points* with respect to U_H.

Choosing another coefficient $a_i \neq 0$ in the equation of H and proceeding in a similar way, we obtain a different affine chart.

Being an "improper point" is of course a relative concept: a point may lie at infinity with respect to a chart and be proper with respect to another one; even more, every point P can be considered as an improper point simply by choosing a chart U_H determined by a hyperplane H containing P.

If $\mathbb{P}(V) = \mathbb{P}^n(\mathbb{K})$ we will denote by $U_i = \mathbb{P}^n(\mathbb{K}) \setminus H_i$ the chart determined by the fundamental hyperplane $H_i = \{x_i = 0\}$; proceeding as above it turns out that the map $j_i \colon \mathbb{K}^n \to U_i$

$$j_i(y_1, \ldots, y_n) = [y_1, \ldots, y_{i-1}, 1, y_{i+1}, \ldots, y_n]$$

is a bijection with inverse

$$j_i^{-1}([x_0, \ldots, x_n]) = \left(\dfrac{x_0}{x_i}, \ldots, \dfrac{x_{i-1}}{x_i}, \dfrac{x_{i+1}}{x_i}, \ldots, \dfrac{x_n}{x_i}\right).$$

If $i = 0$ we recover the map j_0 defined at the beginning of this subsection.

Observe that $U_0 \cup \cdots \cup U_n = \mathbb{P}^n(\mathbb{K})$; the family $\{(U_i, j_i^{-1})\}_{i=0,\ldots,n}$ is called *standard atlas* of $\mathbb{P}^n(\mathbb{K})$.

The composition of j_i with the natural inclusion of U_i in $\mathbb{P}^n(\mathbb{K})$ gives an embedding of $\mathbb{K}^n$ in $\mathbb{P}^n(\mathbb{K})$; thus we can think of $\mathbb{P}^n(\mathbb{K})$ as the *completion* of $\mathbb{K}^n$ obtained by adding the hyperplane at infinity H_i.

A projectivity f of $\mathbb{P}(V)$ leaves an affine chart $U_H = \mathbb{P}(V) \setminus H$ invariant (i.e. $f(U_H) = U_H$) if and only if $f(H) = H$. The projectivities that leave an affine chart invariant form a subgroup of $\mathbb{PGL}(V)$. Up to a change of homogeneous coordinates we may assume that H has equation $x_0 = 0$; if A is a matrix associated to f in this system of coordinates, then f leaves H invariant if and only if, up to non-zero scalars, A is a block matrix of the form

$$A = \left(\begin{array}{c|ccc} 1 & 0 & \ldots & 0 \\ \hline B & & C & \end{array} \right),$$

where $B \in \mathbb{K}^n$ and C is an invertible square matrix of order n. Therefore, in affine coordinates f acts on the chart U_H mapping $X \in \mathbb{K}^n$ to the vector $CX + B$, that is, the restriction of f to the chart is an affinity. In other words the subgroup of $\mathbb{P}GL(V)$ of the projectivities leaving a hyperplane invariant is isomorphic to the group of affinities of $\mathbb{K}^n$.

1.3.9 Projective Closure of an Affine Subspace

Let W be an affine subspace of $\mathbb{K}^n$ defined by the equations

$$\begin{cases} a_{1,1}x_1 + \cdots + a_{1,n}x_n + b_1 = 0 \\ \qquad\vdots \qquad\qquad\qquad \vdots \\ a_{h,1}x_1 + \cdots + a_{h,n}x_n + b_h = 0 \end{cases}$$

and consider, for instance, the embedding of $\mathbb{K}^n$ into $\mathbb{P}^n(\mathbb{K})$ given by the map j_0 defined above. Then j_0 transforms W into the subset of proper points (with respect to the hyperplane H_0) of the projective subspace $\overline{W}$ of $\mathbb{P}^n(\mathbb{K})$ defined by the equations

$$\begin{cases} a_{1,1}x_1 + \cdots + a_{1,n}x_n + b_1x_0 = 0 \\ \qquad\vdots \qquad\qquad\qquad \vdots \\ a_{h,1}x_1 + \cdots + a_{h,n}x_n + b_hx_0 = 0 \end{cases}.$$

The subspace $\overline{W}$ coincides with the smallest projective subspace of $\mathbb{P}^n(\mathbb{K})$ containing W, and is called the *projective closure* of W (with respect to j_0).

1.3.10 Projective Transformations and Change of Coordinates

Let $f \colon \mathbb{P}(V) \to \mathbb{P}(W)$ be a projective transformation, with $\dim \mathbb{P}(V) = n$ and $\dim \mathbb{P}(W) = m$. Assume that $\mathcal{R}_1, \mathcal{R}_2$ are projective frames in $\mathbb{P}(V)$ and that $\mathcal{S}_1, \mathcal{S}_2$ are projective frames in $\mathbb{P}(W)$. For any $P \in \mathbb{P}(V)$, denote by X (respectively X') a column vector whose entries are homogeneous coordinates of P with respect to $\mathcal{R}_1$ (respectively $\mathcal{R}_2$). Moreover, denote by Y (resp. Y') a column vector whose entries

are homogeneous coordinates of the point $f(P) \in \mathbb{P}(W)$ with respect to $\mathcal{S}_1$ (resp. $\mathcal{S}_2$).
If A is a matrix associated to f with respect to the frames $\mathcal{R}_1$ and $\mathcal{S}_1$, then $Y = kAX$
for some $k \in \mathbb{K}^*$. On the other hand if we denote by N (resp. M) the change of
coordinates between $\mathcal{R}_1, \mathcal{R}_2$ (resp. between $\mathcal{S}_1, \mathcal{S}_2$), then $X' = \alpha NX$ and $Y' = \beta MY$
for some $\alpha, \beta \in \mathbb{K}^*$. Therefore

$$Y' = \beta MY = \beta kMAX = (\beta k\alpha^{-1})MAN^{-1}X',$$

and hence f is represented by the matrix MAN^{-1} with respect to the frames $\mathcal{R}_2$ and
$\mathcal{S}_2$.

In particular, if $f \in \mathbb{P}GL(V)$ and we use the same frame of $\mathbb{P}(V)$ both in the
source and in the target of f, then the matrices A, B representing f in two different
systems of homogeneous coordinates are *similar* up to a non-zero scalar, i.e. there
exist $M \in GL(n+1, \mathbb{K})$ and $\lambda \in \mathbb{K}^*$ such that $B = \lambda MAM^{-1}$. For instance, if $\mathbb{K}$ is
algebraically closed, the projectivity f is represented by a Jordan matrix in a suitable
frame.

Two projectivities f, g of $\mathbb{P}(V)$ are said to be *conjugate* if there exists a projectivity
$h \in \mathbb{P}GL(V)$ such that $g = h^{-1} \circ f \circ h$. Therefore f, g are conjugate if and only if
the matrices representing f and g in a given system of homogeneous coordinates are
similar up to a non-zero scalar.

1.4 Dual Projective Space and Duality

1.4.1 Dual Projective Space

Let V be a $\mathbb{K}$-vector space and denote by V^* its dual space. The projective space
$\mathbb{P}(V^*)$, also denoted by $\mathbb{P}(V)^*$, is called *dual projective space*. If $\dim V = n+1$,
then $\mathbb{P}(V)^*$ has dimension n and hence it is projectively isomorphic to $\mathbb{P}(V)$.

If $\mathcal{B}$ is a linear basis of V (obtained for instance from a projective frame of $\mathbb{P}(V)$),
then the dual basis $\mathcal{B}^*$ of V^* can be used to induce on $\mathbb{P}(V)^*$ a system of *dual
homogeneous coordinates*. If $[L] \in \mathbb{P}(V)^*$ is the equivalence class of a non-zero
linear functional $L \in V^*$ and if $L(x_0, \ldots, x_n) = a_0 x_0 + \cdots + a_n x_n$ in the system of
coordinates induced by $\mathcal{B}$, then L has coordinates $(a_0, \ldots, a_n)$ with respect to the
basis $\mathcal{B}^*$ of V^* and $[L]$ has homogeneous coordinates $[a_0, \ldots, a_n]$.

Since $a_0 x_0 + \cdots + a_n x_n = 0$ represents a hyperplane of $\mathbb{P}(V)$, the space $\mathbb{P}(V)^*$
can be identified in a natural way with the set of projective hyperplanes of $\mathbb{P}(V)$, and
hence it has a structure of projective space. Accordingly we say that given hyperplanes
are independent (respectively, in general position) if the corresponding points of
$\mathbb{P}(V)^*$ are projectively independent (resp. in general position). We call *homogeneous
coordinates of a hyperplane* the homogeneous coordinates of the corresponding
point of $\mathbb{P}(V)^*$; in this way the hyperplane of equation $a_0 x_0 + \cdots + a_n x_n = 0$ has
homogeneous coordinates $[a_0, \ldots, a_n]$.

1.4.2 Duality Correspondence

Let $S = \mathbb{P}(W)$ be a projective subspace of $\mathbb{P}(V)$ of dimension k. Then the annihilator $\text{Ann}(W) = \{L \in V^* \mid L|_W \equiv 0\}$ is a linear subspace of V^* of dimension $n - k$. Consider the map δ, having the set of subspaces of $\mathbb{P}(V)$ of dimension k as domain and the set of subspaces of $\mathbb{P}(V)^*$ of dimension $n - k - 1$ as target, which associates to the subspace $S = \mathbb{P}(W)$ the subspace $\mathbb{P}(\text{Ann}(W))$. The map δ, called *duality correspondence*, is a bijection that reverses inclusions and satisfies

$$\delta(S_1 \cap S_2) = L(\delta(S_1), \delta(S_2)) \qquad \delta(L(S_1, S_2)) = \delta(S_1) \cap \delta(S_2).$$

When $k = n - 1$ we recover the one-to-one correspondence between hyperplanes of $\mathbb{P}(V)$ and points of $\mathbb{P}(V)^*$; when $k = 0$ we get a one-to-one correspondence between points of $\mathbb{P}(V)$ and hyperplanes of $\mathbb{P}(V)^*$.

1.4.3 Linear Systems of Hyperplanes

Every projective subspace of $\mathbb{P}(V)^*$ is called a *linear system*; as seen in Sect. 1.4.2, the points of a linear system are hyperplanes of $\mathbb{P}(V)$. If $\mathcal{L}$ is a linear system, the intersection S of all hyperplanes in $\mathcal{L}$ is called the *centre of the linear system* and $\mathcal{L}$ coincides with the set of all hyperplanes of $\mathbb{P}(V)$ containing S. This is why the linear system $\mathcal{L}$ is also called the *linear system of hyperplanes with centre S* and denoted $\Lambda_1(S)$.

Since δ reverses inclusions, a hyperplane H of $\mathbb{P}(V)$ contains S if and only if the point $\delta(H) \in \mathbb{P}(V)^*$ belongs to $\delta(S)$; therefore $\Lambda_1(S) = \delta(S)$ and hence, if $\dim S = k$, it follows that $\dim \Lambda_1(S) = n - k - 1$.

More explicitly, if we consider a Cartesian representation of S by means of a minimal set of $n - k$ homogeneous linear equations, we may view S as the intersection of $n - k$ hyperplanes $H_1, \ldots, H_{n-k}$ of $\mathbb{P}(V)$ such that $\delta(H_1), \ldots, \delta(H_{n-k})$ are projectively independent points in $\mathbb{P}(V)^*$. On the other hand, as we saw above,

$$\Lambda_1(S) = \delta(S) = \delta(H_1 \cap \cdots \cap H_{n-k}) = L(\delta(H_1), \ldots, \delta(H_{n-k}))$$

which shows again that $\dim \Lambda_1(S) = n - k - 1$.

When $k = n - 2$ the linear system $\Lambda_1(S)$ has dimension 1 and is called the *pencil of hyperplanes with centre S*. For instance if $n = 2$ and S contains only one point P, then $\Lambda_1(S)$ is the *pencil of lines* of the projective plane $\mathbb{P}(V)$ with centre P; if $n = 3$ and S is a line, then $\Lambda_1(S)$ is the *pencil of planes* of the projective space $\mathbb{P}(V)$ with centre the line S.

If L is a projective hyperplane of $\mathbb{P}(V)$, we denote by $\Lambda_1(S) \cap U_L$ the set of intersections of the projective hyperplanes of the linear system $\Lambda_1(S)$ with the chart $U_L = \mathbb{P}(V) \setminus L$; this set consists of affine hyperplanes and is called *linear system of affine hyperplanes.* When the centre S is not contained in L (and hence $L \notin \Lambda_1(S)$), the affine hyperplanes of $\Lambda_1(S) \cap U_L$ meet in the affine subspace $S \cap U_L$ and $\Lambda_1(S) \cap U_L$ is called a *proper linear system of affine hyperplanes.* If $S \subset L$, instead, it is called an *improper linear system of affine hyperplanes.*

In particular, if $\Lambda_1(S)$ is a pencil (i.e. it has dimension 1), then $\Lambda_1(S) \cap U_L$ is called a *proper pencil* (or an *improper pencil*) of affine hyperplanes; any two hyperplanes of an improper pencil are parallel. When $n = 2$ we thus recover the notion of proper pencil of lines (which meet at a point) and that of improper pencil of lines (consisting of all lines parallel to a fixed line).

In the case of a proper linear system, mapping a hyperplane $H \in \Lambda_1(S)$ to the intersection $H \cap U_L$ defines a bijection between the projective linear system $\Lambda_1(S)$ and the proper affine linear system $\Lambda_1(S) \cap U_L$. If instead $S \subset L$, the hyperplane L belongs to $\Lambda_1(S)$ but its intersection with U_L is empty.

1.4.4 Duality Principle

Since $\mathbb{P}(V)^*$ is projectively isomorphic to $\mathbb{P}(V)$, the bijection δ defined in Sect. 1.4.2 can be seen as a map which transforms subspaces of dimension k of $\mathbb{P}(V)$ into subspaces of "dual dimension" $n - k - 1$ of $\mathbb{P}(V)$. The properties of δ recalled above imply the following:

Theorem 1.4.1 (Duality principle) *Let $\mathbb{P}(V)$ be a projective space of dimension n. Let $\mathcal{P}$ be an assertion about subspaces of $\mathbb{P}(V)$, their intersections, joins, and dimensions. Denote by $\mathcal{P}^*$ the "dual" assertion obtained from $\mathcal{P}$ by replacing the words "intersection, join, contained, containing, dimension" by "join, intersection, containing, contained, dual dimension" respectively (the "dual dimension" is $n - k - 1$ if the dimension is k). Then $\mathcal{P}$ holds if and only if $\mathcal{P}^*$ holds.*

A proposition is said to be *self-dual* if it coincides with its dual statement. An example of a self-dual proposition is the following:

Theorem 1.4.2 (Desargues' Theorem) *Let $\mathbb{P}(V)$ be a projective plane and $A_1, A_2, A_3, B_1, B_2, B_3$ points of $\mathbb{P}(V)$ in general position. Consider the triangles T_1 and T_2 of $\mathbb{P}(V)$ with vertices A_1, A_2, A_3 and B_1, B_2, B_3; one says that T_1 and T_2 are in central perspective if there exists a point O in the plane, distinct from the A_i and B_i, such that all lines $L(A_i, B_i)$ pass through O. Then T_1 and T_2 are in central perspective if and only if the points $P_1 = L(A_2, A_3) \cap L(B_2, B_3)$, $P_2 = L(A_3, A_1) \cap L(B_3, B_1)$ and $P_3 = L(A_1, A_2) \cap L(B_1, B_2)$ are collinear.*

For a proof of Desargues' Theorem see Exercise 12.

1.4.5 Dual Projectivity

Let f be a projectivity of $\mathbb{P}(V)$ represented by a matrix $A \in \mathrm{GL}(n+1, \mathbb{K})$ in a system of homogeneous coordinates. The map f sends the hyperplane H of equation ${}^tCX = 0$ to the hyperplane $f(H)$ of equation ${}^tC'X = 0$ where $C' = {}^tA^{-1}C$. In fact,

$$ {}^tCX = 0 \iff {}^tCA^{-1}AX = 0 \iff {}^t({}^tA^{-1}C)AX = 0 \iff {}^tC'AX = 0. $$

Then the map $f_*\colon \mathbb{P}(V)^* \to \mathbb{P}(V)^*$ defined by $f_*(H) = f(H)$ is represented by the matrix ${}^tA^{-1}$ in the dual system of homogeneous coordinates in $\mathbb{P}(V)^*$. So f_* is a projectivity of $\mathbb{P}(V)^*$, called *dual projectivity*.

   A hyperplane H of equation ${}^tCX = 0$ turns out to be invariant under f if and only if there exists $\lambda \neq 0$ such that $C = \lambda {}^tA^{-1}C$, i.e. ${}^tAC = \lambda C$, and therefore if and only if C is an eigenvector of the matrix ${}^tA$.

1.5 Projective Spaces of Dimension 1

1.5.1 Cross-Ratio

Given four points P_1, P_2, P_3, P_4 of a projective line $\mathbb{P}(V)$, where P_1, P_2, P_3 are distinct, the *cross-ratio* $\beta(P_1, P_2, P_3, P_4)$ is defined as the number $\frac{y_1}{y_0} \in \mathbb{K} \cup \{\infty\}$, where $[y_0, y_1]$ are homogeneous coordinates of P_4 in the projective frame $\{P_1, P_2, P_3\}$ of $\mathbb{P}(V)$.

   In particular one has $\beta(P_1, P_2, P_3, P_1) = 0$, $\beta(P_1, P_2, P_3, P_2) = \infty$ and $\beta(P_1, P_2, P_3, P_3) = 1$. In this way for every $P_4 \neq P_2$ the number $\beta(P_1, P_2, P_3, P_4)$ is the affine coordinate of P_4 in the affine chart $\mathbb{P}(V) \setminus \{P_2\}$ with respect to the affine frame $\{P_1, P_3\}$.

   If $[\lambda_i, \mu_i]$ are homogeneous coordinates of P_i for $i = 1, \ldots, 4$ in a given system of homogeneous coordinates, we have that

$$ \beta(P_1, P_2, P_3, P_4) = \frac{(\lambda_1\mu_4 - \lambda_4\mu_1)\,(\lambda_3\mu_2 - \lambda_2\mu_3)}{(\lambda_1\mu_3 - \lambda_3\mu_1)\,(\lambda_4\mu_2 - \lambda_2\mu_4)} $$

(setting as usual $\frac{1}{0} = \infty$). By means of this formula we can extend the notion of cross-ratio to the case when three of the four points, but not necessarily the first three, are pairwise distinct.

   If $\lambda_i \neq 0$ for each i and we consider the affine coordinates $z_i = \frac{\mu_i}{\lambda_i}$ of the points P_i, the cross-ratio is given by

$$ \beta(P_1, P_2, P_3, P_4) = \frac{(z_4 - z_1)(z_3 - z_2)}{(z_3 - z_1)(z_4 - z_2)}. $$

Observe that, if P_1, P_2, P_3 are distinct points of a projective line, then for every $k \in \mathbb{K} \cup \{\infty\}$ there exists a unique point Q such that $\beta(P_1, P_2, P_3, Q) = k$.

Let $\mathcal{F}_O$ denote the pencil of lines passing through a point O in a projective plane $\mathbb{P}(V)$. The image of the pencil under the duality correspondence is a line in $\mathbb{P}(V)^*$, and therefore $\mathcal{F}_O$ has a natural structure of 1-dimensional projective space. Thus, given four lines in a projective plane $\mathbb{P}(V)$ passing through a point O, three of which are pairwise distinct, we can define the *cross-ratio of the four lines* as the cross-ratio of the four collinear points of $\mathbb{P}(V)^*$ corresponding to them.

Similarly, given a subspace S of codimension 2 in a projective space $\mathbb{P}(V)$, we denote by $\mathcal{F}_S$ the pencil of hyperplanes centred at S, i.e. the set of hyperplanes of $\mathbb{P}(V)$ containing S. Since $\mathcal{F}_S$ is a 1-dimensional subspace of the dual projective space $\mathbb{P}(V)^*$, given hyperplanes $H_1, H_2, H_3, H_4 \in \mathcal{F}_S$ such that at least three of them are distinct, the cross-ratio $\beta(H_1, H_2, H_3, H_4)$ is well defined.

The main property of the cross-ratio is its invariance under projective isomorphisms:

Theorem 1.5.1 *Let P_1, P_2, P_3, P_4 be points of a projective line $\mathbb{P}(V)$, with P_1, P_2, P_3 distinct, and let Q_1, Q_2, Q_3, Q_4 be points of a projective line $\mathbb{P}(W)$, with Q_1, Q_2, Q_3 distinct. Then there exists a projective isomorphism $f : \mathbb{P}(V) \rightarrow \mathbb{P}(W)$ such that $f(P_i) = Q_i$ for $i = 1, \ldots, 4$ if and only if $\beta(P_1, P_2, P_3, P_4) = \beta(Q_1, Q_2, Q_3, Q_4)$.*

This property can also be used to construct a projective isomorphism f between two projective lines $\mathbb{P}(V)$ and $\mathbb{P}(W)$ sending three distinct points P_1, P_2, P_3 of $\mathbb{P}(V)$ to three distinct points Q_1, Q_2, Q_3 of $\mathbb{P}(W)$. Namely, for every $X \in \mathbb{P}(V)$, $f(X)$ is necessarily the unique point in $\mathbb{P}(W)$ such that $\beta(P_1, P_2, P_3, X) = \beta(Q_1, Q_2, Q_3, f(X))$.

1.5.2   Symmetries of the Cross-Ratio

Since the cross-ratio of four points of a projective line depends on the order of the points and since there are 24 ways of ordering four distinct points, a priori there are 24 possible values that the cross-ratio of an unordered set of four points can take. From the analytic expression of the cross-ratio we deduce that it is invariant under the following permutations

$$\text{Id,} \quad (1\ 2)(3\ 4), \quad (1\ 3)(2\ 4), \quad (1\ 4)(2\ 3),$$

so that

$$\beta(P_1, P_2, P_3, P_4) = \beta(P_2, P_1, P_4, P_3) = \beta(P_3, P_4, P_1, P_2) = \beta(P_4, P_3, P_2, P_1).$$

As a consequence, in order to describe the values of the cross-ratio relative to the 24 ordered quadruples obtained by arranging the points P_1, P_2, P_3, P_4, it suffices to consider the ordered quadruples where P_1 appears in the first position. If we set $\beta(P_1, P_2, P_3, P_4) = k$, it turns out that

$$\beta(P_1, P_2, P_4, P_3) = \frac{1}{k}, \qquad \beta(P_1, P_4, P_3, P_2) = 1 - k,$$

$$\beta(P_1, P_4, P_2, P_3) = \frac{1}{1 - k}, \qquad \beta(P_1, P_3, P_4, P_2) = \frac{k - 1}{k},$$

$$\beta(P_1, P_3, P_2, P_4) = \frac{k}{k - 1}.$$

Therefore, by permuting the four points in all possible ways, the cross-ratio can take at most 6 distincts values, namely

$$k, \ \frac{1}{k}, \ 1 - k, \ \frac{1}{1 - k}, \ \frac{k - 1}{k}, \ \frac{k}{k - 1}.$$

It follows that two sets $\{P_1, P_2, P_3, P_4\}$ and $\{Q_1, Q_2, Q_3, Q_4\}$, each consisting of four distinct points of a projective line $\mathbb{P}(V)$, are projectively equivalent if and only if, letting $k = \beta(P_1, P_2, P_3, P_4)$, we have $\beta(Q_1, Q_2, Q_3, Q_4) \in \left\{ k, \frac{1}{k}, 1 - k, \frac{1}{1 - k}, \frac{k - 1}{k}, \frac{k}{k - 1} \right\}$.

However for certain special values of k the distinct values of the cross-ratio of four points are in fact fewer than 6 (cf. Exercise 21).

If $\beta(P_1, P_2, P_3, P_4) = -1$ we say that the quadruple (P_1, P_2, P_3, P_4) forms a *harmonic quadruple*.

1.5.3   Classification of the Projectivities of $\mathbb{P}^1(\mathbb{C})$ and of $\mathbb{P}^1(\mathbb{R})$

Let f be a projectivity of $\mathbb{P}^1(\mathbb{K})$ with either $\mathbb{K} = \mathbb{C}$ or $\mathbb{K} = \mathbb{R}$.

If $\mathbb{K} = \mathbb{C}$, then there exists a system of homogeneous coordinates in $\mathbb{P}^1(\mathbb{C})$ where f is represented by one of the following matrices:

(a) $\begin{pmatrix} \lambda & 0 \\ 0 & \mu \end{pmatrix}$ with $\lambda, \mu \in \mathbb{K}^*$, $\lambda \neq \mu$;

(b) $\begin{pmatrix} \lambda & 0 \\ 0 & \lambda \end{pmatrix}$ with $\lambda \in \mathbb{K}^*$;

(c) $\begin{pmatrix} \lambda & 1 \\ 0 & \lambda \end{pmatrix}$ with $\lambda \in \mathbb{K}^*$.

Recall that the matrix representing a projectivity is determined up to multiplication by a non-zero scalar and that the matrix $\begin{pmatrix} 1 & \frac{1}{\lambda} \\ 0 & 1 \end{pmatrix}$ is similar to the matrix $\begin{pmatrix} 1 & 1 \\ 0 & 1 \end{pmatrix}$ for any $\lambda \in \mathbb{K}^*$. Therefore we may assume $\lambda = 1$ in the matrices listed above.

If $\mathbb{K} = \mathbb{R}$ and the eigenvalues of one (and hence of every) matrix associated to f are real, then in a suitable system of homogeneous coordinates of $\mathbb{P}^1(\mathbb{R})$ the projectivity f is represented by one of the matrices (a), (b), (c). However it may happen that the eigenvalues of the matrix associated to f are not real, but rather complex conjugate: $a + ib$ and $a - ib$, with $b \neq 0$. If so, f is represented, in a suitable system of homogeneous coordinates, by a matrix of type:

(d) $\begin{pmatrix} a & -b \\ b & a \end{pmatrix}$ with $a \in \mathbb{R}, b \in \mathbb{R}^*$,

called a "real Jordan matrix". Also in this case, by choosing a suitable multiple of the matrix we may assume, for instance, $a^2 + b^2 = 1$.

If we examine the fixed-point sets in the four cases, we see the following possibilities:

(a) there exist exactly two fixed points, that is the points of coordinates $[1, 0]$ and $[0, 1]$; in this case we say that f is a *hyperbolic projectivity*;
(b) all points are fixed and f is the identity map;
(c) there is a unique fixed point $[1, 0]$; f is then called a *parabolic projectivity*;
(d) there are no fixed points and then we say that f is an *elliptic projectivity*.

Therefore the four cases can be distinguished on the basis of the number of fixed points. In particular the previous matrices give explicit examples of projectivities of $\mathbb{P}^1(\mathbb{C})$ having one or two fixed points and examples of projectivities of $\mathbb{P}^1(\mathbb{R})$ where the fixed-point set has cardinality zero, one or two.

1.5.4   Characteristic of a Projectivity

Assume that $\mathbb{K} = \mathbb{C}$ or $\mathbb{K} = \mathbb{R}$. Let f be a hyperbolic projectivity of $\mathbb{P}^1(\mathbb{K})$ and denote its two distinct fixed points by A and B (cf. Sect. 1.5.3). Let $\mathcal{R}$ be a projective frame in $\mathbb{P}^1(\mathbb{K})$ having A and B as fundamental points (so that $[A]_{\mathcal{R}} = [1, 0]$ and $[B]_{\mathcal{R}} = [0, 1]$). Since A and B are fixed points, f is represented, in the system of homogeneous coordinates induced by $\mathcal{R}$, by a matrix of the form $\begin{pmatrix} \lambda & 0 \\ 0 & \mu \end{pmatrix}$.

If $P \in \mathbb{P}^1(\mathbb{K}) \setminus \{A, B\}$ and $[P]_{\mathcal{R}} = [a, b]$, then $[f(P)]_{\mathcal{R}} = [\lambda a, \mu b]$, so the definition implies immediately that $\beta(A, B, P, f(P)) = \frac{\mu}{\lambda}$ (in particular it does not depend on the point P). The value $\frac{\mu}{\lambda}$ is called the *characteristic of the projectivity*.

Clearly the characteristic depends on the order of the fixed points: if we exchange A and B, then the characteristic takes the value $\frac{\lambda}{\mu}$.

1.6 Conjugation and Complexification

The space $\mathbb{R}^n$ embeds in a natural way in $\mathbb{C}^n$ and can be characterized as the set of points $z = (z_1, \ldots, z_n) \in \mathbb{C}^n$ such that $\sigma(z) = z$, where $\sigma \colon \mathbb{C}^n \to \mathbb{C}^n$ denotes the involution defined by $\sigma(z_1, \ldots, z_n) = (\overline{z_1}, \ldots, \overline{z_n})$.

If $a_1 z_1 + \cdots + a_n z_n = c$ is the equation of an affine hyperplane L of $\mathbb{C}^n$ (where at least one of the complex numbers $a_1, \ldots, a_n$ is non-zero and $c \in \mathbb{C}$), then $\sigma(L)$ is the hyperplane of equation $\overline{a_1} z_1 + \cdots + \overline{a_n} z_n = \overline{c}$ and it is called the *conjugate* of L.

If the numbers $a_1, \ldots, a_n, c$ are real, the equation $a_1 x_1 + \cdots + a_n x_n = c$ defines an affine hyperplane H of $\mathbb{R}^n$; as a matter of fact, the same equation defines in $\mathbb{C}^n$ an affine hyperplane called the *complexification* of H and denoted $H_{\mathbb{C}}$. Proceeding in an analogous way one can define the *complexification* $S_{\mathbb{C}}$ of an affine subspace S of $\mathbb{R}^n$ of any dimension; $S_{\mathbb{C}}$ is an affine subspace of $\mathbb{C}^n$ having the same dimension as S such that $\sigma(S_{\mathbb{C}}) = S_{\mathbb{C}}$ and $S_{\mathbb{C}} \cap \mathbb{R}^n = S$.

If we regard $\mathbb{R}^n$ as embedded in $\mathbb{C}^n$, then any affinity of $\mathbb{R}^n$ extends to an affinity of $\mathbb{C}^n$. The converse is not true: for example, an affinity $T(X) = MX + N$ of $\mathbb{C}^n$ in general does not map points of $\mathbb{R}^n$ to points of $\mathbb{R}^n$. This happens if and only if T transforms $n + 1$ affinely independent points of $\mathbb{R}^n$ into $n + 1$ affinely independent points of $\mathbb{R}^n$, i.e. if and only if M is a matrix of $\mathrm{GL}(n, \mathbb{R})$ and $N \in \mathbb{R}^n$. In this case the restriction of T to $\mathbb{R}^n$ is an affinity of $\mathbb{R}^n$.

Analogous considerations can be made in the projective case. Also $\mathbb{P}^n(\mathbb{R})$ is naturally embedded in $\mathbb{P}^n(\mathbb{C})$ and may be seen as the set of points of $\mathbb{P}^n(\mathbb{C})$ that admit a representative in $\mathbb{R}^{n+1} \setminus \{0\}$. If, for the sake of simplicity, we still denote by σ the involution $\mathbb{P}^n(\mathbb{C}) \to \mathbb{P}^n(\mathbb{C})$ defined by $\sigma([z_0, \ldots, z_n]) = [\overline{z_0}, \ldots, \overline{z_n}]$, then $\mathbb{P}^n(\mathbb{R})$ coincides with the set of points $P \in \mathbb{P}^n(\mathbb{C})$ such that $\sigma(P) = P$. Similarly one can define the conjugate $\sigma(H)$ of a projective hyperplane H of $\mathbb{P}^n(\mathbb{C})$ and the complexification $S_{\mathbb{C}}$ of a projective subspace S of $\mathbb{P}^n(\mathbb{R})$.

Every projectivity of $\mathbb{P}^n(\mathbb{R})$ extends to a projectivity of $\mathbb{P}^n(\mathbb{C})$. Conversely, a projectivity f of $\mathbb{P}^n(\mathbb{C})$ represented by a matrix $A \in \mathrm{GL}(n + 1, \mathbb{C})$ transforms points of $\mathbb{P}^n(\mathbb{R})$ into points of $\mathbb{P}^n(\mathbb{R})$ if and only if there exists $\lambda \in \mathbb{C}^*$ such that $\lambda A \in \mathrm{GL}(n + 1, \mathbb{R})$; in this case the restriction of f to $\mathbb{P}^n(\mathbb{R})$ is a projectivity of $\mathbb{P}^n(\mathbb{R})$.

1.7 Affine and Projective Hypersurfaces

Quadrics, plane algebraic curves and conics are the examples of hypersurfaces we will focus on in what follows. To avoid useless repetitions, in this section we collect notations, definitions and some basic results about hypersurfaces that will be studied in greater detail and applied to the above special cases in subsequent sections.

1.7.1 Homogeneous Polynomials

From now on we assume that $\mathbb{K} = \mathbb{C}$ or $\mathbb{K} = \mathbb{R}$. We denote the ring of polynomials with coefficients in $\mathbb{K}$ in the indeterminates $x_0, \dots, x_n$ by $\mathbb{K}[x_0, \dots, x_n]$ and the degree of the polynomial $F(x_0, \dots, x_n)$ by $\deg F$. We recall (cf. Sect. 1.1) that F_{x_i} denotes the partial derivative of F with respect to x_i; higher-order partial derivatives are denoted in a similar way.

A non-zero polynomial $F \in \mathbb{K}[x_0, \dots, x_n]$ is said to be *homogeneous* if all its monomials have the same degree (homogeneous polynomials are usually denoted by capital letters). Every non-zero polynomial can be written uniquely as a sum of homogeneous polynomials.

Homogeneous polynomials have many important properties; here we just recall that:

(a) a non-zero polynomial $F(x_0, \dots, x_n)$ is homogeneous of degree d if and only if the identity $F(tx_0, \dots, tx_n) = t^d F(x_0, \dots, x_n)$ holds in $\mathbb{K}[x_0, \dots, x_n, t]$;

(b) (*Euler's identity*) if $F(x_0, \dots, x_n)$ is homogeneous, then the equality

$$\sum_{i=0}^{n} x_i F_{x_i} = (\deg F)\, F$$

holds in $\mathbb{K}[x_0, \dots, x_n]$;

(c) a polynomial that divides a homogeneous polynomial is itself homogeneous.

If $F(x_0, \dots, x_n) \in \mathbb{K}[x_0, \dots, x_n]$ is homogeneous, then the polynomial $f(x_1, \dots, x_n) = F(1, x_1, \dots, x_n)$ is called the *dehomogenized polynomial* of F with respect to x_0. If F is homogeneous of degree d and x_0 does not divide F, the dehomogenized polynomial of F still has degree d.

If $f(x_1, \dots, x_n) \in \mathbb{K}[x_1, \dots, x_n]$ is a degree d polynomial, the polynomial

$$F(x_0, \dots, x_n) = x_0^d f\left(\frac{x_1}{x_0}, \dots, \frac{x_n}{x_0}\right)$$

is called the *homogenized polynomial* of f with respect to x_0. It is immediate to check that F is homogeneous of degree d and is not divisible by x_0.

Homogenizing a polynomial $f \in \mathbb{K}[x_1, \dots, x_n]$ and then dehomogenizing the polynomial thus obtained gives f back. On the other hand, if F is homogeneous, then the polynomial obtained by first dehomogenizing it with respect to x_0 and then homogenizing the result coincides with F if and only if x_0 does not divide F. Therefore, there exists a one-to-one correspondence between degree d polynomials of $\mathbb{K}[x_1, \dots, x_n]$ and homogeneous degree d polynomials of $\mathbb{K}[x_0, \dots, x_n]$ that are not divisible by x_0. Analogous definitions can be given for any other variable x_i.

If $F(x_0, \dots, x_n)$ and $f(x_1, \dots, x_n)$ are polynomials that are obtained one from another by homogenizing and dehomogenizing (so that, in particular, x_0 does not divide F), then F is irreducible if and only if f is. More precisely, if F factors as

$F = c F_1^{m_1} \cdot \ldots \cdot F_s^{m_s}$, where the F_i are homogeneous irreducible polynomials and c is a non-zero constant, then f factors as $f = c f_1^{m_1} \cdot \ldots \cdot f_s^{m_s}$, where f_i is the polynomial obtained by dehomogenizing F_i, and conversely.

If we consider homogeneous polynomials in two variables with complex coefficients, one has the following important result, which is the homogeneous version of the Fundamental theorem of algebra:

Theorem 1.7.1 (Factorization of complex homogeneous polynomials in two variables) *Let $F(x_0, x_1) \in \mathbb{C}[x_0, x_1]$ be a homogeneous polynomial of degree $d > 0$. Then there exist $(a_1, b_1), \ldots, (a_d, b_d) \in \mathbb{C}^2 \setminus \{(0, 0)\}$ such that*

$$F(x_0, x_1) = (a_1 x_1 - b_1 x_0) \cdot \ldots \cdot (a_d x_1 - b_d x_0).$$

The pairs (a_i, b_i) are uniquely determined up to the order and non-zero multiplicative constants.

A homogeneous pair $[a, b] \in \mathbb{P}^1(\mathbb{C})$ is called a *root of multiplicity m* of a homogeneous polynomial $F(x_0, x_1) \in \mathbb{C}[x_0, x_1]$ if m is the largest non-negative integer such that $(ax_1 - bx_0)^m$ divides $F(x_0, x_1)$.

Theorem 1.7.1 does not hold for real homogeneous polynomials in two variables: just consider the polynomial $x_0^2 + x_1^2$. However, since the only irreducible real polynomials in one variable are those of degree one and those of degree two with negative discriminant, we have the following real version:

Theorem 1.7.2 (Factorization of real homogeneous polynomials in two variables) *Let $F(x_0, x_1) \in \mathbb{R}[x_0, x_1]$ be a homogeneous polynomial of positive degree. Then there exist pairs $(a_1, b_1), \ldots, (a_h, b_h) \in \mathbb{R}^2 \setminus \{(0, 0)\}$ and triples $(\alpha_1, \beta_1, \gamma_1), \ldots,$ $(\alpha_k, \beta_k, \gamma_k) \in \mathbb{R}^3 \setminus \{(0, 0, 0)\}$ with $\beta_j^2 - 4\alpha_j \gamma_j < 0$ such that*

$$F(x_0, x_1) = \prod_{i=1}^{h} (a_i x_1 - b_i x_0) \cdot \prod_{j=1}^{k} \left(\alpha_j x_1^2 + \beta_j x_0 x_1 + \gamma_j x_0^2 \right).$$

The pairs (a_i, b_i) and the triples $(\alpha_j, \beta_j, \gamma_j)$ are uniquely determined up to the order and non-zero multiplicative constants.

In particular every homogeneous polynomial of odd degree in $\mathbb{R}[x_0, x_1]$ has at least a real linear factor and hence a root in $\mathbb{P}^1(\mathbb{R})$.

1.7.2 Affine and Projective Hypersurfaces

We define an equivalence relation on $\mathbb{K}[x_1, \ldots, x_n]$ by declaring that polynomials $f, g \in \mathbb{K}[x_1, \ldots, x_n]$ are equivalent if and only if there exists $\lambda \in \mathbb{K}^*$ such that $f = \lambda g$; if this is the case, we say that f and g are proportional. An *affine hypersurface of $\mathbb{K}^n$* is defined as a proportionality class of polynomials of $\mathbb{K}[x_1, \ldots, x_n]$ of positive degree; an affine hypersurface is called an *affine curve* when $n = 2$ and an *affine surface* when $n = 3$.

If f is a representative of the hypersurface $\mathcal{I}$, we say that $f(x_1, \ldots, x_n) = 0$ is an *equation* of the hypersurface and that the degree of f is the *degree* of $\mathcal{I}$. If $\mathcal{I} = [f]$ and $\mathcal{J} = [g]$, we denote the hypersurface $[fg]$ by $\mathcal{I} + \mathcal{J}$; moreover, for every positive integer m, we denote $m\mathcal{I} = [f^m]$.

The hypersurface $\mathcal{I} = [f]$ is said to be *irreducible* if f is. If $f = cf_1^{m_1} \cdot \ldots \cdot f_s^{m_s}$ with c a non-zero constant and the f_i distinct irreducible polynomials, the hypersurfaces $\mathcal{I}_i = [f_i]$ are called the *irreducible components* of $\mathcal{I}$ and the integer m_i is called the *multiplicity* of the component $\mathcal{I}_i$. Every irreducible component of multiplicity $m_i > 1$ is called a *multiple component*; hypersurfaces without multiple components are said to be *reduced*.

For any $f \in \mathbb{K}[x_1, \ldots, x_n]$ we set

$$V(f) = \{(x_1, \ldots, x_n) \in \mathbb{K}^n \mid f(x_1, \ldots, x_n) = 0\}.$$

Since $V(\lambda f) = V(f) \ \forall \lambda \in \mathbb{K}^*$, we call the (well-defined) set $V(\mathcal{I}) = V(f)$ the *support* of the hypersurface $\mathcal{I} = [f]$.

As the field $\mathbb{K}$ is infinite, the Identity principle for polynomials ensures that a polynomial f of $\mathbb{K}[x_1, \ldots, x_n]$ satisfies $f(a_1, \ldots, a_n) = 0$ for every $(a_1, \ldots, a_n) \in \mathbb{K}^n$ if and only if $f = 0$. As a consequence, the complement of any hypersurface is non-empty.

While a hypersurface uniquely determines its support, the converse is not true in general: for instance, the hypersurfaces of equations $f = 0$ and $f^m = 0$, although different, have the same supports. Actually, if $\mathbb{K} = \mathbb{C}$ the correspondence between hypersurfaces and supports becomes a bijection in the case of reduced hypersurfaces, while if $\mathbb{K} = \mathbb{R}$ there exist distinct reduced hypersurfaces with the same supports (for example the real plane curves of equations $x_1^2 + x_2^2 = 0$ and $x_1^4 + x_2^4 = 0$). We will sometimes abuse the notation and denote both a hypersurface and its support by the same symbol.

Another remarkable phenomenon is that, while the support of a complex hypersurface contains infinitely many points if $n \geq 2$, if the field is not algebraically closed there exist hypersurfaces whose support is finite (as in the case of the curve of $\mathbb{R}^2$ of equation $x_1^2 + x_2^2 = 0$), or even empty (think of the curve of $\mathbb{R}^2$ of equation $x_1^2 + x_2^2 + 1 = 0$).

An affine hypersurface $\mathcal{I}$ is called a *cone* if there exists a point $P \in \mathcal{I}$ (called *vertex*) such that, for every $Q \in \mathcal{I}$, Q different from P, the line joining P and Q is contained in $\mathcal{I}$. For instance, if f is a homogeneous polynomial then the affine hypersurface $\mathcal{I} = [f]$ is a cone with vertex at the origin. The set of vertices of a cone may be infinite (cf. Exercise 112); for example the vertices of the cone of $\mathbb{R}^3$ of equation $x_1 x_2 = 0$ are precisely the points of the x_3-axis.

Any proportionality class of homogeneous polynomials of positive degree in $\mathbb{K}[x_0, \ldots, x_n]$ is called a *projective hypersurface of* $\mathbb{P}^n(\mathbb{K})$. Similarly to what we did in the affine case one defines the concepts of *equation* and *degree* of a projective hypersurface, of *irreducible hypersurface, irreducible component, multiplicity* of a component, *cone*. A hypersurface is said to be a *plane curve* when $n = 2$ and a *projective surface* when $n = 3$.

The *support* of a projective hypersurface can also be defined in a similar way as the set of points $P = [x_0, \ldots, x_n] \in \mathbb{P}^n(\mathbb{K})$ such that $F(x_0, \ldots, x_n) = 0$. Note that this is a good definition, since it depends neither on the choice of the representative of $\mathcal{I}$ nor on the choice of the homogeneous coordinates of P; in fact, since F is homogeneous, one has $F(kx_0, \ldots, kx_n) = k^d F(x_0, \ldots, x_n)$ for every $k \in \mathbb{K}$, with $d = \deg F$. For instance, the support of a hypersurface of $\mathbb{P}^1(\mathbb{K})$ is a (possibly empty) finite set.

Every hyperplane of $\mathbb{P}^n(\mathbb{K})$ is the support of an irreducible hypersurface of degree 1 (that we shall also call a hyperplane).

Also in the projective case there is a bijective correspondence between reduced projective hypersurfaces and their supports only when $\mathbb{K} = \mathbb{C}$ (cf. Exercise 63 for the case of curves).

The (affine or projective) hypersurfaces of degree $2, 3, 4$ are called *quadrics, cubics, quartics*, respectively. The quadrics of $\mathbb{P}^2(\mathbb{K})$ are called *conics*.

1.7.3 Intersection of a Hypersurface with a Hyperplane

Let $\mathcal{I}$ be a hypersurface of $\mathbb{P}^n(\mathbb{K})$ of equation $F(x_0, \ldots, x_n) = 0$, and let H be a hyperplane not contained in $\mathcal{I}$ of equation $x_i = L(x_0, \ldots, \widehat{x_i}, \ldots, x_n)$, where L is a homogeneous polynomial of degree 1 that does not depend on the variable x_i.

If P_i is the ith point of the standard projective frame of $\mathbb{P}^n(\mathbb{K})$ and H_i is the coordinate hyperplane of equation $x_i = 0$, then P_i does not lie in $H \cup H_i$, and therefore one can define the perspectivity $f : H_i \to H$ centred at P_i (cf. Sect. 1.2.8), i.e. the restriction of the projection $\pi_{P_i} : \mathbb{P}^n(\mathbb{K}) \setminus \{P_i\} \to H$ onto H centred at P_i (cf. Sect. 1.2.7). Recall that the standard homogeneous coordinates of $\mathbb{P}^n(\mathbb{K})$ induce homogeneous coordinates $x_0, \ldots, \widehat{x_i}, \ldots, x_n$ on H_i. In turn, the latter coordinates induce, via the projective isomorphism f, homogeneous coordinates on H that are usually still denoted by $x_0, \ldots, \widehat{x_i}, \ldots, x_n$. It is immediate to check that, with these choices, the point of H of coordinates $[x_0, \ldots, \widehat{x_i}, \ldots, x_n]$ coincides with the point of $\mathbb{P}^n(\mathbb{K})$ of coordinates $[x_0, \ldots, L(x_0, \ldots, \widehat{x_i}, \ldots, x_n), \ldots, x_n]$. The polynomial

$$G(x_0, \ldots, \widehat{x_i}, \ldots, x_n) = F(x_0, \ldots, x_{i-1}, L(x_0, \ldots, \widehat{x_i}, \ldots, x_n), x_{i+1}, \ldots, x_n)$$

is non-zero because H is not contained in $\mathcal{I}$; in the coordinates of H just described, the equation $G(x_0, \ldots, \widehat{x_i}, \ldots, x_n) = 0$ defines a hypersurface of H, that we denote by $\mathcal{I} \cap H$ and that has the same degree as $\mathcal{I}$.

It is easy to check that the point $[a_0, \ldots, a_n]$ of $\mathbb{P}^n(\mathbb{K})$ belongs to $\mathcal{I} \cap H$ if and only if $a_i = L(a_0, \ldots, \widehat{a_i}, \ldots, a_n)$ and $G(a_0, \ldots, \widehat{a_i}, \ldots, a_n) = 0$.

Once a homogeneous coordinate system in H, and thus a projective isomorphism between H and $\mathbb{P}^{n-1}(\mathbb{K})$, has been chosen, the previous considerations essentially reduce the definition of the hypersurface $\mathcal{I} \cap H$ to that of a hypersurface in $\mathbb{P}^{n-1}(\mathbb{K})$ (cf. Sect. 1.7.2).

Another way of fixing a homogeneous coordinate system in H is obtained by using a parametric representation of the hyperplane. Let $P_0, \ldots, P_{n-1}$ be projectively independent points such that $H = L(P_0, \ldots, P_{n-1})$. We fix representatives $(p_{i,0}, \ldots, p_{i,n})$ of the points P_i and denote by $\lambda_0, \ldots, \lambda_{n-1}$ the corresponding coordinate system on H. From now on we will write $\lambda_0 P_0 + \cdots + \lambda_{n-1} P_{n-1}$ instead of

$$[\lambda_0 p_{0,0} + \cdots + \lambda_{n-1} p_{n-1,0}, \ldots, \lambda_0 p_{0,n} + \cdots + \lambda_{n-1} p_{n-1,n}].$$

Then we can represent H parametrically as the set of points $\lambda_0 P_0 + \cdots + \lambda_{n-1} P_{n-1}$ as $[\lambda_0, \ldots, \lambda_{n-1}]$ varies in $\mathbb{P}^{n-1}(\mathbb{K})$.

We will also write $F(\lambda_0 P_0 + \cdots + \lambda_{n-1} P_{n-1})$ instead of

$$F(\lambda_0 p_{0,0} + \cdots + \lambda_{n-1} p_{n-1,0}, \ldots, \lambda_0 p_{0,n} + \cdots + \lambda_{n-1} p_{n-1,n}).$$

This convention will always be tacitly used henceforth.

The polynomial

$$G(\lambda_0, \ldots, \lambda_{n-1}) = F(\lambda_0 P_0 + \cdots + \lambda_{n-1} P_{n-1})$$

(which is non-zero as H is not contained in $\mathcal{I}$) is homogeneous in $\lambda_0, \ldots, \lambda_{n-1}$ of the same degree as F. Therefore, when H is not contained in $\mathcal{I}$, the polynomial $G(\lambda_0, \ldots, \lambda_{n-1})$ defines a hypersurface of H (endowed with the homogenous coordinates $\lambda_0, \ldots, \lambda_{n-1}$) which is denoted by $\mathcal{I} \cap H$ and has the same degree as $\mathcal{I}$.

In a similar way one can see that the intersection of an affine hypersurface with an affine hyperplane not contained in it is a hypersurface of H.

1.7.4   Projective Closure of an Affine Hypersurface

Identify $\mathbb{K}^n$ with the affine chart $U_0 = \mathbb{P}^n(\mathbb{K}) \setminus \{x_0 = 0\}$ by means of the map $j_0 \colon \mathbb{K}^n \to U_0$ defined by $j_0(x_1, \ldots, x_n) = [1, x_1, \ldots, x_n]$.

Let F be a homogeneous polynomial that defines the projective hypersurface $\mathcal{I}$ of $\mathbb{P}^n(\mathbb{K})$ and assume that x_0 is not the only irreducible factor of F. If f denotes the polynomial obtained by dehomogenizing F with respect to x_0, the affine hypersurface defined by f is called the *affine part* of $\mathcal{I}$ in the chart U_0. Its support is the intersection of the support of $\mathcal{I}$ with U_0: hence we denote it by $\mathcal{I} \cap U_0$. The affine part $\mathcal{I} \cap U_0$ has the same degree as $\mathcal{I}$ if and only if x_0 does not divide F.

If $\pi \colon \mathbb{K}^{n+1} \setminus \{0\} \to \mathbb{P}^n(\mathbb{K})$ is the canonical quotient map, the set $\pi^{-1}(\mathcal{I}) \cup \{0\}$ is a cone of $\mathbb{K}^{n+1}$ with vertex 0. The affine part $\mathcal{I} \cap U_0$ can then be regarded as the intersection of the cone $\pi^{-1}(\mathcal{I}) \cup \{0\}$ with the affine hyperplane of $\mathbb{K}^{n+1}$ of equation $x_0 = 1$.

Similarly, for every hyperplane H of $\mathbb{P}^n(\mathbb{K})$ one can define the affine part of $\mathcal{I}$ in the chart $U_H = \mathbb{P}^n(\mathbb{K}) \setminus H$.

Let $\mathcal{I} = [f]$ be an affine hypersurface of $\mathbb{K}^n$. If $F(x_0, \ldots, x_n)$ is the homogenized polynomial of f with respect to x_0, we say that the projective hypersurface $\overline{\mathcal{I}} = [F]$ is the *projective closure* of $\mathcal{I}$. Since by dehomogenizing F with respect to x_0 we get f back, the affine part $\overline{\mathcal{I}} \cap U_0$ of $\overline{\mathcal{I}}$ coincides with $\mathcal{I}$. In addition, since x_0 does not divide F, the intersection of $\overline{\mathcal{I}}$ with the hyperplane $H_0 = \{x_0 = 0\}$ is a hypersurface of H_0 having the same degree as $\mathcal{I}$. For example, if $n = 2$ the projective closure of the curve $\mathcal{I}$ contains only finitely many points of the line $x_0 = 0$, which are called *improper points* or *points at infinity* of $\mathcal{I}$.

We remark that, if $\mathcal{I} = [F]$ is a projective hypersurface and x_0 does not divide F, then $\overline{\mathcal{I} \cap U_0} = \mathcal{I}$.

In addition, if $\mathcal{I}$ is an affine hypersurface and H is an affine hyperplane of $\mathbb{K}^n$ not contained in $\mathcal{I}$, then $\overline{\mathcal{I} \cap H} = \overline{\mathcal{I}} \cap \overline{H}$.

1.7.5 Affine and Projective Equivalence of Hypersurfaces

Recall that two subsets of $\mathbb{K}^n$ are called affinely equivalent if there exists an affinity φ that transforms one into the other. As hypersurfaces are not in general determined by their supports, we introduce a notion of affine equivalence of hypersurfaces based on their equations.

Let $\mathcal{I}$ be an affine hypersurface of $\mathbb{K}^n$ of equation $f(X) = 0$, where $X = (x_1, \ldots, x_n)$, and let $\varphi(X) = AX + B$ be an affinity of $\mathbb{K}^n$, with $A \in \mathrm{GL}(n, \mathbb{K})$ and $B \in \mathbb{K}^n$. We denote by $\varphi(\mathcal{I})$ the affine hypersurface of equation $g(X) = f(\varphi^{-1}(X)) = 0$. This notation is consistent with the fact that φ transforms the support of $\mathcal{I}$ into the support of $\varphi(\mathcal{I})$.

Two affine hypersurfaces $\mathcal{I}$, $\mathcal{J}$ of $\mathbb{K}^n$ are said to be *affinely equivalent* if there exists an affinity φ of $\mathbb{K}^n$ such that $\mathcal{I} = \varphi(\mathcal{J})$. As we mentioned above, the supports of affinely equivalent hypersurfaces are affinely equivalent.

Projective equivalence of projective hypersurfaces can be defined in an analogous way. Let $\mathcal{I}$ be a hypersurface of $\mathbb{P}^n(\mathbb{K})$ of equation $F(x_0, \ldots, x_n) = 0$ and g a projectivity of $\mathbb{P}^n(\mathbb{K})$. If $X = (x_0, \ldots, x_n)$ and $N \in \mathrm{GL}(n + 1, \mathbb{K})$ is a matrix associated to g, then g acts by mapping the point of coordinates X to the point of coordinates NX. We denote by $g(\mathcal{I})$ the projective hypersurface of equation $G(X) = F(N^{-1}X) = 0$.

Two projective hypersurfaces $\mathcal{I}$, $\mathcal{J}$ of $\mathbb{P}^n(\mathbb{K})$ are said to be *projectively equivalent* if there exists a projectivity g of $\mathbb{P}^n(\mathbb{K})$ such that $\mathcal{I} = g(\mathcal{J})$. If this is the case, one can check that g transforms the support of $\mathcal{J}$ into the support of $\mathcal{I}$, and therefore the supports of projectively equivalent hypersurfaces are projectively equivalent sets. Moreover the degree, the number and the multiplicities of the irreducible components of a hypersurface are preserved by projective isomorphisms (hence by changes of homogeneous coordinates of $\mathbb{P}^n(\mathbb{K})$).

Let $\mathcal{I}$ and $\mathcal{J}$ be two hypersurfaces not having $[x_0]$ as an irreducible component and assume that g is a projectivity of $\mathbb{P}^n(\mathbb{K})$ such that $\mathcal{I} = g(\mathcal{J})$. If the affine chart U_0 is invariant under g, then g is represented by a block matrix of the form (cf. Sect. 1.3.8)

$$N = \left(\begin{array}{c|ccc} 1 & 0 & \dots & 0 \\ \hline B & & C & \end{array} \right),$$

with $B \in \mathbb{K}^n$ and C an invertible square matrix of order n. The restriction of g to the chart U_0 is the affinity of $\mathbb{R}^n$ given by $Y \mapsto CY + B$ and transforms $\mathcal{J} \cap U_0$ into $\mathcal{I} \cap U_0$, while the restriction of g to the hyperplane at infinity H_0 is the projectivity of H_0 represented by the matrix C that transforms $\mathcal{J} \cap H_0$ into $\mathcal{I} \cap H_0$.

Analogously, if $\mathcal{I}$ and $\mathcal{J}$ are affinely equivalent hypersurfaces of $\mathbb{K}^n$ and φ is an affinity of $\mathbb{K}^n$ such that $\mathcal{I} = \varphi(\mathcal{J})$, we may regard φ as the restriction to U_0 of a projectivity g of $\mathbb{P}^n(\mathbb{K})$ that fixes U_0 and such that $\overline{\mathcal{I}} = g(\overline{\mathcal{J}})$ and $\overline{\mathcal{I}} \cap H_0 = g|_{H_0}(\overline{\mathcal{J}} \cap H_0)$. Therefore, the projective closures and the parts at infinity of affinely equivalent hypersurfaces are projectively equivalent.

1.7.6   Intersection of a Hypersurface and a Line

Let $\mathcal{I}$ be a projective hypersurface of degree d of $\mathbb{P}^n(\mathbb{K})$ having equation $F(x_0, \dots, x_n) = 0$ and let r be a projective line.

Let R and Q be distinct points of r. In accordance with the considerations and notations established in Sect. 1.7.3, once we have chosen two representatives $(r_0, \dots, r_n)$ and $(q_0, \dots, q_n)$ of R and Q respectively, the line r is the set of points $\lambda R + \mu Q$ as $[\lambda, \mu]$ varies in $\mathbb{P}^1(\mathbb{K})$ and the intersection points of $\mathcal{I}$ and r are obtained by solving the equation

$$G(\lambda, \mu) = F(\lambda R + \mu Q) = F(\lambda r_0 + \mu q_0, \dots, \lambda r_n + \mu q_n) = 0.$$

If r is contained in $\mathcal{I}$, the polynomial $G(\lambda, \mu)$ vanishes identically; otherwise, it is homogeneous of degree d. Therefore, if $\mathbb{K} = \mathbb{C}$, by Theorem 1.7.1 the equation $G(\lambda, \mu) = 0$ has precisely d roots $[\lambda, \mu]$ in $\mathbb{P}^1(\mathbb{C})$, counted with multiplicity. If $\mathbb{K} = \mathbb{R}$, instead, the equation has at most d real roots, corresponding to the linear factors of $G(\lambda, \mu)$.

Looking at the contribution of each root, if $[\lambda_0, \mu_0]$ is a root of multiplicity m of the polynomial $G(\lambda, \mu)$ then we say that $\mathcal{I}$ and r have *multiplicity of intersection* m at the corresponding point $P = \lambda_0 R + \mu_0 Q$, and we write $I(\mathcal{I}, r, P) = m$.

Note that the multiplicity of intersection is well defined, because it does not depend on the choice of the two points on the line r.

We set by definition $I(\mathcal{I}, r, P) = 0$ if $P \notin \mathcal{I} \cap r$ and $I(\mathcal{I}, r, P) = \infty$ if r is contained in $\mathcal{I}$.

Concerning the notion of multiplicity just introduced, one can verify that:

(a) if g is a projectivity of $\mathbb{P}^n(\mathbb{K})$, then $I(\mathcal{I}, r, P) = I(g(\mathcal{I}), g(r), g(P))$, hence in particular the multiplicity of intersection is preserved under projective equivalence;

(b) if $\mathcal{I}$ and $\mathcal{J}$ are projective hypersurfaces of $\mathbb{P}^n(\mathbb{K})$, then

$$I(\mathcal{I} + \mathcal{J}, r, P) = I(\mathcal{I}, r, P) + I(\mathcal{J}, r, P);$$

(c) if the line r is not contained in the projective hypersurface $\mathcal{I}$ of degree d of $\mathbb{P}^n(\mathbb{C})$, then $\sum_{P \in r} I(\mathcal{I}, r, P) = d$ (that is to say, $\mathcal{I}$ and r intersect precisely in d points counted with multiplicity). In the case of real hypersurfaces, even counting intersections with multiplicity, in general we can only say that $\sum_{P \in r} I(\mathcal{I}, r, P) \leq d$.

Proceeding in a similar way, it is possible to define the multiplicity of intersection of a hypersurface and a line at a point in the affine case, as well. Indeed, if $\mathcal{I} = [f]$ is an affine hypersurface of $\mathbb{K}^n$ and r is the line joining two points $R, Q \in \mathbb{K}^n$ parametrized by $t \mapsto (1 - t)R + tQ$, we say that $I(\mathcal{I}, r, P) = m$ if $P = (1 - t_0)R + t_0 Q$ and t_0 is a root of multiplicity m of the polynomial $g(t) = f((1 - t)R + tQ)$.

As in the projective case, it is possible to check that the multiplicity $I(\mathcal{I}, r, P)$ is well defined and does not depend on the choice of the points R, Q. In addition, $I(\mathcal{I}, r, P) = I(\overline{\mathcal{I}}, \overline{r}, P)$ and therefore the multiplicity of intersection of a projective hypersurface and a line at a point can be computed in affine coordinates in any affine chart containing the point.

1.7.7 Tangent Space to a Hypersurface, Singular Points

Let $\mathcal{I} = [F]$ be a projective hypersurface of $\mathbb{P}^n(\mathbb{K})$ of degree d. We say that a projective line r is *tangent* to $\mathcal{I}$ at P if $I(\mathcal{I}, r, P) \geq 2$.

It is possible to check that the union of the tangent lines to the hypersurface $\mathcal{I} = [F]$ at $P = [v]$ coincides with the projective subspace of $\mathbb{P}^n(\mathbb{K})$ defined by

$$F_{x_0}(v)x_0 + \cdots + F_{x_n}(v)x_n = 0;$$

this space is called *tangent space* to $\mathcal{I}$ at P and is denoted by $T_P(\mathcal{I})$. This notion is well defined, i.e. it does not depend on the choice of the representative of P, since every first-order partial derivative of a homogeneous polynomial of degree d is either zero or homogeneous of degree $d - 1$.

A line r contained in $\mathcal{I}$ is tangent to $\mathcal{I}$ at each of its points, and thus $r \subseteq T_P(\mathcal{I})$ for every $P \in r$.

Denoting the usual gradient of F by ∇F, we will write $\nabla F(P) = 0$ if the gradient of F vanishes at any representative of P.

The point $P \in \mathcal{I}$ is said a *singular point* of $\mathcal{I}$ if $\nabla F(P) = 0$; otherwise it is said *non-singular* or *smooth*. We will denote by $\mathrm{Sing}(\mathcal{I})$ the set of singular points of $\mathcal{I}$. If

P is non-singular, then $T_P(\mathcal{I})$ is a hyperplane of $\mathbb{P}^n(\mathbb{K})$, otherwise it coincides with $\mathbb{P}^n(\mathbb{K})$.

A hypersurface is said to be *non-singular* or *smooth* if all its points are non-singular, otherwise it is said to be *singular*.

A projective hyperplane is said to be *tangent* to $\mathcal{I}$ at P if it is contained in $T_P(\mathcal{I})$: if P is non-singular, then the only hyperplane tangent at P is the tangent space $T_P(\mathcal{I})$; if P is singular, then every hyperplane through P is tangent.

In analogy to the projective case, we say that the affine line r is *tangent* to the affine hypersurface $\mathcal{I}$ at P if $I(\mathcal{I}, r, P) \geq 2$.

If $P = (p_1, \ldots, p_n)$, the union of the lines tangent to $\mathcal{I} = [f]$ at P coincides with the affine subspace $T_P(\mathcal{I})$ of $\mathbb{K}^n$ of equation

$$f_{x_1}(P)(x_1 - p_1) + \cdots + f_{x_n}(P)(x_n - p_n) = 0,$$

which is called *tangent space* to $\mathcal{I}$ at P.

The point $P \in \mathcal{I} = [f]$ is called a *singular point* of $\mathcal{I}$ if $\nabla f(P) = 0$, i.e. if all first-order partial derivatives of f vanish at P; otherwise the point is called *non-singular* or *smooth*. We will denote by $\mathrm{Sing}(\mathcal{I})$ the set of singular points of $\mathcal{I}$. If P is non-singular, the tangent space $T_P(\mathcal{I})$ is an affine hyperplane of $\mathbb{K}^n$, otherwise it coincides with $\mathbb{K}^n$. An affine hyperplane is said to be *tangent* to $\mathcal{I}$ at P if it is contained in $T_P(\mathcal{I})$.

Finally we observe that:

(a) P is singular for $\mathcal{I}$ if and only if it is singular for $\overline{\mathcal{I}}$;
(b) an affine line r is tangent to $\mathcal{I}$ at P if and only if $\overline{r}$ is tangent to $\overline{\mathcal{I}}$ at P;
(c) $\overline{T_P(\mathcal{I})} = T_P(\overline{\mathcal{I}})$.

1.7.8 Multiplicity of a Point of a Hypersurface

Let $\mathcal{I}$ be a projective hypersurface of degree d of $\mathbb{P}^n(\mathbb{K})$ and let P be a point in $\mathbb{P}^n(\mathbb{K})$. Letting r vary in the set of lines passing through P, and excluding the lines contained in the hypersurface, if any, the multiplicity of intersection $I(\mathcal{I}, r, P)$ may vary between 0 and d. The integer

$$m_P(\mathcal{I}) = \min_{r \ni P} I(\mathcal{I}, r, P)$$

is called *multiplicity of P for $\mathcal{I}$* (or also multiplicity of $\mathcal{I}$ at P). Since there is always at least one line not contained in the hypersurface, one has that $0 \leq m_P(\mathcal{I}) \leq d$; in addition $m_P(\mathcal{I}) = 0$ if and only if $P \notin \mathcal{I}$.

One can check (cf. Sect. 1.7.6) that:

(a) if g is a projectivity of $\mathbb{P}^n(\mathbb{K})$, then $m_P(\mathcal{I}) = m_{g(P)}(g(\mathcal{I}))$, and therefore the multiplicity of a hypersurface at a point is preserved under projective equivalence;

(b) if $\mathcal{I}$ and $\mathcal{J}$ are projective hypersurfaces of $\mathbb{P}^n(\mathbb{K})$, then

$$m_P(\mathcal{I} + \mathcal{J}) = m_P(\mathcal{I}) + m_P(\mathcal{J})$$

(cf. Exercise 53);

(c) the multiplicity $m_P(\mathcal{I})$ can also be computed in affine coordinates in any affine chart containing P.

Working in an affine chart where P has affine coordinates $(0, \ldots, 0)$, the affine part of $\mathcal{I}$ is defined by an equation $f = f_m + f_{m+1} + \cdots + f_d = 0$, where every f_i is a homogeneous polynomial of degree i in $\mathbb{K}[x_1, \ldots, x_n]$, unless it is the zero polynomial, and $f_m \neq 0$. In this case every line through P has multiplicity of intersection with $\mathcal{I}$ at P greater than or equal to m and the lines r for which $I(\mathcal{I}, r, P) > m$ are precisely those whose affine part is contained in the hypersurface $C_P(\mathcal{I})$ of $\mathbb{K}^n$ of equation $f_m = 0$, which is called the *affine tangent cone* to $\mathcal{I}$ at P (indeed, it is a cone with vertex P). The projective closure $\overline{C_P(\mathcal{I})}$ of $C_P(\mathcal{I})$ is called *projective tangent cone* to $\mathcal{I}$ at P and its support coincides with the union of P and the projective lines through P such that $I(\mathcal{I}, r, P) > m$. For example, the affine tangent cone at $(0, 0)$ to the curve of $\mathbb{R}^2$ of equation $x^2 + y^2 - x^3 = 0$ consists only of the point $(0, 0)$ because $x^2 + y^2 = 0$ contains no lines.

By interpreting $f = f_m + f_{m+1} + \cdots + f_d$ as the Taylor expansion of f centred at $P = (0, \ldots, 0)$, one sees immediately that $P \in \mathcal{I}$ is a point of multiplicity 1 for $\mathcal{I}$ if and only if at least one of the first-order partial derivatives of f does not vanish at P, i.e. if and only if the point is non-singular. In this case sometimes one says that P is a *simple point*. Instead, P is a point of multiplicity $m > 1$ if and only if f and all its partial derivatives of order smaller than m vanish at P but there exists at least one partial derivative of order m that does not vanish at P.

If we do not work in an affine chart but rather we use a homogeneous equation $F = 0$ defining $\mathcal{I}$, by Euler's identity (cf. Sect. 1.7.1) P is a point of multiplicity $m > 1$ if and only if all partial derivatives of F of order $m - 1$ vanish at P and there exists at least one partial derivative of order m that does not vanish at P.

1.7.9 Real Hypersurfaces

Extending what has been done in Sect. 1.6, for any polynomial $f \in \mathbb{C}[x_1, \ldots, x_n]$ we denote by $\sigma(f)$ the polynomial obtained from f by conjugating each coefficient. The hypersurface $\sigma(\mathcal{I}) = [\sigma(f)]$ is called the *conjugate* of the affine hypersurface $\mathcal{I} = [f]$ of $\mathbb{C}^n$. For every affine hypersurface $\mathcal{I}$ of $\mathbb{C}^n$ with support $V(\mathcal{I}) \subseteq \mathbb{C}^n$ we can consider the set $V_{\mathbb{R}}(\mathcal{I})$ of the points of the support that are real; in symbols, $V_{\mathbb{R}}(\mathcal{I}) = V(\mathcal{I}) \cap \mathbb{R}^n$. The points of $V_{\mathbb{R}}(\mathcal{I})$ are called the *real points of the hypersurface*.

On the other hand, starting with any polynomial $f \in \mathbb{R}[x_1, \ldots, x_n]$, we can consider both the equivalence class $[f]_{\mathbb{R}}$ in $\mathbb{R}[x_1, \ldots, x_n]$ and the equivalence class $[f]_{\mathbb{C}}$ in $\mathbb{C}[x_1, \ldots, x_n]$. Therefore, for every affine hypersurface $\mathcal{I} = [f]_{\mathbb{R}}$ of $\mathbb{R}^n$ with support $V(\mathcal{I}) \subseteq \mathbb{R}^n$ we may consider the affine hypersurface $\mathcal{I}_\mathbb{C} = [f]_\mathbb{C}$ of $\mathbb{C}^n$, called the *complexification* of $\mathcal{I}$; then one has $V_\mathbb{R}(\mathcal{I}_\mathbb{C}) = V(\mathcal{I})$.

For example, the support of the affine curve $\mathcal{I} = [x^2 + y^2]_\mathbb{R}$ in $\mathbb{R}^2$ consists only of the point $(0, 0)$, while the support in $\mathbb{C}^2$ of the curve $\mathcal{I}_\mathbb{C} = [x^2 + y^2]_\mathbb{C}$ is the union of the lines $x + iy = 0$ and $x - iy = 0$. Similarly, the complex support of the curve of equation $x^2 + 1 = 0$ is the union of the lines $x + i = 0$ and $x - i = 0$, whereas the real support is empty.

If $g \in \mathbb{C}[x_1, \ldots, x_n]$, it may happen that $[g]_\mathbb{C}$ contains real representatives, i.e. there exist $\alpha \in \mathbb{C}^*$ and $h \in \mathbb{R}[x_1, \ldots, x_n]$ such that $g = \alpha h$. In this case one says that the affine hypersurface $\mathcal{I} = [g]_\mathbb{C}$ of $\mathbb{C}^n$ is a *real hypersurface*; if we denote $\mathcal{I}_\mathbb{R} = [h]_\mathbb{R}$, then $\mathcal{I} = (\mathcal{I}_\mathbb{R})_\mathbb{C}$. Hence there exists a natural bijection between hypersurfaces of $\mathbb{R}^n$ and real hypersurfaces of $\mathbb{C}^n$.

If $g \in \mathbb{C}[x_1, \ldots, x_n]$, the polynomial $g\sigma(g)$ has real coefficients, so the hypersurface $[g] + \sigma([g])$ is real.

If $\eta(X) = MX + N$ is an affinity of $\mathbb{C}^n$ with $M \in \mathrm{GL}(n, \mathbb{R})$ and $N \in \mathbb{R}^n$ (so that η maps points of $\mathbb{R}^n$ to points of $\mathbb{R}^n$, cf. Sect. 1.6), then η transforms every real hypersurface of $\mathbb{C}^n$ into a real hypersurface.

From the point of view of reducibility, we note that both the curve of equation $x^2 + y^2 = 0$ and that of equation $x^2 + 1 = 0$ are reducible when regarded as complex curves, while they are irreducible when regarded as real curves.

In general every polynomial $h \in \mathbb{R}[x_1, \ldots, x_n]$ has a factorization in $\mathbb{C}[x_1, \ldots, x_n]$ of the form

$$h = c\varphi_1^{m_1} \cdot \ldots \cdot \varphi_s^{m_s} \psi_1^{n_1} \cdot \ldots \cdot \psi_t^{n_t} \sigma(\psi_1)^{n_1} \cdot \ldots \cdot \sigma(\psi_t)^{n_t},$$

where c is a real number, $\varphi_1, \ldots, \varphi_s$ are distinct polynomials with real coefficients that are irreducible in $\mathbb{C}[x_1, \ldots, x_n]$, and $\psi_1, \ldots, \psi_t$ are polynomials with complex coefficients with the following properties: they are irreducible, pairwise distinct and not proportional to any polynomial with real coefficients. Therefore, the real hypersurface $\mathcal{I} = [h]_\mathbb{C}$ of $\mathbb{C}^n$ can have both real irreducible components (determined by the real factors $\varphi_1, \ldots, \varphi_s$ of h) and non-real irreducible components; more precisely, if $\mathcal{J}$ is a non-real irreducible component of $\mathcal{I}$ of multiplicity m, then $\sigma(\mathcal{J})$ is also a non-real irreducible component of $\mathcal{I}$ of the same multiplicity. Since $\psi_j\sigma(\psi_j) \in \mathbb{R}[x_1, \ldots, x_n]$ and it is irreducible over the reals, the irreducible components of the hypersurface $\mathcal{I}_\mathbb{R} = [h]_\mathbb{R}$ are given by the real irreducible factors $\varphi_1, \ldots, \varphi_s, \psi_1\sigma(\psi_1), \ldots, \psi_t\sigma(\psi_t)$. In this case, the points of $V(\psi_j) \cap V(\sigma(\psi_j))$, if any, are real points contributing to the support $V_\mathbb{R}(\mathcal{I}_\mathbb{R})$.

In particular, there are only two possibilities for an irreducible hypersurface $\mathcal{I}$ of $\mathbb{R}^n$: either $\mathcal{I}_\mathbb{C}$ is irreducible, or there exists a complex irreducible hypersurface $\mathcal{J}$ such that $\mathcal{I}_\mathbb{C} = \mathcal{J} + \sigma(\mathcal{J})$. In the latter case $\deg \mathcal{I} = \deg \mathcal{I}_\mathbb{C} = 2 \deg \mathcal{J}$, so this situation can occur only in the case of hypersurfaces of even degree. For example, if $\mathcal{I}$ is a real quadric (that is, if $\deg \mathcal{I} = 2$) that is reducible over $\mathbb{C}$, then its irreducible components are either two real hyperplanes or two complex conjugate hyperplanes.

Analogous considerations can be made in the projective case. Thus for every homogeneous polynomial $F \in \mathbb{C}[x_0, \ldots, x_n]$ the hypersurface $\sigma(\mathcal{I}) = [\sigma(F)]$ is called the *conjugate* of the projective hypersurface $\mathcal{I} = [F]$ of $\mathbb{P}^n(\mathbb{C})$.

In addition, for every homogeneous polynomial $F \in \mathbb{R}[x_0, \ldots, x_n]$ the hypersurface $\mathcal{I}_\mathbb{C} = [F]_\mathbb{C}$ of $\mathbb{P}^n(\mathbb{C})$ is called the *complexification* of the hypersurface $\mathcal{I} = [F]_\mathbb{R}$ of $\mathbb{P}^n(\mathbb{R})$; the support of $\mathcal{I}$ in $\mathbb{P}^n(\mathbb{R})$ coincides with the set of real points of the support of $\mathcal{I}_\mathbb{C}$ in $\mathbb{P}^n(\mathbb{C})$. Singular points $P \in \mathbb{P}^n(\mathbb{R})$ for $\mathcal{I}$ are singular for $\mathcal{I}_\mathbb{C}$, too: more precisely, $m_P(\mathcal{I}) = m_P(\mathcal{I}_\mathbb{C})$.

A projective hypersurface $\mathcal{I} = [F]$ of $\mathbb{P}^n(\mathbb{C})$ defined by a homogeneous polynomial $F \in \mathbb{C}[x_0, \ldots, x_n]$ is called a *real projective hypersurface* if F contains a real representative $G \in \mathbb{R}[x_0, \ldots, x_n]$. If f is a projectivity of $\mathbb{P}^n(\mathbb{C})$ that maps points of $\mathbb{P}^n(\mathbb{R})$ to points of $\mathbb{P}^n(\mathbb{R})$ (so that there exists a matrix $A \in \mathrm{GL}(n+1, \mathbb{R})$ representing it, cf. Sect. 1.6), then f transforms every real hypersurface of $\mathbb{P}^n(\mathbb{C})$ into a real hypersurface.

In view of the previous discussion, by combining the inclusion of $\mathbb{R}^n$ in $\mathbb{P}^n(\mathbb{R})$ and $\mathbb{P}^n(\mathbb{R})$ in $\mathbb{P}^n(\mathbb{C})$, for every polynomial $f \in \mathbb{R}[x_1, \ldots, x_n]$ it can be useful to relate the geometric properties (such as, for instance, support and irreducibility) of the real affine hypersurface $\mathcal{I} = [f]_\mathbb{R}$ with those of its projective closure $\overline{\mathcal{I}}$ in $\mathbb{P}^n(\mathbb{R})$ and of the complexification $(\overline{\mathcal{I}})_\mathbb{C}$ in $\mathbb{P}^n(\mathbb{C})$. For example, the polynomial $x_1^2 + x_2^2 - 1 \in \mathbb{R}[x_1, x_2]$ defines an irreducible conic $\mathcal{I}$ in $\mathbb{R}^2$ whose projective closure $\overline{\mathcal{I}}$ in $\mathbb{P}^2(\mathbb{R})$ does not intersect the line at infinity in any real point; this happens because the complexification $(\overline{\mathcal{I}})_\mathbb{C}$ in $\mathbb{P}^2(\mathbb{C})$ intersects H_0 in the hypersurface $x_1^2 + x_2^2 = 0$ whose support consists of the complex points $[1, i]$ and $[1, -i]$ which are not real.

1.8 Quadrics

1.8.1 First Notions and Projective Classification

A projective hypersurface of $\mathbb{P}^n(\mathbb{K})$ of degree 2 is called a *quadric*; a quadric of $\mathbb{P}^2(\mathbb{K})$ is called a *conic*.

If $F(x_0, \ldots, x_n) = 0$ is the equation of a quadric $\mathcal{Q}$ of $\mathbb{P}^n(\mathbb{K})$, there exists a unique symmetric matrix A of order $n + 1$ such that

$$F(x_0, \ldots, x_n) = {}^t X A X,$$

where ${}^t X = (x_0\ x_1\ \ldots\ x_n)$. In this case we say that the quadric is represented by the symmetric matrix A.

The quadric $\mathcal{Q}$ of equation ${}^t X A X = 0$ is said to be *non-degenerate* if the matrix A is invertible; the rank of A is called *rank* of the quadric (this notion is well defined since A is determined by $\mathcal{Q}$ up to a non-zero multiplicative scalar). For instance, the quadrics of rank 1 are hyperplanes counted twice.

If $P = [Y]$ is a point of $Q = [F]$ where $F(X) = {}^tXAX$, if we compute the gradient of F at Y we immediately see that $\nabla F(Y) = 2\,{}^tYA$. Therefore the tangent space $T_P(Q)$ is defined by ${}^tYAX = 0$ and $\mathrm{Sing}(Q) = \{P = [Y] \in \mathbb{P}^n(\mathbb{K}) \mid AY = 0\}$. Hence the quadric is singular if and only if $\det A = 0$, that is if and only if Q is degenerate; furthermore the singular locus of Q is a projective subspace of $\mathbb{P}^n(\mathbb{K})$.

The quadric Q is reducible (that is, the polynomial F is reducible) if and only if A has rank 1 or 2. In particular, if Q is reducible and $n \geq 2$, then it is degenerate and hence singular; on the other hand, there exist quadrics that are singular but irreducible, such as the quadric of $\mathbb{P}^3(\mathbb{C})$ of equation $x_1^2 + x_2^2 - x_3{}^2 = 0$, for instance.

If we intersect a quadric Q with a hyperplane H not contained in Q, we obtain a quadric of H which is singular at a point P if and only if the hyperplane is tangent to Q at P (cf. Exercise 58). More precisely, when the quadric Q is non-degenerate and the hyperplane H is tangent to Q at P, the quadric $Q \cap H$ has rank $n - 1$ and $\mathrm{Sing}(Q \cap H) = \{P\}$ (for a proof of this result see Exercise 167).

Recall that any projectivity of $\mathbb{P}^n(\mathbb{K})$ transforms a quadric into a quadric (cf. Sect. 1.7.5). More precisely, if Q has equation ${}^tXAX = 0$ and the projectivity g is represented by an invertible matrix N, then $g(Q)$ has equation ${}^tX\,{}^tN^{-1}AN^{-1}X = 0$. Since the matrices A and ${}^tN^{-1}AN^{-1}$ have equal ranks, g transforms Q into a quadric of the same rank.

Two quadrics Q and Q' of $\mathbb{P}^n(\mathbb{K})$ of equations ${}^tXAX = 0$ and ${}^tXA'X = 0$, respectively, are *projectively equivalent* (cf. Sect. 1.7.5) if and only if there exists $\lambda \in \mathbb{K}^*$ such that the matrices A and $\lambda A'$ are congruent, i.e. there exists $M \in \mathrm{GL}(n + 1, \mathbb{K})$ such that $\lambda A' = {}^tMAM$. If $\mathbb{K} = \mathbb{C}$ this latter fact occurs if and only if A and A' are congruent, while if $K = \mathbb{R}$ it occurs if and only if A is congruent to $\pm A'$.

As a consequence, it is possible to classify quadrics up to projective equivalence by using the classification of symmetric matrices up to congruence.

If $\mathbb{K} = \mathbb{C}$, the quadrics Q and Q' are projectively equivalent if and only if A and A' have the same rank.

If $\mathbb{K} = \mathbb{R}$, instead, A and A' are congruent if and only if they have the same signature (by signature of A we mean the pair $\mathrm{sign}(A) = (i_+(A), i_-(A))$ where $i_+(A)$ denotes the positivity index of the matrix A and $i_-(A)$ denotes its negativity index; recall that $i_+(A) + i_-(A)$ coincides with the rank of A). Since $\mathrm{sign}(-A) = (i_-(A), i_+(A))$, up to exchanging A with $-A$ ($\pm A$ define the same quadric) we shall assume $i_+(A) \geq i_-(A)$ from now on. Having agreed on this convention, the real quadrics Q and Q', represented by the symmetric matrices A and A' respectively, are projectively equivalent if and only if $\mathrm{sign}(A) = \mathrm{sign}(A')$.

From the previous considerations we obtain the following:

Theorem 1.8.1 (Projective classification of quadrics in $\mathbb{P}^n(\mathbb{K})$)

(a) *Any quadric of $\mathbb{P}^n(\mathbb{C})$ of rank r is projectively equivalent to the quadric of equation*

$$\sum_{i=0}^{r-1} x_i^2 = 0.$$

(b) *Any quadric of $\mathbb{P}^n(\mathbb{R})$ of signature $(p, r - p)$, with $p \geq r - p$, is projectively equivalent to the quadric of equation*

$$\sum_{i=0}^{p-1} x_i^2 - \sum_{i=p}^{r-1} x_i^2 = 0.$$

1.8.2 Polarity with Respect to a Quadric

Let $\mathcal{Q}$ be a quadric of $\mathbb{P}^n(\mathbb{K})$ of equation ${}^tXAX = 0$ with A a symmetric matrix.

If $P = [Y] \in \mathbb{P}^n(\mathbb{K})$, the equation ${}^tYAX = 0$ defines a subspace of $\mathbb{P}^n(\mathbb{K})$ which does not depend on the choice of the representative Y of P; therefore we will write ${}^tPAX = 0$ henceforth, without specifying which representative of P has been chosen. Similarly, if $P = [Y]$, AP denotes the point having the vector AY as a representative.

For every $P \in \mathbb{P}^n(\mathbb{K})$ the subspace of $\mathbb{P}^n(\mathbb{K})$ of equation ${}^tPAX = 0$ is denoted by $\mathrm{pol}_\mathcal{Q}(P)$ and called the *polar space of P with respect to $\mathcal{Q}$.*

If $P \in \mathcal{Q}$, then $\mathrm{pol}_\mathcal{Q}(P) = T_P(\mathcal{Q})$ (cf. Sect. 1.8.1); in particular, if P is singular for $\mathcal{Q}$, then $\mathrm{pol}_\mathcal{Q}(P) = \mathbb{P}^n(\mathbb{K})$.

If $P \in \mathbb{P}^n(\mathbb{K}) \setminus \mathrm{Sing}(\mathcal{Q})$, the equation ${}^tPAX = 0$ defines a hyperplane of $\mathbb{P}^n(\mathbb{K})$, which is called the *polar hyperplane of P with respect to $\mathcal{Q}$.* This hyperplane, that will often be denoted simply by $\mathrm{pol}(P)$, corresponds to the point of coordinates AP in the dual space $\mathbb{P}^n(\mathbb{K})^*$. Therefore we have a map

$$\mathrm{pol} \colon \mathbb{P}^n(\mathbb{K}) \setminus \mathrm{Sing}(\mathcal{Q}) \to \mathbb{P}^n(\mathbb{K})^*$$

that associates the point $AP \in \mathbb{P}^n(\mathbb{K})^*$ to the point $P \in \mathbb{P}^n(\mathbb{K}) \setminus \mathrm{Sing}(\mathcal{Q})$.

For instance, if the fundamental point $P_i = [0, \ldots, 1, \ldots, 0]$ is non-singular for $\mathcal{Q}$, the polar hyperplane of P_i has equation $a_{i,0}x_0 + \cdots + a_{i,n}x_n = 0$.

If $\mathcal{Q}$ is a non-degenerate quadric (i.e. it is non-singular), the matrix A is invertible and therefore the map $\mathrm{pol} \colon \mathbb{P}^n(\mathbb{K}) \to \mathbb{P}^n(\mathbb{K})^*$ is a projective isomorphism. In this case for every hyperplane H of $\mathbb{P}^n(\mathbb{K})$ there exists a unique point, called the *pole* of H, having H as polar hyperplane with respect to $\mathcal{Q}$. In particular the pole of the ith fundamental hyperplane $H_i = \{x_i = 0\}$ is the point $[A_{i,0}, \ldots, A_{i,n}]$ (where $A_{i,j} = (-1)^{i+j} \det(c_{i,j}(A))$, cf. Sect. 1.1).

Let us recall the main properties of polarity:

(a) (reciprocity) $P \in \mathrm{pol}(R) \iff R \in \mathrm{pol}(P)$
   (in particular for every $P \in \mathbb{P}^n(\mathbb{K})$ and for every $R \in \mathrm{Sing}(\mathcal{Q})$ one has that $P \in \mathrm{pol}(R)$ and hence, by reciprocity, $R \in \mathrm{pol}(P)$, that is, $\mathrm{pol}(P)$ contains the singular locus of the quadric);
(b) $P \in \mathrm{pol}(P) \iff P \in \mathcal{Q}$;
(c) if $P \notin \mathcal{Q}$, $\mathrm{pol}(P)$ is a hyperplane which intersects $\mathcal{Q}$ in the locus of points of intersection between $\mathcal{Q}$ and the lines passing through P and tangent to the quadric.

Property (c) immediately implies that there are at most two tangents to a non-degenerate conic passing through a given point P of the projective plane $\mathbb{P}^2(\mathbb{K})$. In particular, if there are exactly two tangents passing through P, the polar of P is the line joining the two tangency points on the conic.

Of course in the real case the polar of a point not lying on the quadric might not intersect the quadric at all (for instance the polar of $[0, 0, 1]$ with respect to the real conic $x_0^2 + x_1^2 - x_2^2 = 0$ is the line $x_2 = 0$ which does not meet the conic in any real point).

Two points P, R in $\mathbb{P}^n(\mathbb{K})$ are said to be *conjugate with respect to the quadric* Q if ${}^tPAR = 0$. In particular property (b) implies that the support of the quadric can be seen as the set of points of $\mathbb{P}^n(\mathbb{K})$ which are self-conjugate with respect to Q.

We say that $n + 1$ projectively independent points $P_0, \ldots, P_n$ of $\mathbb{P}^n(\mathbb{K})$ are the vertices of a *self-polar* $(n + 1)$-*hedron* for Q if for any $i \neq j$ the points P_i and P_j are conjugate with respect to Q (if $n = 2$ we speak of a *self-polar triangle*). In this case if $P_i \notin \mathrm{Sing}(Q)$ it turns out that $\mathrm{pol}(P_i) = L(P_0, \ldots, P_{i-1}, P_{i+1}, \ldots, P_n)$; this fact occurs for each P_i if Q is non-degenerate, which justifies the terminology "self-polar".

If $P_i = [v_i]$ for $i = 0, \ldots, n$ and A is a symmetric matrix representing the quadric with respect to the basis $\{v_0, \ldots, v_n\}$, then the points $P_0, \ldots, P_n$ are the vertices of a self-polar $(n + 1)$-hedron for Q if and only if $\{v_0, \ldots, v_n\}$ is an orthogonal basis for the inner product on $\mathbb{K}^{n+1}$ associated to the matrix A. As a consequence, representing Q by means of a diagonal matrix is equivalent to choosing a projective frame in $\mathbb{P}^n(\mathbb{K})$ whose fundamental points are the vertices of a self-polar $(n+1)$-hedron. In particular the theorem of projective classification of quadrics (cf. Theorem 1.8.1) implies that every quadric has a self-polar $(n + 1)$-hedron; if the quadric has rank r, it contains exactly r points not lying on the quadric and $n - r + 1$ points belonging to $\mathrm{Sing}(Q)$.

If Q is a non-degenerate quadric represented by an invertible matrix A, the image of Q under the projective isomorphism pol_Q is a non-degenerate quadric, called *dual quadric* and denoted by Q^*, which is associated to the symmetric matrix A^{-1} with respect to the dual coordinates. We can thus regard the support of the dual quadric as the set of the hyperplanes tangent to Q. By means of the usual identification with the bidual, one has that $Q^{**} = Q$.

If Q is non-degenerate, the set of matrices representing Q^* contains in particular the adjoint matrix $A^* = (\det A)A^{-1}$. Recall that A^* is defined also when $\det A = 0$ (since it is the matrix whose (i, j) entry is the value $A_{j,i} \in \mathbb{K}$ defined in Sect. 1.1) and A^* is non-zero if $\mathrm{rk}\, A = n$. If Q is a quadric of $\mathbb{P}^n(\mathbb{K})$ of rank n, we can therefore extend the previous notion by defining the dual quadric of Q as the quadric represented by the matrix A^*. In this case Q^* has rank 1 and its support is the image of the degenerate projective transformation pol.

1.8.3 Intersection of a Quadric with a Line

Assume that Q is a quadric of $\mathbb{P}^n(\mathbb{K})$ of equation ${}^tXAX = 0$ and r is a line. As already observed (cf. Sect. 1.7.6), if r is not contained in Q, then $Q \cap r$ consists of at most two points, possibly coinciding when the line is tangent. If r is not tangent to Q (in particular it is not contained in the quadric), we say that the line r is *secant* to Q if $Q \cap r$ consists of two distinct points and that r is *external* to Q if the support of $Q \cap r$ is empty.

More precisely, if $\mathbb{K} = \mathbb{C}$ no line can be external and, if r is not tangent, then $Q \cap r$ consists of exactly two distinct points. If Q and r are real and $Q \cap r$ contains at least one real point, then $Q \cap r$ consists of two real points, possibly coinciding; moreover, if a real line is tangent to a real quadric, then the point of tangency is real.

Explicitly, if P, R are two distinct points of r and we regard the line as the set of points $\lambda P + \mu R$ as $[\lambda, \mu]$ varies in $\mathbb{P}^1(\mathbb{K})$, then

$$\lambda P + \mu R \in Q \quad \Longleftrightarrow \quad {}^tPAP\,\lambda^2 + 2\,{}^tPAR\,\lambda\mu + {}^tRAR\,\mu^2 = 0. \qquad (1.1)$$

In particular if $P, R \in Q$ are conjugate with respect to Q, then $r \subseteq Q$. Moreover, if $P \in \mathrm{Sing}(Q)$ and $R \in Q$, then r is contained in Q; therefore any quadric with a singular point P is a cone with vertex P.

1.8.4 Projective Quadrics in $\mathbb{P}^2(\mathbb{K})$ and in $\mathbb{P}^3(\mathbb{K})$

In this section we study quadrics in projective spaces of low dimension in a more detailed way.

As already said, quadrics in the projective plane are called conics. A conic is defined by an equation of type

$$a_{0,0}x_0^2 + a_{1,1}x_1^2 + a_{2,2}x_2^2 + 2a_{0,1}x_0x_1 + 2a_{0,2}x_0x_2 + 2a_{1,2}x_1x_2 = 0,$$

that is ${}^tXAX = 0$ with

$$A = \begin{pmatrix} a_{0,0} & a_{0,1} & a_{0,2} \\ a_{0,1} & a_{1,1} & a_{1,2} \\ a_{0,2} & a_{1,2} & a_{2,2} \end{pmatrix} \quad \text{and} \quad X = \begin{pmatrix} x_0 \\ x_1 \\ x_2 \end{pmatrix}.$$

The conic is non-degenerate if the symmetric matrix A is invertible; it is called *simply degenerate* if A has rank 2 and *doubly degenerate* if A has rank 1.

In contrast to what happens for quadrics of $\mathbb{P}^n(\mathbb{C})$ when $n \geq 3$, a conic of $\mathbb{P}^2(\mathbb{C})$ is reducible if and only if it is degenerate; in this case its irreducible components are two lines (which coincide when the conic is doubly degenerate). If $\mathbb{K} = \mathbb{R}$ the previous equivalence fails: for instance the conic of $\mathbb{P}^2(\mathbb{R})$ of equation $x_0^2 + x_1^2 = 0$ is degenerate, yet the polynomial $x_0^2 + x_1^2$ is irreducible in $\mathbb{R}[x_0, x_1, x_2]$.

We can list the distinct projective canonical forms of conics of $\mathbb{P}^2(\mathbb{K})$ by specializing the result of Theorem 1.8.1 for $n = 2$:

Theorem 1.8.2 (Projective classification of conics of $\mathbb{P}^2(\mathbb{K})$)

(a) *Every conic of $\mathbb{P}^2(\mathbb{C})$ is projectively equivalent to exactly one conic in the following list:*

   (C1) $x_0^2 + x_1^2 + x_2^2 = 0$
   (C2) $x_0^2 + x_1^2 = 0$
   (C3) $x_0^2 = 0$.

(b) *Every conic of $\mathbb{P}^2(\mathbb{R})$ is projectively equivalent to exactly one conic in the following list:*

   (R1) $x_0^2 + x_1^2 + x_2^2 = 0$
   (R2) $x_0^2 + x_1^2 - x_2^2 = 0$
   (R3) $x_0^2 + x_1^2 = 0$
   (R4) $x_0^2 - x_1^2 = 0$
   (R5) $x_0^2 = 0$.

As expected, in the complex case there is one model for each value of the rank; in the real case, instead, we find: two non-degenerate models ((R1) whose support is empty and (R2) whose support contains infinitely many real points), two simply degenerate models ((R3) whose real support consists of a single point and (R4) which is the union of two distinct real lines), and one model (R5) of rank 1.

The conic C of equation ${}^t\!XAX = 0$ is singular if and only if $\det A = 0$, i.e. if and only if C is degenerate. If $\mathbb{K} = \mathbb{C}$ and C is simply degenerate, the only singular point is the point where the two irreducible components of the conic meet; if C is doubly degenerate, all points of the curve are singular.

If $\mathbb{K} = \mathbb{R}$ and C is simply degenerate, C may be reducible, with two distinct lines as irreducible components, or irreducible, in which case the complexified conic $C_{\mathbb{C}}$ has two conjugate complex lines as irreducible components (cf. Sect. 1.7.9). In either case the two lines meet at a real point which is the only singular point of C. If $\mathbb{K} = \mathbb{R}$ and C has rank 1, the support is a line, all of whose points are singular for C.

One obtains equations in canonical form by choosing in $\mathbb{P}^2(\mathbb{K})$ a projective frame whose fundamental points P_1, P_2, P_3 are the vertices of a self-polar triangle. A triangle of this type always exists (cf. Sect. 1.8.2 or Exercise 170) and its construction is very simple when $n = 2$.

To do that, if the conic is non-degenerate, we can choose a point P_1 not belonging to C and consider the line $r_1 = \mathrm{pol}(P_1)$ (in particular $P_1 \notin r_1$). Then we choose $P_2 \in r_1 \setminus C$ and denote $r_2 = \mathrm{pol}(P_2)$; by reciprocity $P_1 \in r_2$ and therefore $r_1 \neq r_2$. Finally we take $P_3 = r_1 \cap r_2$ and denote $r_3 = \mathrm{pol}(P_3)$; then $r_3 = L(P_1, P_2)$. By construction P_1, P_2, P_3 are vertices of a self-polar triangle in which each edge is the polar of the opposite vertex.

If the conic is simply degenerate, the polar of any point passes through the only singular point Q; if we choose a point P_1 not on the conic and a point $P_2 \neq Q$

on pol(P_1), then P_1, P_2, Q are vertices of a self-polar triangle. If the conic is doubly degenerate and its support is a line r (counted twice), in order to construct a self-polar triangle it suffices to take two distinct points on r and another point not in r.

The list of projective models of quadrics of the three-dimensional projective space is longer:

Theorem 1.8.3 (Projective classification of quadrics of $\mathbb{P}^3(\mathbb{K})$)

(a) *Every quadric of $\mathbb{P}^3(\mathbb{C})$ is projectively equivalent to exactly one quadric in the following list:*

(C1) $\quad x_0^2 + x_1^2 + x_2^2 + x_3{}^2 = 0$

(C2) $\quad x_0^2 + x_1^2 + x_2^2 = 0$

(C3) $\quad x_0^2 + x_1^2 = 0$

(C4) $\quad x_0^2 = 0.$

(b) *Every quadric of $\mathbb{P}^3(\mathbb{R})$ is projectively equivalent to exactly one quadric in the following list:*

(R1) $\quad x_0^2 + x_1^2 + x_2^2 + x_3{}^2 = 0$

(R2) $\quad x_0^2 + x_1^2 + x_2^2 - x_3{}^2 = 0$

(R3) $\quad x_0^2 + x_1^2 - x_2^2 - x_3{}^2 = 0$

(R4) $\quad x_0^2 + x_1^2 + x_2^2 = 0$

(R5) $\quad x_0^2 + x_1^2 - x_2^2 = 0$

(R6) $\quad x_0^2 + x_1^2 = 0$

(R7) $\quad x_0^2 - x_1^2 = 0$

(R8) $\quad x_0^2 = 0.$

Observe that in the real case there are 3 distinct projective types of non-degenerate quadrics of $\mathbb{P}^3(\mathbb{R})$ (i.e. the models (R1), (R2) and (R3) in Theorem 1.8.3), two types of rank 3 (the models (R4) and (R5)), two types of rank 2 (the models (R6) and (R7)) and only one type (R8) of rank 1.

If A is a symmetric matrix of order 4 representing Q, we know that Q is singular if and only if $\det A = 0$. More precisely:

(a) if $\mathrm{rk}\, A = 3$, Q has a unique singular point P and it is a cone with vertex P (the support may contain only the point P if $\mathbb{K} = \mathbb{R}$, as for the model (R4) of Theorem 1.8.3);

(b) if $\mathrm{rk}\, A = 2$, Q contains a line r of singular points; in the complex case Q is reducible and its support consists of two distinct planes intersecting in r; if $\mathbb{K} = \mathbb{R}$ the quadric Q can be either reducible with two distinct planes intersecting in r as irreducible components, or irreducible, in which case the irreducible components of the complexified quadric $Q_{\mathbb{C}}$ are two conjugate complex planes intersecting in the real line r (cf. Sect. 1.7.9);

(c) if $\mathrm{rk}\, A = 1$, all points of Q are singular, the quadric is reducible and its support consists of a plane counted twice.

Concerning the intersection of $\mathcal{Q}$ with a plane H of $\mathbb{P}^3(\mathbb{K})$, recall that H is *tangent* to $\mathcal{Q}$ if $H \subseteq T_P(\mathcal{Q})$ for some $P \in \mathcal{Q}$.

If not tangent, H is said:

(a) *external* to $\mathcal{Q}$ if the conic $\mathcal{Q} \cap H$ has an empty support (this can occur only if $\mathbb{K} = \mathbb{R}$);
(b) *secant* to $\mathcal{Q}$ if the conic $\mathcal{Q} \cap H$ is not empty.

In particular, if we look at the intersection of $\mathcal{Q}$ with the tangent plane $T_P(\mathcal{Q})$ at a non-singular point P, we observe that, if $T_P(\mathcal{Q}) \not\subseteq \mathcal{Q}$, then $\mathcal{Q} \cap T_P(\mathcal{Q})$ is a degenerate conic which is singular at P (cf. Sect. 1.8.1). As a consequence $\mathcal{Q} \cap T_P(\mathcal{Q})$ is a cone with vertex P (cf. Sect. 1.8.3).

Therefore for any smooth point P of $\mathcal{Q}$ we can have one of the following situations:

(a) $\mathcal{Q} \cap T_P(\mathcal{Q}) = T_P(\mathcal{Q})$; in this case $\mathcal{Q}$ is reducible, that is it consists of two planes; in particular $\mathcal{Q} \cap T_R(\mathcal{Q}) = T_R(\mathcal{Q})$ for every smooth point R of $\mathcal{Q}$;
(b) $\mathcal{Q} \cap T_P(\mathcal{Q}) = \{P\}$ (this can happen only if $\mathbb{K} = \mathbb{R}$, when the complexified conic of $\mathcal{Q} \cap T_P(\mathcal{Q})$ is the union of two complex conjugate lines meeting at P);
(c) $\mathcal{Q} \cap T_P(\mathcal{Q})$ is a degenerate conic which is the union of two (possibly coinciding) lines (for instance the intersection of the real cone of equation $x_0^2 + x_1^2 - x_2^2 = 0$ with the tangent plane at any smooth point is a line counted twice).

If $\mathcal{Q}$ is irreducible, case (a) in the latter list cannot occur; then we say that

(a) P is an *elliptic point* if $\mathcal{Q} \cap T_P(\mathcal{Q}) = \{P\}$;
(b) P is a *parabolic point* if $\mathcal{Q} \cap T_P(\mathcal{Q})$ is a line counted twice;
(c) P is a *hyperbolic point* if $\mathcal{Q} \cap T_P(\mathcal{Q})$ consists of two distinct lines.

It can be checked (cf. Exercise 172) that in a quadric $\mathcal{Q}$ of $\mathbb{P}^3(\mathbb{K})$ which is non-degenerate and non-empty either all points are hyperbolic or they are all elliptic (the latter case may occur only if $\mathbb{K} = \mathbb{R}$); in this case we say that $\mathcal{Q}$ is a *hyperbolic quadric* or an *elliptic quadric*, respectively. It can also be checked (cf. Exercise 173) that every non-singular point of an irreducible degenerate quadric of $\mathbb{P}^3(\mathbb{K})$ is parabolic: in this case, if there is at least a non-singular point, we say that $\mathcal{Q}$ is a *parabolic quadric*.

In the complex case, in which no elliptic point can exist, all non-degenerate quadrics are hyperbolic, while all quadrics of rank 3 are parabolic. If $\mathbb{K} = \mathbb{R}$ there are irreducible examples of each of the three types (cf. Exercise 174).

Since the notion of elliptic, hyperbolic or parabolic point is projective, meaning it is invariant under projectivities, one can investigate the nature of the points of an irreducible quadric by using its projective model, or a model that is projectively equivalent to it and admits a simple equation (cf. Exercise 177). For instance the quadric of $\mathbb{P}^3(\mathbb{R})$ of equation $x_0 x_3 - x_1 x_2 = 0$ is projectively equivalent to the (hyperbolic) quadric $x_0^2 + x_1^2 - x_2^2 - x_3^2 = 0$.

1.8.5 Quadrics in $\mathbb{R}^n$

An *affine quadric* Q of $\mathbb{R}^n$, that is an affine hypersurface of $\mathbb{R}^n$ of degree 2, is defined by an equation of the form

$$^tXAX + 2\,^tBX + c = 0,$$

where A is a non-zero symmetric matrix of order n, B is a vector of $\mathbb{R}^n$, $c \in \mathbb{R}$ and $^tX = (x_1 \ldots x_n)$. If we identify $\mathbb{R}^n$ with the affine chart $U_0 = \mathbb{P}^n(\mathbb{R}) \setminus \{x_0 = 0\}$ by means of the map $j_0 \colon \mathbb{R}^n \to U_0$ defined by $j_0(x_1, \ldots, x_n) = [1, x_1, \ldots, x_n]$, the projective closure $\overline{Q}$ of the affine quadric Q is represented by the block matrix

$$\overline{A} = \left(\begin{array}{c|c} c & ^tB \\ \hline B & A \end{array} \right),$$

while the quadric at infinity $Q_\infty = \overline{Q} \cap H_0$ is represented by the matrix A. We say that the affine quadric Q is *degenerate* if its projective closure is degenerate, and we call *rank* of Q the rank of $\overline{Q}$, that is the rank of the matrix $\overline{A}$.

If Q has rank n, $\overline{Q}$ has only one singular point P and it is a cone of vertex P (cf. Sect. 1.8.3). In particular we say that Q is an *affine cone* if $P \in U_0$ and that it is an *affine cylinder* if $P \in H_0$.

Two affine quadrics Q and Q' of $\mathbb{R}^n$ are affinely equivalent if there exists an affinity φ of $\mathbb{R}^n$ such that $Q = \varphi(Q')$ (cf. Sect. 1.7.5). If $\varphi(X) = MX + N$, with M an invertible matrix of order n and $N \in \mathbb{R}^n$, then φ is the restriction to the chart U_0 of the projectivity of $\mathbb{P}^n(\mathbb{R})$ represented by the block matrix

$$\overline{M}_N = \left(\begin{array}{c|c} 1 & 0 \ldots 0 \\ \hline N & M \end{array} \right)$$

(cf. Sect. 1.3.8). If φ transforms Q into Q', then $\overline{M}_N$ represents a projectivity of $\mathbb{P}^n(\mathbb{R})$ that transforms $\overline{Q}$ into $\overline{Q}'$ and M represents a projectivity of H_0 that transforms Q_∞ into Q'_∞; therefore if two affine quadrics are affinely equivalent, their projective closures and their quadrics at infinity are projectively equivalent (the converse implication holds too, as we will see later on in this section).

The map $\varphi(X) = MX + N$ is an isometry of $\mathbb{R}^n$ if and only if $M \in O(n)$; the quadrics Q and Q' of $\mathbb{R}^n$ are said *metrically equivalent* if there exists an isometry $\varphi(X) = MX + N$ such that $\varphi(Q) = Q'$.

For every $X = (x_1, \ldots, x_n) \in \mathbb{R}^n$ let $\widetilde{X} = \begin{pmatrix} 1 \\ x_1 \\ \vdots \\ x_n \end{pmatrix} = \begin{pmatrix} 1 \\ X \end{pmatrix}$, that is denote by $\widetilde{X}$

the vector of $\mathbb{R}^{n+1}$ obtained from X by appending 1 to it as first coordinate. Then the equation of Q can be written in the form $\widetilde{X}\overline{A}\widetilde{X} = 0$; moreover $\widetilde{\varphi(X)} = \begin{pmatrix} 1 \\ \varphi(X) \end{pmatrix} =$

$\begin{pmatrix} 1 \\ MX + N \end{pmatrix} = \overline{M}_N\widetilde{X}$. This is why from now on we will say that the affine quadric Q of $\mathbb{R}^n$ is represented by the symmetric matrix $\overline{A}$ of order $n + 1$ and that the affinity φ of $\mathbb{R}^n$ is represented by the matrix $\overline{M}_N$.

So the quadric $\varphi^{-1}(Q)$ is represented by the matrix

$$
{}^t\overline{M}_N\overline{A}\,\overline{M}_N = \begin{pmatrix} \begin{array}{c|c} 1 & {}^tN \\ \hline 0 & \\ \vdots & {}^tM \\ 0 & \end{array} \end{pmatrix} \begin{pmatrix} c & {}^tB \\ \hline B & A \end{pmatrix} \begin{pmatrix} 1 & 0 \ldots 0 \\ \hline N & M \end{pmatrix} =
$$

$$
= \begin{pmatrix} {}^tNAN + 2\,{}^tBN + c & {}^t({}^tM(AN + B)) \\ \hline {}^tM(AN + B) & {}^tMAM \end{pmatrix}.
$$

By using an affinity $\overline{A}$ and A transform to congruent matrices, so their signatures remain unchanged and in particular their ranks are preserved. Now, the matrix of a quadric of $\mathbb{R}^n$ is determined up to a scalar $\alpha \neq 0$, hence either positive or negative. Since for every symmetric real matrix S we have $\text{sign}(\alpha S) = (i_+(S), i_-(S))$ if $\alpha > 0$, while $\text{sign}(\alpha S) = (i_-(S), i_+(S))$ if $\alpha < 0$, up to multiplying the quadric's equation by -1 (i.e. modifying $\overline{A}$ to $-\overline{A}$ and A to $-A$) we may assume $i_+(A) \geq i_-(A)$ and $i_+(\overline{A}) \geq i_-(\overline{A})$. Henceforth we shall always adopt this convention, with the effect that the above signatures now depend only on the quadric but not on the particular equation, and the pair $(\text{sign}(\overline{A}), \text{sign}(A))$ becomes an affine invariant of the quadric.

The quadric Q of equation $F(X) = {}^tXAX + 2\,{}^tBX + c = 0$ is said to have *centre* at $C \in \mathbb{R}^n$ if $F(C + X) = F(C - X)$ for every $X \in \mathbb{R}^n$, equivalently if the polynomials $F(C + X)$ and $F(C - X)$ coincide. Clearly the origin O of $\mathbb{R}^n$ is a centre for Q if and only if $B = O$. If C is a centre for Q, the translation $\tau(X) = X + C$ satisfies $\tau(O) = C$ and O becomes a centre for the translated quadric $Q' = \tau^{-1}(Q)$. Therefore for Q' all degree one coefficients are zero, i.e. $AC + B = 0$, making C a solution of the system $AX = -B$. Conversely, if the system $AX = -B$ admits a solution C, the point C must be a centre for the quadric.

It can be checked that, if φ is an affinity, C is a centre for Q if and only if $\varphi^{-1}(C)$ is a centre for the quadric $\varphi^{-1}(Q)$. Therefore the property of having a centre is affine (that is any quadric which is affinely equivalent to a quadric with centre has itself a centre).

Quadrics without centre are called *paraboloids*.

It turns out that $\text{rk}\,A \leq \text{rk}\,\overline{A} \leq \text{rk}\,A + 2$ and that the system $AX = -B$ admits solutions (that is, the quadric has a centre) if and only if $\text{rk}\,\overline{A} \leq \text{rk}\,A + 1$, while Q is a paraboloid if and only if $\text{rk}\,\overline{A} = \text{rk}\,A + 2$. In particular Q is a non-degenerate paraboloid if and only if $\det\overline{A} \neq 0$ and $\det A = 0$.

Recall (cf. Sect. 1.8.2) that, if Q is non-degenerate, the pole of the hyperplane at infinity H_0 with respect to $\overline{Q}$ is the point $C = [\overline{A}_{0,0}, \overline{A}_{0,1} \ldots, \overline{A}_{0,n}] \in \mathbb{P}^n(\mathbb{R})$.

If Q is a non-degenerate paraboloid, then $\overline{A}_{0,0} = \det A = 0$; hence the pole of the hyperplane at infinity H_0 with respect to $\overline{Q}$ is an improper point, H_0 is tangent to $\overline{Q}$ and therefore (cf. Sect. 1.8.1) $\overline{Q} \cap H_0$ is degenerate (of rank $n - 1$). The converse implication holds too: if Q is non-degenerate and $\overline{Q} \cap H_0$ is degenerate, then rk $\overline{A} = n+1$ and $\det A = 0$, so that rk $A = n-1 = $ rk $\overline{A} - 2$ and hence Q is a non-degenerate paraboloid.

If Q is a non-degenerate quadric with centre, then rk $\overline{A} = n + 1$ and rk $A = n$, hence the system $AX = -B$ has only one solution (that is, there exists only one centre) which, by using Cramer's rule, is given by the point $C = \left(\dfrac{\overline{A}_{0,1}}{\overline{A}_{0,0}}, \ldots, \dfrac{\overline{A}_{0,n}}{\overline{A}_{0,0}} \right) \in \mathbb{R}^n$. If we regard C as a point of $\mathbb{P}^n(\mathbb{R})$, we have that $C = [\overline{A}_{0,0}, \overline{A}_{0,1} \ldots, \overline{A}_{0,n}]$. Therefore the centre C coincides with the pole of the hyperplane at infinity H_0 with respect to $\overline{Q}$ and this is a proper point (because $\overline{A}_{0,0} = \det A \neq 0$). Since $\overline{Q} \cap H_0$ is non-degenerate, we may distinguish cases according to the nature of the intersection between $\overline{Q}$ and the hyperplane at infinity:

(a) Q is called *hyperboloid* if non-degenerate, with centre and $\overline{Q} \cap H_0$ is a non-degenerate quadric with non-empty support;

(b) Q is called *ellipsoid* if non-degenerate, with centre and $\overline{Q} \cap H_0$ is a non-degenerate quadric with empty support.

By taking into account the property of being a quadric with centre or not, and using classical results of linear algebra, it is possible to determine distinct canonical forms up to affine equivalence and up to metric equivalence for quadrics in $\mathbb{R}^n$, including degenerate ones. If we restrict ourselves to using isometries,

(a) the Spectral theorem ensures that, after applying a linear isometry $X \mapsto MX$ of $\mathbb{R}^n$ (with $M \in O(n)$), we can assume that the symmetric matrix A is diagonal with $a_{i,i} = 0$ for each $i = $ rk $A + 1, \ldots, n$;

(b) if Q has a centre, by translating the origin to a centre we may assume $B = 0$;

(c) if Q has no centre, after applying an isometry (and multiplying the equation by a non-zero constant, if necessary) we may assume $B = (0, \ldots, 0, -1)$ and $c = 0$.

Therefore we obtain:

Theorem 1.8.4 (Metric classification of quadrics of $\mathbb{R}^n$) *Assume that Q is a quadric of $\mathbb{R}^n$ of equation $\widetilde{X}\overline{A}\widetilde{X} = 0$ with $p = i_+(A) \geq i_-(A)$ and $i_+(\overline{A}) \geq i_-(\overline{A})$. Let $r = $ rk A and $\overline{r} = $ rk $\overline{A}$. Then:*

(a) *there exist $\lambda_1, \ldots, \lambda_r \in \mathbb{R}$ with $0 < \lambda_1 \leq \cdots \leq \lambda_p$ and $0 < \lambda_{p+1} \leq \cdots \leq \lambda_r$ such that Q is metrically equivalent to the quadric defined by one of the following equations:*

(m1) $x_1^2 + \lambda_2 x_2^2 + \cdots + \lambda_p x_p^2 - \lambda_{p+1} x_{p+1}^2 - \cdots - \lambda_r x_r^2 = 0$
in this case we have $\lambda_1 = 1$, $\overline{r} = r$ and $\mathrm{sign}(\overline{A}) = \mathrm{sign}(A)$;

(m2) $\lambda_1 x_1^2 + \cdots + \lambda_p x_p^2 - \lambda_{p+1} x_{p+1}^2 - \cdots - \lambda_r x_r^2 - 1 = 0$

  in this case we have $\mathrm{sign}(\overline{A}) = (p, i_-(A) + 1)$ *(in particular* $\overline{r} = r + 1$*);*

(m3) $\lambda_1 x_1^2 + \cdots + \lambda_p x_p^2 - \lambda_{p+1} x_{p+1}^2 - \cdots - \lambda_r x_r^2 + 1 = 0$

  in this case we have $\mathrm{sign}(\overline{A}) = (p + 1, i_-(A))$ *(in particular* $\overline{r} = r + 1$*);*

(m4) $\lambda_1 x_1^2 + \cdots + \lambda_p x_p^2 - \lambda_{p+1} x_{p+1}^2 - \cdots - \lambda_r x_r^2 - 2x_n = 0$

  in this case we have $\overline{r} = r + 2$*.*

(b) In cases (m2), (m3), (m4) *the numbers* $\lambda_1, \ldots, \lambda_r$ *are uniquely determined by* $\mathcal{Q}$*.*

For quadrics of type (m1) the numbers $\lambda_2, \ldots, \lambda_r$ are not uniquely determined: for instance the quadric of equation $x_1^2 - 2x_2^2 = 0$ is metrically equivalent to the quadric of equation $x_1^2 - \frac{1}{2}x_2^2 = 0$ (it suffices to consider the isometry that exchanges x_1 with x_2 and multiply the equation thus obtained by a suitable scalar).

We arrive at a metric canonical form of type (m1), (m2) or (m3) if $\mathcal{Q}$ has a centre, of type (m4) if the quadric is a paraboloid.

If we do not restrict ourselves to using isometries in the process of simplifying the equation of the quadric, by applying a suitable affinity after the metric reduction described above, it is possible to "normalize" the coefficients, so that the coefficients of the equation belong to the set $\{1, -1, 0\}$. Therefore we have:

Theorem 1.8.5 (Affine classification of quadrics of $\mathbb{R}^n$) *Assume that* $\mathcal{Q}$ *is a quadric of* $\mathbb{R}^n$ *of equation* ${}^t\widetilde{X}\overline{A}\widetilde{X} = 0$ *with* $p = i_+(A) \geq i_-(A)$ *and* $i_+(\overline{A}) \geq i_-(\overline{A})$*. Set* $r = \mathrm{rk}\,A$ *and* $\overline{r} = \mathrm{rk}\,\overline{A}$*. Then* $\mathcal{Q}$ *is affinely equivalent to exactly one among the following quadrics:*

(a1) $x_1^2 + \cdots + x_p^2 - x_{p+1}^2 - \cdots - x_r^2 = 0,$

  in this case we have $\overline{r} = r$ *and* $\mathrm{sign}(\overline{A}) = \mathrm{sign}(A)$*;*

(a2) $x_1^2 + \cdots + x_p^2 - x_{p+1}^2 - \cdots - x_r^2 - 1 = 0$

  in this case we have $\mathrm{sign}(\overline{A}) = (p, i_-(A) + 1)$ *(in particular* $\overline{r} = r + 1$*);*

(a3) $x_1^2 + \cdots + x_p^2 - x_{p+1}^2 - \cdots - x_r^2 + 1 = 0$

  in this case we have $\mathrm{sign}(\overline{A}) = (p + 1, i_-(A))$ *(in particular* $\overline{r} = r + 1$*);*

(a4) $x_1^2 + \cdots + x_p^2 - x_{p+1}^2 - \cdots - x_r^2 - 2x_n = 0$

  in this case we have $\overline{r} = r + 2$*.*

We arrive at one of the first three models in the case of quadrics with centre, at the fourth model in the case of a paraboloid.

For each model in Theorem 1.8.5 the pair $(\mathrm{sign}(\overline{A}), \mathrm{sign}(A))$ is different and therefore it allows us to distinguish the different affine types, so that two affine quadrics $\mathcal{Q}$ and $\mathcal{Q}'$ of equations ${}^t\widetilde{X}\overline{A}\widetilde{X} = 0$ and ${}^t\widetilde{X}\overline{A}'\widetilde{X} = 0$, respectively, are affinely equivalent if and only if $(\mathrm{sign}(\overline{A}), \mathrm{sign}(A)) = (\mathrm{sign}(\overline{A}'), \mathrm{sign}(A'))$. We express this property by saying that the pair $(\mathrm{sign}(\overline{A}), \mathrm{sign}(A))$ is a complete system of affine invariants.

So, if $\overline{\mathcal{Q}}$ is projectively equivalent to $\overline{\mathcal{Q}'}$ and $\mathcal{Q}_\infty$ is projectively equivalent to $\mathcal{Q}'_\infty$, then $\mathrm{sign}(\overline{A}) = \mathrm{sign}(\overline{A}')$ and $\mathrm{sign}(A) = \mathrm{sign}(A')$ and hence $\mathcal{Q}$ and $\mathcal{Q}'$ are affinely equivalent. In other words, if the projective closures and the quadrics at infinity of two affine quadrics Q and Q' are projectively equivalent, then Q and Q' are affinely equivalent.

In order to decide which is the affine canonical form of a quadric it is therefore sufficient to compute the signatures of A and $\overline{A}$. As a matter of fact, Theorem 1.8.5 implies that it is not always necessary to compute both signatures. One can start by computing the signature of A (hence $\operatorname{rk} A$) and the rank of $\overline{A}$: if $\operatorname{rk} \overline{A} = \operatorname{rk} A$ or if $\operatorname{rk} \overline{A} = \operatorname{rk} A + 2$ the equation of the affine model is determined. If instead $\operatorname{rk} \overline{A} = \operatorname{rk} A + 1$, it is necessary to compute the signature of $\overline{A}$ too.

Recall that the positivity index of a symmetric matrix coincides with the number of positive eigenvalues of the matrix and hence with the number of positive roots of its characteristic polynomial. Since all the roots of the characteristic polynomial of a real symmetric matrix are real, the computation of the signatures of A and of $\overline{A}$ from their characteristic polynomials is immediate as a consequence of the following result:

Theorem 1.8.6 (Descartes' criterion) *If the roots of a polynomial $p(x)$ with real coefficients are all real, then $p(x)$ has as many positive roots, counted with multiplicity, as the number of sign changes in the sequence of non-zero coefficients of the polynomial.*

1.8.6 Diametral Hyperplanes, Axes, Vertices

In this section we shall only consider non-degenerate quadrics of $\mathbb{R}^n$.

In Sect. 1.8.5 we saw that, if Q is a non-degenerate quadric with centre, the centre coincides with the pole of the hyperplane at infinity H_0 with respect to $\overline{Q}$. So when working with non-degenerate quadrics we can extend the notion of centre by calling *centre of a non-degenerate quadric* Q the pole of the hyperplane at infinity H_0 with respect to $\overline{Q}$. In particular the notion of centre can be used also for non-centred quadrics. By using the terminology thus introduced, it turns out that:

(a) the pole of H_0 is proper, that is it lies in $\mathbb{R}^n$, if and only if Q is a non-degenerate quadric with centre (in this case Q has as centre precisely the pole of H_0);
(b) the pole of H_0 is improper, that is it belongs to the hyperplane at infinity H_0, if and only if Q is a non-degenerate paraboloid (and $\overline{Q}$ is tangent to H_0 at the centre of Q).

An affine hyperplane is called a *diametral hyperplane* of a non-degenerate quadric Q (*diametral plane* if $n = 3$, *diameter* if $n = 2$) if its projective closure is the polar hyperplane of an improper point. If R is the centre of the quadric, by reciprocity a hyperplane is diametral if and only if its projective closure passes through the centre R. For instance if Q is a non-degenerate paraboloid, all diametral hyperplanes are parallel to one line, because their projective closures pass through the improper centre of Q.

If Q is a non-degenerate quadric of $\mathbb{R}^n$ and if the diametral hyperplane $\text{pol}_{\overline{Q}}(P) \cap \mathbb{R}^n$ relative to a point $P \in H_0$ is orthogonal to the direction l_P determined by the point at infinity P, then $\text{pol}_{\overline{Q}}(P)$ is called *principal hyperplane* of Q.

Moreover, we call *axis* of a non-degenerate quadric Q every line of $\mathbb{R}^n$ which is the intersection of principal hyperplanes (if $n = 2$ the notions of principal hyperplane and axis coincide). We call *vertex* of Q every point where Q intersects an axis (no confusion is possible between this notion of vertex and that of vertex of a cone, because cones are degenerate quadrics).

It can be checked that, for every vertex V of a non-degenerate quadric Q of $\mathbb{R}^n$, the tangent hyperplane to Q at V is orthogonal to the axis through V (for a proof see Exercise 191).

The hyperplane $\text{pol}_{\overline{Q}}(P) \cap \mathbb{R}^n = \{{}^t P \overline{A} \widetilde{X} = 0\}$ relative to $P = [0, p_1, \ldots, p_n]$ is principal if and only if there exists $\lambda \neq 0$ such that $(p_1 \ldots p_n) A = \lambda (p_1 \ldots p_n)$, that is if and only if the vector $(p_1, \ldots, p_n)$ is an eigenvector of A relative to a non-zero eigenvalue. The eigenvectors $v_0 \in \mathbb{R}^n$ of A relative to the eigenvalue 0, if any, do not correspond to any principal hyperplane, because in this case the polar hyperplane determined by the point $P = [(0, v_0)]$ is the hyperplane at infinity H_0, which is not the projective closure of any affine hyperplane.

If the quadric Q is non-degenerate and with centre, it can be proved (cf. Exercise 190) that there exist n pairwise orthogonal axes of the quadric that meet in the centre of Q. Therefore the quadric can be represented in metric canonical form by choosing in $\mathbb{R}^n$ a Cartesian coordinate system having the centre of Q as origin and n pairwise orthogonal axes of the quadric as coordinate axes.

If Q is a non-degenerate paraboloid, it can be proved (cf. Exercise 190) that Q has only one axis (which is the intersection of $n - 1$ pairwise orthogonal principal hyperplanes of Q) and only one vertex. Thus we arrive at the metric canonical form of the equation of Q by choosing in $\mathbb{R}^n$ a Cartesian coordinate system having as origin the vertex V of Q and as coordinate axes pairwise orthogonal lines passing through V in such a way that the nth coordinate hyperplane is the tangent hyperplane to Q at V and the remaining $n - 1$ coordinate hyperplanes are pairwise orthogonal principal hyperplanes of Q.

Given a point $P = (a_1, \ldots, a_n) \in \mathbb{R}^n$ and a real number $\eta \geq 0$, the set of points of $\mathbb{R}^n$ whose distance from P is equal to η is a quadric of equation

$$(x_1 - a_1)^2 + \cdots + (x_n - a_n)^2 = \eta. \tag{1.2}$$

More generally, we call *sphere* any quadric of $\mathbb{R}^n$ defined by Eq. (1.2) with $\eta \in \mathbb{R}$. This sphere is an ellipsoid with non-empty support if $\eta > 0$, an ellipsoid with empty support if $\eta < 0$; if $\eta = 0$ it is a degenerate quadric with support consisting of only one point.

A sphere is a quadric with exactly one centre; all hyperplanes through the centre are principal, all lines through the centre are axes and all points in the support are vertices (cf. Exercise 190).

The image of a sphere under an isometry is a sphere; in particular the property of being a sphere is invariant under isometric coordinate changes.

1.8.7   Conics of $\mathbb{R}^2$

In this section, using the notation introduced so far, we examine in detail conics of $\mathbb{R}^2$ both in the affine and in the metric setting, as we will do later for quadrics of $\mathbb{R}^3$. Since $\mathbb{R}^2 \subset \mathbb{C}^2 \subset \mathbb{P}^2(\mathbb{C})$, it is useful to consider a conic C of $\mathbb{R}^2$ not only as embedded in its projective closure in $\mathbb{P}^2(\mathbb{R})$, but also as embedded in the complexified conic $C_\mathbb{C}$ in $\mathbb{C}^2$ and in the projective closure of $C_\mathbb{C}$ in $\mathbb{P}^2(\mathbb{C})$ (cf. Sect. 1.7.9).

A doubly degenerate conic has as support a line consisting entirely of singular points.

A simply degenerate conic C can be either reducible (with two distinct lines as irreducible components) or irreducible (in this case the complexified conic $C_\mathbb{C}$ has two complex conjugate lines as irreducible components, cf. Sect. 1.7.9). In both cases the two lines meet at a real point, which is the only singular point of C. If the intersection point of these two lines is proper, the conic is an affine cone; if it is improper, the conic is an affine cylinder (cf. Sect. 1.8.5). More precisely, the conic is called a *real cone* or a *real cylinder* if the support is a union of lines, an *imaginary cone* if the support reduces to the unique singular point, an *imaginary cylinder* if the support is empty.

A non-degenerate conic C is called:

(a) *parabola* if it has no centre (in this case H_0 is tangent to $\overline{C}$),
(b) *hyperbola* if it has a centre and $\overline{C}$ intersects H_0 in two distinct points,
(c) *ellipse* if it has a centre and the line H_0 is external to $\overline{C}$ (more precisely, it is called a *real ellipse* if the support of C is not empty, an *imaginary ellipse* otherwise).

In $\mathbb{R}^2$ the classification obtained in Theorem 1.8.5 reads:

Theorem 1.8.7 (Affine classification of conics of $\mathbb{R}^2$) *Every conic of $\mathbb{R}^2_{(x,y)}$ is affinely equivalent to exactly one among the following conics:*
(a1) $x^2 + y^2 - 1 = 0$ (*real ellipse*);
(a2) $x^2 + y^2 + 1 = 0$ (*imaginary ellipse*);
(a3) $x^2 - y^2 + 1 = 0$ (*hyperbola*);
(a4) $x^2 - 2y = 0$ (*parabola*);
(a5) $x^2 - y^2 = 0$ (*pair of distinct incident real lines*);
(a6) $x^2 + y^2 = 0$ (*pair of complex conjugate incident lines*);
(a7) $x^2 - 1 = 0$ (*pair of distinct parallel real lines*);
(a8) $x^2 + 1 = 0$ (*pair of distinct parallel complex conjugate lines*);
(a9) $x^2 = 0$ (*pair of coincident real lines*).

Table 1.1 Affine invariants for conics of $\mathbb{R}^2$

Model	sign(A)	sign($\overline{A}$)
(a1)	(2, 0)	(2, 1)
(a2)	(2, 0)	(3, 0)
(a3)	(1, 1)	(2, 1)
(a4)	(1, 0)	(2, 1)
(a5)	(1, 1)	(1, 1)
(a6)	(2, 0)	(2, 0)
(a7)	(1, 0)	(1, 1)
(a8)	(1, 0)	(2, 0)
(a9)	(1, 0)	(1, 0)

Observe that:

(a) for canonical forms (a1), (a2) or (a3) the origin is the centre of the conic C and the axes x and y are diameters with points at infinity which are conjugate with respect to $\overline{C}$;

(b) for the parabola C of equation (a4) the origin O is a point of the conic, the axis x is the tangent line at O and the axis y is the line passing through O and through the (improper) centre of C;

(c) the model (a5) is a real cone, (a6) is an imaginary cone, (a7) is a real cylinder, (a8) is an imaginary cylinder; the doubly degenerate model (a9) is a cone for which all points of the support are vertices.

An affine line r is called an *asymptote* of a non-degenerate conic C if its projective closure $\overline{r}$ is tangent to $\overline{C}$ at one of its improper points (that is, if $\overline{r}$ is the polar of an improper point of the conic with respect to $\overline{C}$). A real ellipse has no asymptotes because it has no improper points, a parabola has no asymptotes because the tangent at the unique point at infinity is the line at infinity, while a hyperbola has two asymptotes. The asymptotes of a hyperbola, being the affine parts of the polars of the improper points, are diameters and they meet in the centre of the hyperbola.

By using the convention previously adopted (cf. Sect. 1.8.5) of choosing equations such that $i_+(A) \geq i_-(A)$ and $i_+(\overline{A}) \geq i_-(\overline{A})$ and of denoting $\widetilde{X} = \begin{pmatrix} 1 \\ x \\ y \end{pmatrix}$, we know that two conics of $\mathbb{R}^2$ of equations $\widetilde{X}\overline{A}\widetilde{X} = 0$ and $\widetilde{X}\overline{A'}\widetilde{X} = 0$ are affinely equivalent if and only if sign($\overline{A}$) = sign($\overline{A'}$) and sign(A) = sign(A').

Table 1.1 shows the values of these invariants for the affine models listed in Theorem 1.8.7.

We call *circle* every sphere of $\mathbb{R}^2$, that is every conic of equation $(x - x_0)^2 + (y - y_0)^2 = \eta$. This circle is a real ellipse if $\eta > 0$ and an imaginary ellipse if $\eta < 0$; if $\eta = 0$ it is a degenerate conic whose support consists of only one point. The image of a circle under an isometry is a circle; in particular being a circle is an invariant property under isometric coordinate changes.

The complexification of a circle has $I_1 = [0, 1, i]$ and $I_2 = [0, 1, -i]$ as improper points; they are called *cyclic points* of the Euclidean plane. One can easily check that any conic of $\mathbb{R}^2$ whose complexification has the cyclic points as improper points is a circle.

The cyclic points are closely related to metric properties. For instance by associating to every line through a point $P \in \mathbb{R}^2$ the orthogonal line through P we define an involution in the pencil $\mathcal{F}_P$ of lines through P. In an affine coordinate system where $P = (0, 0)$ we thus associate the line of equation $bx + ay = 0$ to the line of equation $ax - by = 0$. Since the projective closures of these two lines meet the line at infinity $x_0 = 0$ at the points $[0, b, a]$ and $[0, a, -b]$, respectively, the involution given by the orthogonality relation in $\mathcal{F}_P$ induces the so-called *absolute involution* $[0, b, a] \mapsto [0, a, -b]$ on the improper line. This latter map has no real fixed points, while the corresponding involution of $(H_0)_{\mathbb{C}}$ (which is projectively isomorphic to $\mathbb{P}^1(\mathbb{C})$) has the cyclic points I_1 and I_2 as fixed points.

The name "absolute involution" originates from the relationship between this involution and the polarity with respect to the quadric of H_0 of equation $x_1^2 + x_2^2 = 0$, classically called the *absolute conic*. Namely this polarity associates to the point $[b, a]$ the hyperplane of H_0 of equation $bx_1 + ax_2 = 0$, whose support consists of the point $[a, -b]$ only.

A line of $\mathbb{C}^2$ having a cyclic point as improper point is called an *isotropic line*; exactly two isotropic lines pass through every point of $\mathbb{C}^2$ (and hence through every point of $\mathbb{R}^2$).

A point F of $\mathbb{R}^2$ is called a *focus* of a non-degenerate conic $\mathcal{C}$ if the projective closures of the isotropic lines passing through F are tangent to the projective closure of the complexification of $\mathcal{C}$. A non-degenerate conic has one or two foci (cf. Exercise 194).

A line of $\mathbb{R}^2$ is called a *directrix* of $\mathcal{C}$ if its projective closure is the polar line of a focus with respect to $\overline{\mathcal{C}}$.

In the case of $\mathbb{R}^2$, restricting our attention only to non-degenerate conics, Theorem 1.8.4 yields:

Theorem 1.8.8 (Metric classification of conics of $\mathbb{R}^2$) *Every non-degenerate conic of $\mathbb{R}^2_{(x,y)}$ is metrically equivalent to exactly one among the following conics:*

(m1) $\dfrac{x^2}{a^2} + \dfrac{y^2}{b^2} - 1 = 0$ *with* $a \geq b > 0$ (*real ellipse*);

(m2) $\dfrac{x^2}{a^2} + \dfrac{y^2}{b^2} + 1 = 0$ *with* $a \geq b > 0$ (*imaginary ellipse*);

(m3) $\dfrac{x^2}{a^2} - \dfrac{y^2}{b^2} + 1 = 0$ *with* $a > 0, b > 0$ (*hyperbola*);

(m4) $x^2 - 2cy = 0$ *with* $c > 0$ (*parabola*).

Using the canonical equations of non-degenerate conics listed in Theorem 1.8.8, we can compute explicit formulas for vertices, foci, axes and directrices of the non-degenerate metric models listed above (cf. Exercise 196).

Starting from the canonical metric equations it is also easy to check (cf. Exercise 197) that non-degenerate conics can be characterized in terms of metric conditions, as we already did for the circle:

(a) a parabola is the locus of points of the plane such that the distance to a given point (the focus) equals the distance to a given line not passing through that point (the directrix);
(b) an ellipse is the locus of points of the plane such that the sum of the distances to two given distinct points of the plane (the foci) is constant;
(c) a hyperbola is the locus of points of the plane such that the absolute value of the difference of the distances to two given distinct points of the plane (the foci) is constant.

For another metric characterization of non-degenerate conics see Exercise 198 and the Remark following it.

1.8.8   Quadrics of $\mathbb{R}^3$

Recall that a quadric Q of $\mathbb{R}^3$ is non-degenerate if and only if a matrix $\overline{A}$ representing it is invertible. Moreover, Q is reducible if and only if $\overline{A}$ has rank 1 or 2.

If $\operatorname{rk}\overline{A} = 4$, Q can be either a paraboloid (if it has no centre, i.e. if H_0 is tangent to $\overline{Q}$), or a hyperboloid (if Q has a centre and H_0 is secant to $\overline{Q}$) or an ellipsoid (if Q has a centre and H_0 is external to $\overline{Q}$). We can further refine this terminology according to the nature of points of $\overline{Q}$ (recall that points of non-degenerate quadrics can be only elliptic or hyperbolic, cf. Sect. 1.8.4). Hence we speak of

(a) *elliptic hyperboloid*, or *two-sheeted hyperboloid*, if all its points are elliptic;
(b) *hyperbolic hyperboloid*, or *one-sheeted hyperboloid*, if all its points are hyperbolic;
(c) *elliptic paraboloid* if all its points are elliptic;
(d) *hyperbolic paraboloid*, or *saddle*, if all its points are hyperbolic.

Note that all points of an ellipsoid Q are necessarily elliptic: if there were a hyperbolic point P for Q, then $Q \cap T_P(Q)$ would be the union of two lines so that $\overline{Q} \cap H_0$ would contain real points (cf. Exercise 175 too).

We also say that an ellipsoid is *real* (resp. *imaginary*) if its support is non-empty (resp. empty).

If $\operatorname{rk}\overline{A} = 3$, we know that the projective closure $\overline{Q}$ is a cone having only one singular point S. Therefore the quadric Q is either an affine cone (if S is a proper point, which happens if and only if $\det A \neq 0$) or an affine cylinder (if S is an improper point, which happens if and only if $\det A = 0$). More precisely, we speak of a *real cone* and of a *real cylinder* if the support is a union of lines, of an *imaginary cone* if the support only contains the unique singular point, of an *imaginary cylinder* if the support is empty. For a real cylinder, the terminology is sometimes further refined according to the intersection of $\overline{Q}$ with the plane H_0. Thus if Q is a real cylinder, we say that $\mathcal{Q}$ is a *hyperbolic cylinder* if $\overline{\mathcal{Q}} \cap H_0$ is a pair of distinct real lines, an *elliptic cylinder* if $\overline{\mathcal{Q}} \cap H_0$ is a pair of complex conjugate lines, a *parabolic cylinder* if $\overline{\mathcal{Q}} \cap H_0$ is a line counted twice. Note how these names have nothing to do with the nature of points on the quadric since all non-singular points on a cylinder are parabolic, the quadric being degenerate and irreducible (cf. Exercise 173). The terminology actually refers to the fact that non-degenerate plane sections are hyperbolas, ellipses and parabolas, respectively.

In $\mathbb{R}^3$ the classification of Theorem 1.8.5 translates into:

Theorem 1.8.9 (Affine classification of quadrics of $\mathbb{R}^3$) *Every quadric of $\mathbb{R}^3_{(x,y,z)}$ is affinely equivalent to exactly one among the following quadrics:*

(a1)	$x^2 + y^2 + z^2 - 1 = 0$	*(real ellipsoid)*;
(a2)	$x^2 + y^2 + z^2 + 1 = 0$	*(imaginary ellipsoid)*;
(a3)	$x^2 + y^2 - z^2 - 1 = 0$	*(hyperbolic hyperboloid)*;
(a4)	$x^2 + y^2 - z^2 + 1 = 0$	*(elliptic hyperboloid)*;
(a5)	$x^2 + y^2 - 2z = 0$	*(elliptic paraboloid)*;
(a6)	$x^2 - y^2 - 2z = 0$	*(hyperbolic paraboloid)*;
(a7)	$x^2 + y^2 + z^2 = 0$	*(imaginary cone)*;
(a8)	$x^2 + y^2 - z^2 = 0$	*(real cone)*;
(a9)	$x^2 + y^2 + 1 = 0$	*(imaginary cylinder)*;
(a10)	$x^2 + y^2 - 1 = 0$	*(elliptic cylinder)*;
(a11)	$x^2 - y^2 + 1 = 0$	*(hyperbolic cylinder)*;
(a12)	$x^2 - 2z = 0$	*(parabolic cylinder)*;
(a13)	$x^2 + y^2 = 0$	*(pair of incident complex planes)*;
(a14)	$x^2 - y^2 = 0$	*(pair of distinct incident real planes)*;
(a15)	$x^2 + 1 = 0$	*(pair of parallel complex planes)*;
(a16)	$x^2 - 1 = 0$	*(pair of distinct parallel real planes)*;
(a17)	$x^2 = 0$	*(pair of coincident real planes)*.

As already observed (cf. Sect. 1.8.5), by the convention of choosing equations such that $i_+(A) \geq i_-(A)$ and $i_+(\overline{A}) \geq i_-(\overline{A})$, two quadrics of $\mathbb{R}^3$ of equations $\widetilde{X}A X = 0$ and $\widetilde{X}A'X = 0$ are affinely equivalent if and only if $\text{sign}(\overline{A}) = \text{sign}(\overline{A'})$ and $\text{sign}(A) = \text{sign}(A')$. Useful, although partial, information can be obtained from two additional affine invariants, such as the vanishing of the determinant of A and the sign of the determinant of $\overline{A}$ (since $\overline{A}$ is a matrix of even order).

Table 1.2 shows the values of these invariants for the affine models listed in Theorem 1.8.9. The pictures of some of these models appear in Fig. 1.1.

When considering only non-degenerate quadrics, in order to distinguish the affine type of $\mathcal{Q}$ it suffices to compute the determinants of A and $\overline{A}$ and the signature of A. In fact,

(a) $\mathcal{Q}$ is a real ellipsoid $\iff \det A \neq 0$, $\det \overline{A} < 0$ and A is positive definite;

(b) $\mathcal{Q}$ is an imaginary ellipsoid $\iff \det A \neq 0$, $\det \overline{A} > 0$ and A is positive definite;

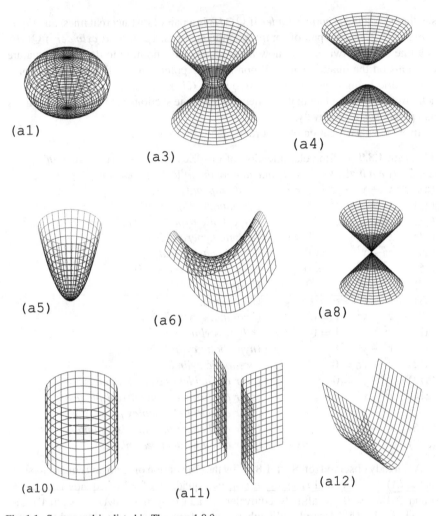

Fig. 1.1 Some quadrics listed in Theorem 1.8.9

(c) $\mathcal{Q}$ is a hyperbolic hyperboloid $\Longleftrightarrow$ $\det A \neq 0$, $\det \overline{A} > 0$ and A is indefinite;

(d) $\mathcal{Q}$ is an elliptic hyperboloid $\Longleftrightarrow$ $\det A \neq 0$, $\det \overline{A} < 0$ and A is indefinite;

(e) $\mathcal{Q}$ is an elliptic paraboloid $\Longleftrightarrow$ $\det A = 0$ and $\det \overline{A} < 0$;

(f) $\mathcal{Q}$ is a hyperbolic paraboloid $\Longleftrightarrow$ $\det A = 0$ and $\det \overline{A} > 0$.

In the case of $\mathbb{R}^3$ the classification obtained in Theorem 1.8.4 yields metric canonical equations for quadrics of $\mathbb{R}^3$.

Table 1.2 Affine invariants for quadrics of $\mathbb{R}^3$

Model	$\det(A)$	$\det(\overline{A})$	$\text{sign}(A)$	$\text{sign}(\overline{A})$
(a1)	$\neq 0$	<0	$(3, 0)$	$(3, 1)$
(a2)	$\neq 0$	>0	$(3, 0)$	$(4, 0)$
(a3)	$\neq 0$	>0	$(2, 1)$	$(2, 2)$
(a4)	$\neq 0$	<0	$(2, 1)$	$(3, 1)$
(a5)	0	<0	$(2, 0)$	$(3, 1)$
(a6)	0	>0	$(1, 1)$	$(2, 2)$
(a7)	$\neq 0$	0	$(3, 0)$	$(3, 0)$
(a8)	$\neq 0$	0	$(2, 1)$	$(2, 1)$
(a9)	0	0	$(2, 0)$	$(3, 0)$
(a10)	0	0	$(2, 0)$	$(2, 1)$
(a11)	0	0	$(1, 1)$	$(2, 1)$
(a12)	0	0	$(1, 0)$	$(2, 1)$
(a13)	0	0	$(2, 0)$	$(2, 0)$
(a14)	0	0	$(1, 1)$	$(1, 1)$
(a15)	0	0	$(1, 0)$	$(2, 0)$
(a16)	0	0	$(1, 0)$	$(1, 1)$
(a17)	0	0	$(1, 0)$	$(1, 0)$

1.9 Plane Algebraic Curves

The definitions and considerations about affine and projective hypersurfaces presented in Sect. 1.7 apply in particular to hypersurfaces of the plane (i.e. to the case $n = 2$), namely to *plane algebraic curves*. These will sometimes be called simply "curves" from now on. In particular we have the notions of projective closure of an affine curve, affine and projective equivalence of curves, multiplicity of a point of a curve, tangent line, singular curve, and so on.

Therefore, in this section we just highlight some results about the local study of a curve at one of its points and the intersection of two curves.

1.9.1 Local Study of a Plane Algebraic Curve

Let $\mathcal{C}$ be a projective curve of degree d of $\mathbb{P}^2(\mathbb{K})$, where $\mathbb{K} = \mathbb{C}$ or $\mathbb{K} = \mathbb{R}$, and let P be a point of $\mathbb{P}^2(\mathbb{K})$.

If $F(x_0, x_1, x_2) = 0$ is a homogeneous equation of $\mathcal{C}$, then P is a simple point of the curve if and only if not every first-order partial derivative of F vanishes at P (cf. Sect. 1.7.8). In this case the tangent line to $\mathcal{C}$ at P has equation

$$F_{x_0}(P)x_0 + F_{x_1}(P)x_1 + F_{x_2}(P)x_2 = 0.$$

A singular point $P \in C$ (i.e. a point such that $m_P(C) > 1$) is called a *double point* if $m_P(C) = 2$, a *triple point* if $m_P(C) = 3$, an *m-tuple* point if $m_P(C) = m$. The point P is an m-tuple point of C if and only if all $(m-1)$th derivatives of F vanish at P and there is at least one mth derivative of F that does not vanish at P.

All points of a multiple irreducible component turn out to be singular. If C is a reduced curve and $C_1, \ldots, C_m$ are its irreducible components, then $\mathrm{Sing}(C)$ is a finite set and, more precisely (cf. Exercises 54 and 56),

$$\mathrm{Sing}(C) = \bigcup_{j=1}^{m} \mathrm{Sing}(C_j) \cup \bigcup_{i \neq j}(C_i \cap C_j).$$

A line r is called a *principal tangent* to C at P if $I(C, r, P) > m_P(C)$.

At a simple point, the notions of tangent line and of principal tangent coincide. If P is a singular point of multiplicity m, any line through P is a tangent, while the principal tangents are those contained in the tangent cone to C at P (cf. Sect. 1.7.8). The set of principal tangents to a curve $C + D$ at P consists of the principal tangents to C at P together with the principal tangents to D at P (cf. Exercise 53).

In order to study the curve locally at a point P, it can be useful to choose a chart in which $P = (0, 0)$. In such a chart, if f is a polynomial that defines the affine part of C and if we write f as a sum of homogeneous terms, the multiplicity at the origin coincides with the degree m of the non-zero homogeneous part f_m of f of minimal degree. So the affine parts of the principal tangents at P are defined by the linear factors of f_m. In particular, there are at most $m = m_P(C)$ principal tangents at P. More precisely, if $\mathbb{K} = \mathbb{C}$ by Theorem 1.7.1 there are m principal tangents (counted with multiplicity); if $\mathbb{K} = \mathbb{R}$, instead, by Theorem 1.7.2 there are at most m principal tangents, but there may be none. For example, the point $[1, 0, 0]$ is a double point of the real curve of equation $x_0 x_1^2 + x_0 x_2^2 - x_1^3 = 0$ at which there are no principal tangents.

If $\mathbb{K} = \mathbb{C}$, the singular point P is said to be *ordinary* if there are precisely $m_P(C)$ distinct principal tangents at P. A double point is called a *node* if it is ordinary, a *cusp* if it is non-ordinary; more precisely, one speaks of an *ordinary cusp* if the only principal tangent at P has multiplicity of intersection exactly 3 with the curve at the point.

If $\mathbb{K} = \mathbb{R}$, a singular point P is said to be *ordinary* if it is ordinary for the complexified curve $C_{\mathbb{C}}$; in that case, since $m_P(C_{\mathbb{C}}) = m_P(C)$, the complexified curve $C_{\mathbb{C}}$ has $m_P(C)$ distinct principal tangent lines at P but it is possible that the number of the real ones is strictly smaller than $m_P(C)$.

If, in some affine chart, the affine part of C has equation $f(x, y) = 0$ and $P = (a, b)$ is non-singular, the line of equation

$$f_x(P)(x - a) + f_y(P)(y - b) = 0$$

is the affine part of the (principal) tangent to C at P and is called the *affine tangent* at P.

Finally, generalizing the notion already introduced for non-degenerate conics, an affine line r is called an *asymptote* for an affine curve $\mathcal{D}$ if its projective closure $\bar{r}$ is a principal tangent to $\bar{\mathcal{D}}$ at one of its improper points.

1.9.2 The Resultant of Two Polynomials

Let D be a unique factorization domain (we are interested in the cases $D = \mathbb{K}$ and $D = \mathbb{K}[x_1, \ldots, x_n]$). Let us consider two polynomials $f, g \in D[x]$ of positive degrees m and p, respectively,

$$f(x) = a_0 + a_1 x + \cdots + a_m x^m \qquad a_m \neq 0$$

$$g(x) = b_0 + b_1 x + \cdots + b_p x^p \qquad b_p \neq 0.$$

We call *Sylvester matrix* of f and g the square matrix of order $m + p$

$$S(f, g) = \begin{pmatrix} a_0 & a_1 & \ldots\ldots & a_m & 0 & \ldots & 0 \\ 0 & a_0 & a_1 & \ldots\ldots & a_m & 0 & \ldots \\ & & \ddots & \ddots & & & \ddots \\ 0 & \ldots & 0 & a_0 & a_1 & \ldots\ldots & a_m \\ b_0 & b_1 & \ldots\ldots & b_p & 0 & \ldots & 0 \\ 0 & b_0 & b_1 & \ldots\ldots & b_p & 0 & \ldots \\ & & \ddots & \ddots & & & \ddots \\ 0 & \ldots & 0 & b_0 & b_1 & \ldots\ldots & b_p \end{pmatrix}$$

whose first p lines are determined by the coefficients of f, while the following ones by the coefficients of g. The determinant of $S(f, g)$ is called the *resultant* of f and g; in symbols, $\mathrm{Ris}(f, g) = \det S(f, g)$.

The main property of the resultant is that f and g have a common factor of positive degree in $D[x]$ if and only if $\mathrm{Ris}(f, g) = 0$. If $D = \mathbb{C}$, then $\mathrm{Ris}(f, g) = 0$ if and only if f and g have a common root.

If $f, g \in \mathbb{K}[x_1, \ldots, x_n]$, we may choose an indeterminate, say x_n, and regard f and g as polynomials in x_n whose coefficients are polynomials in $x_1, \ldots, x_{n-1}$, that is, $f, g \in D[x_n]$ with $D = \mathbb{K}[x_1, \ldots, x_{n-1}]$. Denote by $S(f, g, x_n)$ the Sylvester matrix and by $\mathrm{Ris}(f, g, x_n)$ the resultant of f and g regarded as elements of $D[x_n]$, so that $\mathrm{Ris}(f, g, x_n) \in \mathbb{K}[x_1, \ldots, x_{n-1}]$. In addition, we denote by $\deg_{x_n} f$ the degree of f with respect to the variable x_n.

Among the many important properties of the resultant, we recall only those that are essential for our purposes.

Specialization property: Setting $X = (x_1, \ldots, x_{n-1})$, let

$$f(x_1, \ldots, x_n) = a_0(X) + a_1(X)x_n + \cdots + a_m(X)x_n^m \qquad a_m(X) \neq 0$$

$$g(x_1, \ldots, x_n) = b_0(X) + b_1(X)x_n + \cdots + b_p(X)x_n^p \qquad b_p(X) \neq 0$$

and let $R(x_1, \ldots, x_{n-1}) = \mathrm{Ris}(f, g, x_n)$.

For any $c = (c_1, \ldots, c_{n-1}) \in \mathbb{K}^{n-1}$, one can alternatively:

(a) evaluate the polynomials f and g at $X = c$ first, thus obtaining the polynomials in one variable $f_c(x_n) = f(c, x_n)$ and $g_c(x_n) = g(c, x_n)$, and then compute the resultant of $f_c(x_n), g_c(x_n) \in \mathbb{K}[x_n]$;

(b) compute the resultant $R(x_1, \ldots, x_{n-1})$ first, and then evaluate it at $X = c$.

If for instance $a_m(c) \neq 0$ and $b_p(c) \neq 0$, then $\deg_{x_n} f(x_1, \ldots, x_n) = \deg f_c(x_n)$ and $\deg_{x_n} g(x_1, \ldots, x_n) = \deg g_c(x_n)$, so that the Sylvester matrix of $f_c(x_n)$ and $g_c(x_n)$ coincides with the matrix $S(f, g, x_n)$ evaluated at $X = c$. As a consequence,

$$\mathrm{Ris}(f(c, x_n), g(c, x_n)) = R(c),$$

which is to say that specializing at $X = c$ commutes with computing the resultant.

Homogeneity property: Let $F(x_1, \ldots, x_n)$ and $G(x_1, \ldots, x_n)$ be homogeneous polynomials of degrees m and p, respectively. Set $X = (x_1, \ldots, x_{n-1})$ and let

$$F(x_1, \ldots, x_n) = A_0(X) + A_1(X)x_n + \cdots + A_m x_n^m$$

$$G(x_1, \ldots, x_n) = B_0(X) + B_1(X)x_n + \cdots + B_p x_n^p,$$

where every non-zero A_i (resp. B_i) is a homogeneous polynomial of degree $m - i$ (resp. $p - i$) in $\mathbb{K}[x_1, \ldots, x_{n-1}]$.

If $A_m \neq 0$ and $B_p \neq 0$, the polynomial $\mathrm{Ris}(F, G, x_n)$ is either homogeneous of degree mp in the variables $x_1, \ldots, x_{n-1}$ or zero.

1.9.3   Intersection of Two Curves

One can generalize the notion of multiplicity of intersection of two curves at a point, which was previously defined when one of the curves is a line. There are several ways of defining this concept; below we shall recall the one that suits better the effective computation of the multiplicity.

Let $\mathcal{C}$ and $\mathcal{D}$ be two projective curves of $\mathbb{P}^2(\mathbb{K})$ of degrees m and d, respectively, and without common components, and let $P \in \mathcal{C} \cap \mathcal{D}$. Choose a homogeneous coordinate system $[x_0, x_1, x_2]$ such that $[0, 0, 1] \notin \mathcal{C} \cup \mathcal{D}$ and such that P is the only point of $\mathcal{C} \cap \mathcal{D}$ on the line joining P and $[0, 0, 1]$.

If $C = [F]$ and $\mathcal{D} = [G]$, denote by $\mathrm{Ris}(F, G, x_2)$ the resultant of F and G with respect to the variable x_2. Since $F(0, 0, 1) \neq 0$ and $G(0, 0, 1) \neq 0$, one has $\deg_{x_2} F = \deg F$, $\deg_{x_2} G = \deg G$; in addition, since C and $\mathcal{D}$ have no common component, the polynomial $R(x_0, x_1) = \mathrm{Ris}(F, G, x_2)$ is homogeneous of degree md (cf. Sect. 1.9.2, homogeneity property).

If $P = [c_0, c_1, c_2]$, the multiplicity of $[c_0, c_1]$ as root of the polynomial $\mathrm{Ris}(F, G, x_2)$ is called the *multiplicity of intersection of the curves C and $\mathcal{D}$ at P* and we denote it by $I(C, \mathcal{D}, P)$.

One can check either directly or indirectly, in any case with some effort, that this definition does not depend on the homogeneous coordinate system. One can also verify that, in case one of the curves is a line, the multiplicity of intersection just defined coincides with the one defined earlier.

In case C and $\mathcal{D}$ are curves in $\mathbb{P}^2(\mathbb{R})$ and P is a point of $\mathbb{P}^2(\mathbb{R})$, it is apparent that $I(C, \mathcal{D}, P) = I(C_\mathbb{C}, \mathcal{D}_\mathbb{C}, P)$.

Let f and g be the polynomials obtained by dehomogenizing F and G with respect to x_0, which define the affine parts of the curves C and $\mathcal{D}$ in the chart U_0. Since $[0, 0, 1] \notin C \cup \mathcal{D}$, specializing to $x_0 = 1$ does not lower the degrees of F and G and, by the specialization property of the resultant, one has $\mathrm{Ris}(F, G, x_2)(1, x_1) = \mathrm{Ris}(f, g, x_2)(x_1)$; hence the multiplicity of intersection of C and $\mathcal{D}$ at P can also be computed from the equations of their affine parts.

Theorem 1.9.1 *Let C and $\mathcal{D}$ be projective curves of $\mathbb{P}^2(\mathbb{K})$ without common components. Then*

$$I(C, \mathcal{D}, P) \geq m_P(C) \cdot m_P(\mathcal{D})$$

for every point $P \in \mathbb{P}^2(\mathbb{K})$.

See Exercise 110 for a proof of Theorem 1.9.1.

We say that the curves C and $\mathcal{D}$ are *tangent* at P if $I(C, \mathcal{D}, P) \geq 2$. Note that the notion of curves tangent at a point can be characterized in terms of the tangent lines to the curves at the point, namely $I(C, \mathcal{D}, P) \geq 2$ if and only if the curves have at least one common tangent at P. Indeed, if the two curves are not singular at P, then $I(C, \mathcal{D}, P) \geq 2$ if and only if the tangent lines to the two curves at P coincide (cf. Exercise 109). In case at least one of the curves, say C, is singular at P, then all lines through P are tangent to C and at least one of them is tangent also to $\mathcal{D}$; in fact in this case $I(C, \mathcal{D}, P) \geq m_P(C) \cdot m_P(\mathcal{D}) \geq 2$.

Concerning the global intersection of two projective curves, we have the following fundamental result:

Theorem 1.9.2 (Bézout's Theorem) *Let C and $\mathcal{D}$ be projective curves of $\mathbb{P}^2(\mathbb{C})$ of degrees m and d, respectively. If C and $\mathcal{D}$ have no common component, they have exactly md common points, counted with the corresponding multiplicity of intersection (in particular $C \cap \mathcal{D} \neq \emptyset$).*

In the real case Bézout's Theorem does not hold, and it may even happen that two projective curves of $\mathbb{P}^2(\mathbb{R})$ do not intersect at all (just think of the two distinct real conics of equations $x_0^2 + x_1^2 - x_2^2 = 0$ and $x_0^2 + x_1^2 - 2x_2^2 = 0$). However one obtains the following "weak" form of Bézout's Theorem as an immediate consequence of Theorem 1.9.2:

Theorem 1.9.3 (real Bézout's Theorem) *Let C and D be projective curves of $\mathbb{P}^2(\mathbb{R})$ of degrees m and d, respectively. If C and D have more than md common points, they necessarily have a common irreducible component.*

1.9.4 Inflection Points

A point P of a projective curve C is called an *inflection point*, or a *flex*, if it is simple and $I(C, \tau, P) \geq 3$, where τ denotes the (principal) tangent to C at P. In case $I(C, \tau, P) = 3$ one has an *ordinary inflection point*. For example, all points of a line are inflection points, irreducible conics have no inflection points and all inflection points of an irreducible cubic are ordinary.

Inflection points on a projective curve C of equation $F(X) = 0$ are characterized (and hence can be determined) as the non-singular points P such that $H_F(P) = 0$, where $H_F(X) = \det(F_{x_i x_j}(X))$. If F has degree $d \geq 2$, the polynomial $H_F(X)$ is either homogeneous of degree $3(d-2)$ or identically zero (for example if $F = x_1 x_2(x_1 - x_2)$, then $H_F(X) = 0$).

The symmetric matrix $\mathrm{Hess}_F(X) = (F_{x_i x_j}(X))_{0 \leq i,j \leq 2}$ is called the *Hessian matrix* of F and, if $H_F(X) = \det(\mathrm{Hess}_F(X))$ is non-zero, the curve of equation $H_F(X) = 0$ is called the *Hessian curve* of $F(X) = 0$ and is denoted by $H(C)$.

So, when the field is $\mathbb{C}$, the number of inflection points of a projective curve of degree $d \geq 3$ is either infinite or, by Bézout's Theorem, at most $3d(d - 2)$; if non-singular, such a curve must anyhow have at least one inflection point.

We recall that, if C is a curve of equation $F(X) = 0$ and $g(X) = MX$ is a projectivity of $\mathbb{P}^2(\mathbb{K})$, then the curve $D = g^{-1}(C)$ has equation $G(X) = F(MX) = 0$. One checks that the Hessian matrices of F and G are related by the equality $\mathrm{Hess}_G(X) = {}^t\!M\,\mathrm{Hess}_F(MX)\,M$, so that $H_G(X) = (\det M)^2 H_F(MX)$. Therefore, the polynomial $H_G(X)$ is non-zero if and only if $H_F(X)$ is non-zero and, if so, $H(D) = g^{-1}(H(C))$.

1.9.5 Linear Systems, Pencils

For every $d \geq 1$ we denote by Λ_d the set of projective curves of degree d of $\mathbb{P}^2(\mathbb{K})$. From the definition of curve it follows that $\Lambda_d = \mathbb{P}(\mathbb{K}[X]_d)$, where $\mathbb{K}[X]_d = \mathbb{K}[x_0, x_1, x_2]_d$ denotes the vector space consisting of the zero polynomial and all homogeneous polynomials of degree d in x_0, x_1, x_2. A basis of this vector space consists of the monomials $x_0^i x_1^j x_2^{d-i-j}$ of degree d, so that $\mathbb{K}[x_0, x_1, x_2]_d$ has dimension $\dfrac{(d+1)(d+2)}{2}$. As a consequence, Λ_d is a projective space of dimension

$$N = \frac{(d+1)(d+2)}{2} - 1 = \frac{d(d+3)}{2}.$$

Hence lines form a projective space of dimension 2, conics a space of dimension 5, cubics of dimension 9 and so on. In the homogeneous coordinate system induced by the basis of monomials, the coordinates of a curve $C \in \Lambda_d$ are the coefficients of one of the polynomials defining it.

A projective subspace of Λ_d is called a *linear system of curves of degree d*; a linear system of curves of dimension 1 is called a *pencil*. A linear system of curves of dimension r can be represented parametrically starting with $r + 1$ independent points in it, $[F_0], \ldots, [F_r]$, so every curve of the linear system has equation $\sum_{i=0}^{r} \lambda_i F_i(X) = 0$. Alternatively, it can be represented in Cartesian form as an intersection of hyperplanes.

Points belonging to all curves of a linear system are called *base points* of the linear system. For pencils of curves the base points are determined by intersecting any two distinct curves $F_0(X) = 0$ and $F_1(X) = 0$ of the pencil, since every curve of the pencil has an equation of the form $\lambda F_0(X) + \mu F_1(X) = 0$ with $[\lambda, \mu] \in \mathbb{P}^1(\mathbb{K})$. If Q is not a base point of the pencil, there exists a unique curve of the pencil passing through Q, because by imposing that a curve pass through Q one obtains an equation in λ, μ that has a unique solution $[\lambda_0, \mu_0] \in \mathbb{P}^1(\mathbb{K})$.

If two curves of degree d are tangent to the same line l at a point P, then all curves of the pencil spanned by the two curves are tangent to l at P.

Finally we recall that, if f is a projectivity of $\mathbb{P}^2(\mathbb{K})$ and C is a curve of degree d, then the curve $f(C)$ has also degree d, so that f induces a map $\Lambda_d \to \Lambda_d$ that is in fact a projectivity. In particular, f transforms every linear system of curves into a linear system of the same dimension; for instance, f transforms the pencil generated by two curves C_1 and C_2 of degree d into the pencil generated by the curves $f(C_1)$ and $f(C_2)$.

1.9.6 Linear Conditions

The equation of a hyperplane of Λ_d is a homogeneous linear equation in the coordinates of Λ_d, namely in the coefficients of the generic curve of Λ_d. Such an equation is called a *linear condition* on the curves of Λ_d.

For instance, imposing that a curve pass through a given point P is a linear condition. Since N hyperplanes of $\mathbb{P}^N(\mathbb{K})$ have always at least one point in common, there is always at least one curve of degree d that satisfies up to $N = \frac{d(d+3)}{2}$ linear conditions.

In particular, since $\dim \Lambda_2 = 5$, there is at least one conic passing through 5 given points in $\mathbb{P}^2(\mathbb{K})$. And if no 4 of the 5 points are collinear, there is exactly one such conic. In fact, two distinct conics in $\mathbb{P}^2(\mathbb{K})$ either have no common irreducible component or they do. In the former case Bézout's Theorem (cf. Theorem 1.9.2 or Theorem 1.9.3) tells that they meet at 4 distinct points at most. In the latter case

(both conics are degenerate) they share a line plus a point not on the line. This means
two conics can meet at 5 points, no 4 of which are collinear, only if they coincide.
Hence the system of conics passing through 4 non-collinear points is a pencil. If the
5 points satisfy the stronger condition that no 3 are collinear, then the only conic
through them is necessarily non-degenerate.

Imposing that a point P have multiplicity at least $r \geq 1$ for a curve of Λ_d is
equivalent to imposing that all derivatives of order $r - 1$ of a polynomial defining
the curve vanish at P. This corresponds to imposing $\dfrac{r(r+1)}{2}$ linear conditions.

Of course, by imposing k linear conditions one obtains in general a linear system
of dimension greater than or equal to $N - k$; the dimension is exactly $N - k$ only if the
linear conditions that one imposes are independent. For example, the linear system
of projective conics that pass through 4 points lying on the same line r consists of
the reducible conics that are the union of the line r and of another line; this set is in
one-to-one correspondence with the space of lines of the projective plane and is a
linear system of dimension 2 (cf. Exercise 96).

Other examples of linear conditions arise from tangency conditions at given points
to the curves of Λ_d. For example, if r is a line of equation ${}^tRX = 0$ and $P \in r$, then
a curve $\mathcal{C}$ of Λ_d of equation $F(X) = 0$ is tangent to r at P if and only if the vectors
$R = (r_0, r_1, r_2)$ and $(F_{x_0}(P), F_{x_1}(P), F_{x_2}(P))$ are proportional, i.e. if and only if the
matrix

$$M = \begin{pmatrix} r_0 & r_1 & r_2 \\ F_{x_0}(P) & F_{x_1}(P) & F_{x_2}(P) \end{pmatrix}$$

has rank 1. This is equivalent to the vanishing of the determinants of two 2×2
submatrices of M, and therefore to two independent linear conditions (cf. Exercise 95). Note that if M has rank 1, and hence there exists $k \in \mathbb{K}$ such that
$(F_{x_0}(P), F_{x_1}(P), F_{x_2}(P)) = k(r_0, r_1, r_2)$, then automatically $\mathcal{C}$ passes through the
point $P = [p_0, p_1, p_2]$ since

$$(\deg F)F(P) = p_0 F_{x_0}(P) + p_1 F_{x_1}(P) + p_2 F_{x_2}(P) = k(p_0 r_0 + p_1 r_1 + p_2 r_2) = 0.$$

A case in which the two linear conditions are satisfied is when the second row of
M is zero, namely when $\mathcal{C}$ is singular at P; this agrees with the fact that every line
passing through a singular point of a curve is tangent to the curve at that point.

1.9.7 Pencils of Conics

A *pencil of conics* is a linear system of conics of dimension 1. If $\mathcal{C}_1 = [F_1]$ and $\mathcal{C}_2 = [F_2]$ are two distinct conics of the pencil, then the conics of the pencil have equation
$\lambda F_1 + \mu F_2 = 0$ as $[\lambda, \mu]$ varies in $\mathbb{P}^1(\mathbb{K})$. If $F_1(X) = {}^tXA_1X$ and $F_2(X) = {}^tXA_2X$ with
A_1, A_2 symmetric 3×3 matrices, the generic conic $\mathcal{C}_{\lambda,\mu}$ of the pencil has equation
${}^tX(\lambda A_1 + \mu A_2)X = 0$.

The conic $C_{\lambda,\mu}$ is degenerate if and only if $D(\lambda, \mu) = \det(\lambda A_1 + \mu A_2) = 0$. Therefore, if the homogeneous polynomial of degree three $D(\lambda, \mu)$ is non-zero, the pencil contains three degenerate conics at most. If $D(\lambda, \mu)$ is the zero polynomial, instead, all conics of the pencil are degenerate; this happens, for instance, when the conics of the pencil consist of a given line and an arbitrary other line passing through a given point.

However there exist pencils all of whose conics are degenerate without having a common component: just think of the pencil spanned by $x_0^2 = 0$ and by $x_1^2 = 0$.

Recall that the base points of a pencil, i.e. the points common to all conics of the pencil, can be determined by intersecting any two distinct conics C_1, C_2 of the pencil. Since every pencil contains at least one degenerate conic, we may assume that one of the conics spanning the pencil is degenerate, and therefore is the sum of two lines; this simplifies the computation of the intersection points of C_1 and C_2.

For instance, let us consider the pencil of conics passing through 4 non-collinear points (cf. Sect. 1.9.6): if the 4 points are in general position, then the set of base points of the pencil consists exactly of the 4 points; if 3 of the 4 points belong to a line r, then the set of base points is the union of the line r and of the fourth point not lying on r (in particular, it is an infinite set).

As remarked above, it is possible that the locus of base points of a pencil is an infinite set (when all conics of the pencil have a common component) and in this case all conics of the pencil are degenerate. If the pencil contains at least one non-degenerate conic, then the set of base points is finite.

We now describe the types of pencils of conics containing at least one non-degenerate conic: let $\mathcal{F}$ be such a pencil and let C_1 and C_2 be two distinct conics of $\mathcal{F}$, with C_1 degenerate.

If $\mathbb{K} = \mathbb{C}$ then the conics C_1 and C_2 intersect at 4, possibly non-distinct, points. We will denote the quadruple of intersection points by writing any point P as many times as the value of $I(C_1, C_2, P)$. Hence, according to the number of distinct base points and their multiplicities, we obtain the following types of pencils containing at least one non-degenerate conic:

(a) $C_1 \cap C_2 = \{A, B, C, D\}$: the points A, B, C, D turn out to be in general position; the pencil contains 3 degenerate conics, namely $L(A, B) + L(C, D)$, $L(A, C) + L(B, D)$ and $L(A, D) + L(B, C)$;

(b) $C_1 \cap C_2 = \{A, A, B, C\}$: in this case the points A, B, C are not collinear; C_1 and C_2 (hence all conics of the pencil) are tangent at A to a line t_A that contains neither B nor C; the pencil contains 2 degenerate conics, namely $L(A, B) + L(A, C)$ and $t_A + L(B, C)$ (see also Exercise 121);

(c) $C_1 \cap C_2 = \{A, A, B, B\}$: all conics of the pencil are tangent at A to a line t_A not passing through B and at B to a line t_B not passing through A; the pencil contains 2 degenerate conics, namely $t_A + t_B$ and $2L(A, B)$ (see also Exercise 125);

(d) $C_1 \cap C_2 = \{A, A, A, B\}$: in this case C_1 and C_2 intersect at A with multiplicity of intersection 3 and are tangent at A to a line t_A not passing through B; the pencil contains only one degenerate conic, namely $t_A + L(A, B)$;

(e) $C_1 \cap C_2 = \{A, A, A, A\}$: this case occurs when C_1 and C_2 intersect at A with multiplicity of intersection 4 and are tangent to a line t_A; the pencil contains only one degenerate (actually, doubly degenerate) conic: $2t_A$.

If $\mathbb{K} = \mathbb{R}$, it is possible that the intersection points of the conics C_1 and C_2, even if counted with multiplicity, are fewer than 4, and possibly none (just think, for example, of the conics of equations $x^2 + y^2 = 1$ and $x^2 - 4 = 0$). Another case in which there are no base points is when C_1 is a conic whose support consists of just one point P (this happens when C_1 is irreducible and $(C_1)_{\mathbb{C}}$ has two complex conjugate lines meeting at P as irreducible components) and C_2 does not pass through P.

On the other hand the complexified curves $(C_1)_{\mathbb{C}}$ and $(C_2)_{\mathbb{C}}$ do intersect at four, possibly not distinct, points, that turn out to be the complex base points of $\mathcal{F}$. Since $(C_1)_{\mathbb{C}}$ and $(C_2)_{\mathbb{C}}$ are real curves, if $A = [a_0, a_1, a_2]$ is a complex base point of the pencil, then the point $\sigma(A) = [\overline{a_0}, \overline{a_1}, \overline{a_2}]$ is a base point as well. With regard to this, note that the line $L(A, \sigma(A))$ admits an equation with real coefficients, i.e. it is a real line. If instead B and C are any two points of $\mathbb{P}^2(\mathbb{C})$ (so in general $L(B, C)$ does not admit a real equation), then an equation of the line $L(\sigma(B), \sigma(C))$ is obtained from an equation of $L(B, C)$ by conjugating its coefficients; as a consequence, the conic $L(B, C) + L(\sigma(B), \sigma(C))$ is a real conic.

Therefore, in the real case, besides the five types of pencils containing at least one non-degenerate conic listed above, there exist also the following additional types:

(f) $(C_1)_{\mathbb{C}} \cap (C_2)_{\mathbb{C}} = \{A, \sigma(A), B, C\}$ where $A \neq \sigma(A)$ and B, C are real points: the pencil $\mathcal{F}$ has two distinct real base points and contains only one degenerate conic, namely $L(A, \sigma(A)) + L(B, C)$;

(g) $(C_1)_{\mathbb{C}} \cap (C_2)_{\mathbb{C}} = \{A, \sigma(A), B, B\}$ where $A \neq \sigma(A)$ and B is a real point: the pencil has only one real base point, C_1 and C_2 (hence all conics of the pencil) are tangent at B to a real line t_B; the pencil contains two degenerate conics, namely $L(A, B) + L(\sigma(A), B)$ and $L(A, \sigma(A)) + t_B$;

(h) $(C_1)_{\mathbb{C}} \cap (C_2)_{\mathbb{C}} = \{A, \sigma(A), B, \sigma(B)\}$ where $A \neq \sigma(A)$ and $B \neq \sigma(B)$: the pencil has no real base point and contains three degenerate conics; one of them, namely $L(A, \sigma(A)) + L(B, \sigma(B))$, is a pair of lines; the other two, namely $L(A, \sigma(B)) + L(\sigma(A), B)$ and $L(A, B) + L(\sigma(A), \sigma(B))$, are conics having only one point each as support;

(i) $(C_1)_{\mathbb{C}} \cap (C_2)_{\mathbb{C}} = \{A, \sigma(A), A, \sigma(A)\}$ where $A \neq \sigma(A)$: all complexifications of the conics of the pencil are tangent at A and $\sigma(A)$, respectively, to two complex conjugate lines t_A and $t_{\sigma(A)}$; the pencil contains two degenerate conics, namely $t_A + t_{\sigma(A)}$ and $2L(A, \sigma(A))$.

Chapter 2
Exercises on Projective Spaces

Projective spaces and projective subspaces. Projective frames and homogeneous coordinates. Projective transformations and projectivities. Linear systems of hyperplanes and duality. The projective line. Cross-ratio.

Abstract Solved problems on projective spaces and subspaces, projective transformations, the projective line and the cross-ratio, linear systems of hyperplanes and duality.

Notation: Throughout the whole chapter, the symbol $\mathbb{K}$ denotes a subfield of $\mathbb{C}$.

Exercise 1. Show that the points

$$\left[\frac{1}{2}, 1, 1\right], \quad \left[1, \frac{1}{3}, \frac{4}{3}\right], \quad [2, -1, 2]$$

of the real projective plane are collinear, and find an equation of the line containing them.

Solution. The points $\left[\frac{1}{2}, 1, 1\right], \left[1, \frac{1}{3}, \frac{4}{3}\right]$ are distinct, so the point $[x_0, x_1, x_2]$ is collinear with them if and only if the vectors $(1, 2, 2), (3, 1, 4), (x_0, x_1, x_2)$ are linearly dependent, i.e., if and only if

$$0 = \det \begin{pmatrix} 1 & 3 & x_0 \\ 2 & 1 & x_1 \\ 2 & 4 & x_2 \end{pmatrix} = 6x_0 + 2x_1 - 5x_2.$$

Therefore, an equation of the line containing $\left[\frac{1}{2}, 1, 1\right], \left[1, \frac{1}{3}, \frac{4}{3}\right]$ is given by $6x_0 + 2x_1 - 5x_2 = 0$, and this equation is satisfied by the point $[2, -1, 2]$.

© Springer International Publishing Switzerland 2016
E. Fortuna et al., *Projective Geometry*, UNITEXT - La Matematica per il 3+2 104,
DOI 10.1007/978-3-319-42824-6_2

Exercise 2. Find the values $a \in \mathbb{C}$ for which the lines of equations

$$ax_1 - x_2 + 3ix_0 = 0, \quad -iax_0 + x_1 - ix_2 = 0, \quad 3ix_2 + 5x_0 + x_1 = 0$$

of $\mathbb{P}^2(\mathbb{C})$ are concurrent.

Solution (1). If

$$A = \begin{pmatrix} 3i & a & -1 \\ -ia & 1 & -i \\ 5 & 1 & 3i \end{pmatrix},$$

then the given lines all intersect if and only if the homogeneous linear system $AX = 0$ admits a non-trivial solution. This occurs if and only if $0 = \det A = -3a^2 - 4ia - 7$, i.e., if and only if either $a = i$ or $a = -\frac{7}{3}i$.

Solution (2). Via the duality correspondence (see Sect. 1.4.2), the given lines determine three points in the space $\mathbb{P}^2(\mathbb{C})^*$, and the coordinates of these points with respect to the frame induced by the standard basis of $(\mathbb{C}^2)^*$ are given by $[3i, a, -1]$, $[-ia, 1, -i]$, $[5, 1, 3i]$. An easy application of the Duality principle shows that these points are collinear if and only if the given lines are concurrent. Finally, the points $[3i, a, -1]$, $[-ia, 1, -i]$, $[5, 1, 3i]$ are collinear if and only if the determinant of the matrix A introduced above vanishes (see Exercise 1).

Exercise 3. In $\mathbb{P}^3(\mathbb{R})$ consider the points

$$P_1 = [1, 0, 1, 2], \quad P_2 = [0, 1, 1, 1], \quad P_3 = [2, 1, 2, 2], \quad P_4 = [1, 1, 2, 3].$$

(a) Determine whether P_1, P_2, P_3, P_4 are in general position.
(b) Compute the dimension of the subspace $L(P_1, P_2, P_3, P_4)$, and find Cartesian equations of $L(P_1, P_2, P_3, P_4)$.
(c) If possible, complete the set $\{P_1, P_2, P_3\}$ to a projective frame of $\mathbb{P}^3(\mathbb{R})$.

Solution. (a) Let $v_1 = (1, 0, 1, 2)$, $v_2 = (0, 1, 1, 1)$, $v_3 = (2, 1, 2, 2)$, $v_4 = (1, 1, 2, 3)$ be vectors in $\mathbb{R}^4$ such that $P_i = [v_i]$ for every i, and set

$$A = \begin{pmatrix} 1 & 0 & 2 & 1 \\ 0 & 1 & 1 & 1 \\ 1 & 1 & 2 & 2 \\ 2 & 1 & 2 & 3 \end{pmatrix}.$$

It is easy to check that $\det A = 0$, so v_1, v_2, v_3, v_4 are linearly dependent. Therefore, the points P_1, P_2, P_3, P_4 are not in general position.

(b) The determinant of the submatrix given by the first three lines and the first three columns of A is equal to -1, so v_1, v_2, v_3 are linearly independent. Therefore, by point (a) the dimension of the linear subspace spanned by $v_1, \ldots, v_4$ is equal to 3, so $L(P_1, P_2, P_3, P_4) = L(P_1, P_2, P_3)$ and $\dim L(P_1, P_2, P_3, P_4) = 2$. Moreover, the same argument as in the solution of Exercise 1 shows that a Cartesian equation of $L(P_1, P_2, P_3, P_4) = L(P_1, P_2, P_3)$ is given by

$$0 = \det \begin{pmatrix} 1 & 0 & 2 & x_0 \\ 0 & 1 & 1 & x_1 \\ 1 & 1 & 2 & x_2 \\ 2 & 1 & 2 & x_3 \end{pmatrix} = -x_0 - 2x_1 + 3x_2 - x_3.$$

(c) By point (b), if we replace the last column of A with the vector $(0, 0, 0, 1)$ we obtain an invertible matrix, so the vectors $v_1, v_2, v_3, (0, 0, 0, 1)$ provide a basis of $\mathbb{R}^4$. The projective frame induced by this basis is given by the points $P_1, P_2, P_3, [0, 0, 0, 1], [3, 2, 4, 6]$. Therefore, this 5-tuple of points extends P_1, P_2, P_3 to a projective frame of $\mathbb{P}^3(\mathbb{R})$.

Exercise 4. Let $l \subset \mathbb{P}^2(\mathbb{K})$ be the line of equation $x_0 + x_1 = 0$, set $U = \mathbb{P}^2(\mathbb{K}) \setminus l$ and let $\alpha, \beta \colon U \to \mathbb{K}^2$ be defined as follows:

$$\alpha([x_0, x_1, x_2]) = \left(\frac{x_1}{x_0 + x_1}, \frac{x_2}{x_0 + x_1} \right),$$

$$\beta([x_0, x_1, x_2]) = \left(\frac{x_0}{x_0 + x_1}, \frac{x_2}{x_0 + x_1} \right).$$

Find an explicit formula for the composition $\alpha \circ \beta^{-1}$, and check that this map is an affinity.

Solution. Let us first determine β^{-1}. Let $\beta([x_0, x_1, x_2]) = (u, v)$. Since $x_0 + x_1 \neq 0$ on U, we may suppose $x_0 + x_1 = 1$, so that

$$u = \frac{x_0}{x_0 + x_1} = x_0, \quad v = \frac{x_2}{x_0 + x_1} = x_2, \quad x_1 = 1 - x_0 = 1 - u.$$

Therefore, $\beta^{-1}(u, v) = [u, 1 - u, v]$, hence $\alpha(\beta^{-1}(u, v)) = (1 - u, v)$, and $\alpha \circ \beta^{-1}$ is obviously an affinity.

Exercise 5. For $i = 0, 1, 2$, let $j_i \colon \mathbb{K}^2 \to U_i \subseteq \mathbb{P}^2(\mathbb{K})$ be the map introduced in Sect. 1.3.8.

(a) Find two distinct projective lines $r, s \subset \mathbb{P}^2(\mathbb{K})$ such that the affine lines $j_i^{-1}(r \cap U_i), j_i^{-1}(s \cap U_i)$ are parallel for $i = 1, 2$.
(b) Is it possible to find distinct lines $r, s \subset \mathbb{P}^2(\mathbb{K})$ such that the affine lines $j_i^{-1}(r \cap U_i), j_i^{-1}(s \cap U_i)$ are parallel for $i = 0, 1, 2$?

Solution. Let l_i be the projective line of equation $x_i = 0$, $i = 0, 1, 2$. If $r \subset \mathbb{P}^2(\mathbb{K})$ is a projective line, then the set $j_i^{-1}(r \cap U_i)$ is an affine line if and only if $r \neq l_i$. Moreover, for any given distinct projective lines $r \neq l_i$, $s \neq l_i$, the affine lines $j_i^{-1}(r \cap U_i)$ and $j_i^{-1}(s \cap U_i)$ are parallel if and only if the point $s \cap r$ belongs to l_i. Therefore, if r, s are projective lines such that $r \neq s$, $r \notin \{l_1, l_2\}$, $s \notin \{l_1, l_2\}$ and

$r \cap s = [1, 0, 0]$, then r and s satisfy the condition described in (a): for example, one may choose $r = \{x_1 + x_2 = 0\}$, $s = \{x_1 - x_2 = 0\}$.

Moreover, since $l_0 \cap l_1 \cap l_2 = \emptyset$, the condition described in (b) cannot be satisfied by any pair of distinct lines of $\mathbb{P}^2(\mathbb{K})$.

Exercise 6. Let A, B, C, D be points of $\mathbb{P}^2(\mathbb{K})$ in general position, and set

$$P = L(A, B) \cap L(C, D), \quad Q = L(A, C) \cap L(B, D), \quad R = L(A, D) \cap L(B, C).$$

Prove that P, Q, R are not collinear.

Solution. Since A, B, C, D are in general position, we can choose a system of homogeneous coordinates in $\mathbb{P}^2(\mathbb{K})$ where

$$A = [1, 0, 0], \quad B = [0, 1, 0], \quad C = [0, 0, 1], \quad D = [1, 1, 1].$$

An easy computation shows that

$$P = [1, 1, 0], \quad Q = [1, 0, 1], \quad R = [0, 1, 1].$$

Since $\det \begin{pmatrix} 1 & 0 & 1 \\ 1 & 1 & 0 \\ 0 & 1 & 1 \end{pmatrix} \neq 0$, the points P, Q, R are not collinear.

Exercise 7. Let $\mathcal{R} = \{P_0, \ldots, P_{n+1}\}$ be a projective frame of $\mathbb{P}(V)$ and let $0 \leq k < n + 1$. Set $S = L(P_0, P_1, \ldots, P_k)$, $S' = L(P_{k+1}, \ldots, P_{n+1})$.

(a) Show that there exists $W \in \mathbb{P}(V)$ such that $S \cap S' = \{W\}$.
(b) Prove that $\{P_0, \ldots, P_k, W\}$ is a projective frame of S, and that $\{P_{k+1}, \ldots, P_{n+1}, W\}$ is a projective frame of S'.

Solution. (a) By the definition of projective frame, we have $\dim S = k$, $\dim S' = n - k$ and $\dim L(S, S') = n$, so Grassmann's formula implies that $\dim(S \cap S') = \dim S + \dim S' - \dim L(S, S') = 0$, and this proves (a).

(b) In order to show that $\{P_0, \ldots, P_k, W\}$ is a projective frame of S it is sufficient to prove that $\dim L(A) = k$ for every subset $A \subseteq \{P_0, \ldots, P_k, W\}$ containing exactly $k + 1$ points (if this is the case, then clearly $L(A) = S$).

Let A be such a subset. If $W \notin A$, then $\dim L(A) = k$, since the points of $\mathcal{R}$ are in general position. Let us now assume that $W \in A$. The points of $\mathcal{R}$ are in general position, so we have

$$\dim L((A \setminus \{W\}) \cup S') = \dim L((A \setminus \{W\}) \cup \{P_{k+1}, \ldots, P_{n+1}\}) = n,$$
$$\dim L(A \setminus \{W\}) = k - 1.$$

But $\dim S' = n - k$, so

$$\dim(L(A \setminus \{W\}) \cap S') = (k - 1) + (n - k) - n = -1,$$

i.e., $L(A \setminus \{W\}) \cap S' = \emptyset$. Since $W \in S'$, this implies that $W \notin L(A \setminus \{W\})$, so $\dim L(A) = \dim L(A \setminus \{W\}) + 1 = k$. Therefore, $\{P_0, \ldots, P_k, W\}$ is a projective frame of S. In the very same way one can prove that $\{P_{k+1}, \ldots, P_{n+1}, W\}$ is a projective frame of S'.

Exercise 8. Let $r, r' \subset \mathbb{P}^3(\mathbb{K})$ be skew lines, and take $P \in \mathbb{P}^3(\mathbb{K}) \setminus (r \cup r')$. Show that there exists a unique line $l \subset \mathbb{P}^3(\mathbb{K})$ that contains P and meets both r and r'. Compute Cartesian equations for l in the case when $\mathbb{K} = \mathbb{R}$, the line r has equations $x_0 - x_2 + 2x_3 = 2x_0 + x_1 = 0$, the line r' has equations $2x_1 - 3x_2 + x_3 = x_0 + x_3 = 0$, and $P = [0, 1, 0, 1]$.

Solution. Let $S = L(r, P)$, $S' = L(r', P)$. An easy application of Grassmann's formula implies that $\dim S = \dim S' = 2$. Moreover, $S \neq S'$ because otherwise r and r' would be coplanar, hence incident. It follows that $\dim(S \cap S') < 2$. On the other hand, $\dim(S \cap S') = \dim S + \dim S' - \dim L(S, S') \geq 2 + 2 - 3 = 1$, so $l = S \cap S'$ is a line. Since l and r (respectively, r') both lie on S (respectively, S'), we have that $l \cap r \neq \emptyset$ (respectively, $l \cap r' \neq \emptyset$). Therefore l satisfies the required properties.

Let now l' be any line of $\mathbb{P}^3(\mathbb{K})$ containing P and meeting both r and r'. We have $l' \subseteq L(r, P) = S$, $l' \subseteq L(r', P) = S'$, so $l' \subseteq S \cap S' = l$, and $l' = l$ since $\dim l' = \dim l$.

Let us now come to the particular case described in the statement of the exercise. The pencil of planes centred at r has parametric equations

$$\lambda(x_0 - x_2 + 2x_3) + \mu(2x_0 + x_1) = 0, \qquad [\lambda, \mu] \in \mathbb{P}^1(\mathbb{R}).$$

By imposing that the generic plane of the pencil pass through P we obtain $2\lambda + \mu = 0$, so an equation of S is given by $-3x_0 - 2x_1 - x_2 + 2x_3 = 0$. In the same way one proves that S' is described by the equation $-3x_0 + 2x_1 - 3x_2 - 2x_3 = 0$. The system given by the equation of S and the equation of S' provides the required equations for l.

Exercise 9. Let W_1, W_2, W_3 be planes of $\mathbb{P}^4(\mathbb{K})$ such that $W_i \cap W_j$ is a point for every $i \neq j$, and $W_1 \cap W_2 \cap W_3 = \emptyset$. Show that there exists a unique plane W_0 such that $W_0 \cap W_i$ is a projective line for $i = 1, 2, 3$.

Solution. For $i \neq j$ set $P_{ij} = W_i \cap W_j$, and $W_0 = L(P_{12}, P_{13}, P_{23})$ (therefore, $P_{ij} = P_{ji}$ for every $i \neq j$). If P_{12}, P_{13}, P_{23} were not pairwise distinct, then $W_1 \cap W_2 \cap W_3$ would be non-empty, while if P_{12}, P_{13}, P_{23} were pairwise distinct and collinear, then the line containing them would be contained in each W_i. In any case, our hypothesis would be contradicted, so P_{12}, P_{13}, P_{23} are in general position, and W_0 is a plane. Moreover, by construction $W_0 \cap W_i$ contains the line $L(P_{ij}, P_{ik})$, $\{i, j, k\} = \{1, 2, 3\}$. On the other hand, if $\dim(W_0 \cap W_i) > 1$, then $W_0 = W_i$, so $W_i \cap W_j = W_0 \cap W_j$ contains a line for every $j \neq i$, a contradiction. Therefore, $W_0 \cap W_i$ is a line for every $i = 1, 2, 3$.

Let now W_0' be a plane satisfying the properties described in the statement, and let $l_i = W_0' \cap W_i$ for every $i = 1, 2, 3$. Then each l_i is a line, and $W_i \cap W_j \cap W_0' = (W_i \cap W_0') \cap (W_j \cap W_0') = l_i \cap l_j \neq \emptyset$ (the lines l_i, l_j both lie on W_0', so they must intersect). It follows that $P_{ij} \in W_0'$ for every $i, j = 1, 2, 3$, and $W_0 \subseteq W_0'$. Since $\dim W_0' = \dim W_0$, we can conclude that $W_0' = W_0$.

Exercise 10. Let r_1, r_2, r_3 be pairwise skew lines of $\mathbb{P}^4(\mathbb{K})$ and suppose that no hyperplane of $\mathbb{P}^4(\mathbb{K})$ contains $r_1 \cup r_2 \cup r_3$. Prove that there exists a unique line that meets r_i for every $i = 1, 2, 3$.

Solution (1). For every $i, j \in \{1, 2, 3\}, i \neq j$, let $V_{ij} = L(r_i, r_j)$. An easy application of Grassmann's formula implies that $\dim V_{ij} = 3$ for every i, j. Since the lines r_1, r_2, r_3 are not contained in a hyperplane, we have $L(V_{12} \cup V_{13}) = L(r_1, r_2, r_3) = \mathbb{P}^4(\mathbb{K})$, so $\dim(V_{12} \cap V_{13}) = 2$, again by Grassmann's formula. Moreover, if $V_{12} \cap V_{13} \subseteq V_{23}$, then the line r_1 is contained in V_{23}, and this contradicts the fact that r_1, r_2, r_3 are not contained in a hyperplane. It follows that the subspace $l = V_{12} \cap V_{13} \cap V_{23}$ has dimension one, so it is a projective line.

We now check that l meets each r_i, $i = 1, 2, 3$, and that l is the unique line of $\mathbb{P}^4(\mathbb{K})$ with this property. If $\{i, j, k\} = \{1, 2, 3\}$, by construction l lies on the plane $V_{ij} \cap V_{ik}$, which contains also r_i. Since two coplanar projective lines always intersect, we deduce that $l \cap r_i \neq \emptyset$. Moreover, if s is any line meeting each $r_i, i = 1, 2, 3$, then it is easy to show that $s \subseteq V_{ij}$ for every $i, j \in \{1, 2, 3\}$, so $s \subseteq l$. But $\dim s = \dim l = 1$, hence $s = l$.

Solution (2). Let us show how the statement of Exercise 10 can be deduced from the statement of Exercise 8.

If $V_{23} = L(r_2, r_3)$, a direct application of Grassmann's formula implies that $\dim V_{23} = 3$. Moreover, since r_1, r_2, r_3 are not contained in a hyperplane, the line r_1 is not contained in V_{23}, so $r_1 \cap V_{23} = \{P\}$ for some $P \in \mathbb{P}^4(\mathbb{K})$. Now the statement of Exercise 8 implies that there exists a unique line $l \subseteq V_{23}$ that meets r_2 and r_3 and contains P. Therefore, this line meets r_i for every $i = 1, 2, 3$. Moreover, any line that meets both r_2 and r_3 must be contained in V_{23}, so it can intersect r_1 only at P. It follows that l is the unique line of $\mathbb{P}^4(\mathbb{K})$ that satisfies the required properties.

Note. It is not difficult to show that, by duality, the statement of the exercise is equivalent to the following proposition:

Let H_1, H_2, H_3 be planes of $\mathbb{P}^4(\mathbb{K})$ such that $L(H_i, H_j) = \mathbb{P}^4(\mathbb{K})$ for every $i \neq j$, and $H_1 \cap H_2 \cap H_3 = \emptyset$. Then, there exists a unique plane H_0 such that $L(H_0, H_i) \neq \mathbb{P}^4(\mathbb{K})$ for $i = 1, 2, 3$.
On the other hand, an easy application of Grassmann's formula implies that, if S, S' are distinct planes of $\mathbb{P}^4(\mathbb{K})$, then $L(S, S') \neq \mathbb{P}^4(\mathbb{K})$ if and only if $S \cap S'$ is a line, while $L(S, S') = \mathbb{P}^4(\mathbb{K})$ if and only if $S \cap S'$ is a point. It follows that the statements of Exercises 9 and 10 are equivalent.

Exercise 11. Let r and s be distinct lines in $\mathbb{P}^3(\mathbb{K})$ and let f be a projectivity of $\mathbb{P}^3(\mathbb{K})$ such that the fixed-point set of f coincides with $r \cup s$. For every $P \in \mathbb{P}^3(\mathbb{K}) \setminus (r \cup s)$ let l_P be the line joining P and $f(P)$. Prove that the line l_P meets both r and s, for every $P \in \mathbb{P}^3(\mathbb{K}) \setminus (r \cup s)$.

Solution. We first recall that the lines r and s are skew (cf. Sect. 1.2.5). Therefore, the statement of Exercise 8 ensures that, for any given $P \in \mathbb{P}^3(\mathbb{K}) \setminus (r \cup s)$, there exists a line t_P passing through P and meeting both r and s. In order to conclude it is now sufficient to show that $t_P = l_P$.

Let $A = t_P \cap r$, $B = t_P \cap s$. Then $f(P) \in f(L(A, B)) = L(f(A), f(B)) = L(A, B) = t_P$. Therefore, t_P contains both P and $f(P)$, and so it coincides with l_P.

Exercise 12. (*Desargues' Theorem*) Let $\mathbb{P}(V)$ be a projective plane and A_1, A_2, A_3, B_1, B_2, B_3 points of $\mathbb{P}(V)$ in general position. Consider the triangles T_1 and T_2 of $\mathbb{P}(V)$ with vertices A_1, A_2, A_3 and B_1, B_2, B_3; one says that T_1 and T_2 are in central perspective if there exists a point O in the plane, distinct from the A_i and B_i, such that all lines $L(A_i, B_i)$ pass through O.

Prove that T_1 and T_2 are in central perspective if and only if the points $P_1 = L(A_2, A_3) \cap L(B_2, B_3)$, $P_2 = L(A_3, A_1) \cap L(B_3, B_1)$ and $P_3 = L(A_1, A_2) \cap L(B_1, B_2)$ are collinear. (cf. Fig. 2.1).

Solution. It is easy to check that the points P_1, P_2, P_3 satisfy the following properties: they are pairwise distinct, they are distinct from any vertex of T_1 and T_2, and the points A_1, B_1, P_3, P_2 provide a projective frame of $\mathbb{P}(V)$. The point A_2 belongs to the line $L(A_1, P_3)$, so it has coordinates $[1, 0, a_2]$ for some $a_2 \neq 0$, since A_1 and A_2 are distinct from each other. A similar argument shows that:

$$A_3 = [a_3, 1, 1], \quad B_2 = [0, 1, b_2], \quad B_3 = [1, b_3, 1],$$

where $a_3, b_2, b_3 \in \mathbb{K}$, $b_2 \neq 0$, $a_3 \neq 1$ and $b_3 \neq 1$.

Let us set $P_1' = L(A_2, A_3) \cap L(P_2, P_3)$ and $P_1'' = L(B_2, B_3) \cap L(P_2, P_3)$. The points P_1, P_2, P_3 are collinear if and only if $P_1' = P_1''$ (and in this case $P_1 = P_1' = P_1''$). The lines $L(A_2, A_3)$ and $L(B_2, B_3)$ have equations $a_2 x_0 + (1 - a_2 a_3) x_1 - x_2 = 0$ and $(1 - b_2 b_3) x_0 + b_2 x_1 - x_2 = 0$, respectively. So $P_1' = [1, 1, 1 - a_2 a_3 + a_2]$ and

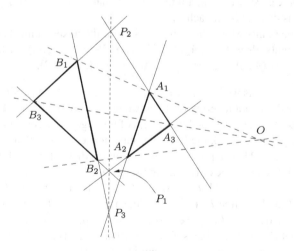

Fig. 2.1 The configuration described in Desargues' Theorem

$P_1'' = [1, 1, 1 - b_2b_3 + b_2]$. Therefore, our previous considerations imply that P_1, P_2 and P_3 are collinear if and only if the following equality holds:

$$a_2(1 - a_3) = b_2(1 - b_3). \tag{2.1}$$

Let us now analyze the condition that T_1 and T_2 are in central perspective. The line $L(A_1, B_1)$ has equation $x_2 = 0$, the line $L(A_2, B_2)$ has equation $a_2x_0 + b_2x_1 - x_2 = 0$, and the line $L(A_3, B_3)$ has equation $(1 - b_3)x_0 + (1 - a_3)x_1 + (a_3b_3 - 1)x_2 = 0$. These three lines are concurrent if and only if the corresponding points of the dual projective plane are collinear, i.e., if and only if

$$\det \begin{pmatrix} 0 & 0 & 1 \\ a_2 & b_2 & -1 \\ 1 - b_3 & 1 - a_3 & a_3b_3 - 1 \end{pmatrix} = 0. \tag{2.2}$$

We observe that, if this condition holds, then the point O belonging to the three lines has coordinates $[b_2, -a_2, 0]$. Since $a_2 \neq 0$ and $b_2 \neq 0$, the point O is distinct from any vertex of T_1 and T_2.

Finally, the conclusion follows from the fact that conditions (2.1) and (2.2) are clearly equivalent.

Note. The solution just described directly proves that the two conditions stated in the exercise are equivalent. We have already noticed in Sect. 1.4.4 that Desargues' Theorem is an example of self-dual proposition, so, in fact, it is sufficient to prove just one implication, since the other one follows from the Duality principle.

Exercise 13. (*Pappus' Theorem*) Let $\mathbb{P}(V)$ be a projective plane, and let $A_1, \ldots, A_6$ be pairwise distinct points such that the lines $L(A_1, A_2)$, $L(A_2, A_3)$,..., $L(A_6, A_1)$ are pairwise distinct. Consider the hexagon of $\mathbb{P}(V)$ with vertices $A_1, \ldots, A_6$, and suppose that there exist two distinct lines r, s such that $A_1, A_3, A_5 \in r$, $A_2, A_4, A_6 \in s$ and $O = r \cap s$ is distinct from each A_i.

Prove that the points where the opposite sides of the hexagon meet are collinear, i.e., that the points $P_1 = L(A_1, A_2) \cap L(A_4, A_5)$, $P_2 = L(A_2, A_3) \cap L(A_5, A_6)$ and $P_3 = L(A_3, A_4) \cap L(A_6, A_1)$ lie on a projective line (cf. Fig. 2.2).

Solution. By hypothesis we have $r = L(A_1, A_3)$ and $s = L(A_2, A_4)$. Since $r \neq s$ and the point $O = r \cap s$ is not a vertex of the hexagon, the points A_1, A_2, A_3, A_4 form a projective frame. In the corresponding system of homogeneous coordinates of $\mathbb{P}(V)$ the line r has equation $x_1 = 0$, the line s has equation $x_0 - x_2 = 0$, and the point O has coordinates $[1, 0, 1]$. The point A_5 lies on r and is distinct from O, from A_1 and from A_2, so it has coordinates $[1, 0, a]$ for some $a \in \mathbb{K} \setminus \{0, 1\}$. In the same way, the point A_6 has coordinates $[1, b, 1]$, where $b \in \mathbb{K} \setminus \{0, 1\}$. The line $L(A_1, A_2)$ has equation $x_2 = 0$ and the line $L(A_4, A_5)$ has equation $ax_0 + (1 - a)x_1 - x_2 = 0$, so the point $P_1 = L(A_1, A_2) \cap L(A_4, A_5)$ has coordinates $[a - 1, a, 0]$. Similar computations show that P_2 has coordinates $[0, b, 1 - a]$ and P_3 has coordinates $[b, b, 1]$. It follows that the points P_1, P_2 and P_3 are collinear, since

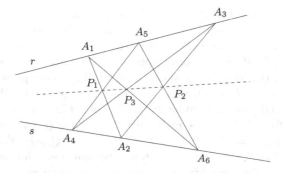

Fig. 2.2 The configuration described in Pappus' Theorem

$$\det \begin{pmatrix} a-1 & a & 0 \\ 0 & b & 1-a \\ b & b & 1 \end{pmatrix} = 0.$$

Exercise 14. Let A, A', B, B' be pairwise distinct non-collinear points of $\mathbb{P}^2(\mathbb{K})$. Prove that A, A', B, B' are in general position if and only if there exists a projectivity $f: \mathbb{P}^2(\mathbb{K}) \to \mathbb{P}^2(\mathbb{K})$ such that $f(A) = B, f(A') = B', f^2 = \mathrm{Id}$.

Solution. Suppose that a projectivity f as in the statement exists. We first observe that $f(B) = f(f(A)) = A$, $f(B') = f(f(A')) = A'$. In particular, the lines $L(A, B)$ and $L(A', B')$ are invariant under f. Moreover, since A, A', B, B' are not collinear, the lines $L(A, B)$ and $L(A', B')$ are distinct and meet at a single point O such that $f(O) = f(L(A, B) \cap L(A', B')) = L(A, B) \cap L(A', B') = O$. It is immediate to check that, if A, A', B, B' were not in general position, than we would have $O \in \{A, A', B, B'\}$, and this would contradict the fact that no point in $\{A, A', B, B'\}$ is a fixed point of f. We have thus shown that A, A', B, B' are in general position, as desired.

Let us now prove the converse implication. Assume that the points A, A', B, B' are in general position. The Fundamental theorem of projective transformations ensures that there exists a (unique) projectivity $f: \mathbb{P}^2(\mathbb{K}) \to \mathbb{P}^2(\mathbb{K})$ such that $f(A) = B$, $f(A') = B', f(B) = A, f(B') = A'$. By construction, f^2 and the identity of $\mathbb{P}^2(\mathbb{K})$ coincide on A, A', B, B', so they coincide on the whole of $\mathbb{P}^2(\mathbb{K})$, again by the Fundamental theorem of projective transformations. Therefore, f satisfies the required properties. We also observe that any projectivity satisfying the conditions described in the statement must coincide with f on A, A', B, B', so f is uniquely determined by such conditions.

Exercise 15. Determine a projectivity $f: \mathbb{P}^2(\mathbb{R}) \to \mathbb{P}^2(\mathbb{R})$ with the following properties: if $P = [1, 2, 1]$, $Q = [1, 1, 1]$ and r, r', s, s' are the lines described by the equations

$$r: x_0 - x_1 = 0, \qquad r': x_0 + x_1 = 0$$
$$s: x_0 + x_1 + x_2 = 0, \ s': x_1 + x_2 = 0,$$

then $f(r) = r', f(s) = s'$, and $f(P) = Q$. Is such a projectivity unique?

Solution. First observe that $P \notin r \cup s$, $Q \notin r' \cup s'$. Let $P_1 = r \cap s = [1, 1, -2]$, $Q_1 = r' \cap s' = [1, -1, 1]$, and let us choose points P_2, P_3 distinct from P on r, s, respectively: for example, we set $P_2 = [0, 0, 1]$, $P_3 = [-1, 1, 0]$. It is easy to check that the points P_1, P_2, P_3, P are in general position. In the same way, if $Q_2 = [0, 0, 1]$, then $Q_2 \in r'$, and Q_1, Q_2, Q are in general position. If Q_3 is any point of s' distinct from Q_1 and from $s' \cap L(Q_2, Q) = [1, 1, -1]$, then the quadruples $\{P_1, P_2, P_3, P\}$ and $\{Q_1, Q_2, Q_3, Q\}$ provide two projective frames of $\mathbb{P}^2(\mathbb{R})$. Therefore, there exists a unique projectivity $f: \mathbb{P}^2(\mathbb{R}) \to \mathbb{P}^2(\mathbb{R})$ such that $f(P_i) = Q_i$ for every $i = 1, 2, 3$ and $f(P) = Q$. Since $r = L(P_1, P_2), s = L(P_1, P_3), r' = L(Q_1, Q_2), s' = L(Q_1, Q_3)$, we also have $f(r) = r', f(s) = s'$. Since Q_3 may be chosen in infinitely many ways, an infinite number of projectivities satisfy the required conditions.

Let us explicitly construct f in the case when $Q_3 = [1, 0, 0]$. A normalized basis associated to the frame $\{P_1, P_2, P_3, P\}$ is given by $\{v_1 = (3, 3, -6), v_2 = (0, 0, 8), v_3 = (-1, 1, 0)\}$, while a normalized basis associated to the frame $\{Q_1, Q_2, Q_3, Q\}$ is given by $\{w_1 = (-1, 1, -1), w_2 = (0, 0, 2), w_3 = (2, 0, 0)\}$. Therefore, f is induced by the unique linear isomorphism $\varphi: \mathbb{R}^3 \to \mathbb{R}^3$ such that $\varphi(v_i) = w_i$ for $i = 1, 2, 3$. A straightforward computation shows that this isomorphism is given, up to a non-zero scalar, by the matrix $\begin{pmatrix} 14 & -10 & 0 \\ -2 & -2 & 0 \\ -1 & -1 & -3 \end{pmatrix}$, so the required projectivity f may be explicitly described by the following formula: $f([x_0, x_1, x_2]) = [14x_0 - 10x_1, -2x_0 - 2x_1, -x_0 - x_1 - 3x_2]$.

Exercise 16. Let r, s, r', s' be lines in $\mathbb{P}^2(\mathbb{K})$ such that $r \neq s$, $r' \neq s'$, and let $g: r \to r', h: s \to s'$ be projective isomorphisms. Find necessary and sufficient conditions for the existence of a projectivity $f: \mathbb{P}^2(\mathbb{K}) \to \mathbb{P}^2(\mathbb{K})$ such that $f|_r = g$ and $f|_s = h$. Show that, when it exists, such an f is unique.

Solution. Let us consider the points $P = r \cap s$, $P' = r' \cap s'$ (which are uniquely defined since $r \neq s$, $r' \neq s'$). Of course, if the required projectivity exists, then we must have $g(P) = h(P)$ (so necessarily $g(P) = h(P) = P'$). Let us show that this condition is also sufficient.

Let P_1, P_2 be pairwise distinct points of $r \setminus \{P\}$, let Q_1, Q_2 be pairwise distinct points of $s \setminus \{P\}$, and set $P_i' = g(P_i)$, $Q_i' = h(Q_i)$, $i = 1, 2$. It is immediate to check that the quadruples $\mathcal{R} = \{P_1, P_2, Q_1, Q_2\}$, $\mathcal{R}' = \{P_1', P_2', Q_1', Q_2'\}$ are in general position, so they define two projective frames of $\mathbb{P}^2(\mathbb{K})$. Therefore, there exists a unique projectivity $f: \mathbb{P}^2(\mathbb{K}) \to \mathbb{P}^2(\mathbb{K})$ such that $f(P_i) = P_i', f(Q_i) = Q_i'$ for $i = 1, 2$. We now show that this projectivity satisfies the required conditions.

We have $f(r) = f(L(P_1, P_2)) = L(P_1', P_2') = r'$, and in the same way $f(s) = s'$. Therefore, $f(P) = f(r \cap s) = r' \cap s' = P'$. So the projective transformations $f|_r$ and g coincide on three pairwise distinct points of r, hence on the whole line r. In the same way one proves that $f|_s = h$.

Finally, the fact that f is unique readily follows from the Fundamental theorem of projective transformations: any projectivity that satisfies the required conditions must coincide with f on $\mathcal{R}$, hence on the whole of $\mathbb{P}^2(\mathbb{K})$.

Exercise 17. Let S, S' be planes of $\mathbb{P}^3(\mathbb{K})$ and r, r' lines of $\mathbb{P}^3(\mathbb{K})$ such that $L(r, S) = L(r', S') = \mathbb{P}^3(\mathbb{K})$, and let $g\colon S \to S'$, $h\colon r \to r'$ be projective isomorphisms. Find necessary and sufficient conditions for the existence of a projectivity $f\colon \mathbb{P}^3(\mathbb{K}) \to \mathbb{P}^3(\mathbb{K})$ such that $f|_S = g$ and $f|_r = h$. Show that, if such an f exists, then it is unique.

Solution. An easy application of Grassmann's formula shows that there exist points $P, P' \in \mathbb{P}^3(\mathbb{K})$ such that $r \cap S = \{P\}, r' \cap S' = \{P'\}$. Of course, if the required projectivity exists, then we must have $g(P) = h(P) = P'$. We now show that this condition is also sufficient.

We extend P to a projective frame $\{P, P_1, P_2, P_3\}$ of S, and we choose distinct points Q_1, Q_2 on r, such that $Q_1 \neq P$, $Q_2 \neq P$. We first show that $\mathcal{R} = \{P_1, P_2, P_3, Q_1, Q_2\}$ is in general position, so it is a projective frame of $\mathbb{P}^3(\mathbb{K})$. To this aim it is sufficient to check that $\mathcal{R}$ does not contain any quadruple of coplanar points. Since $L(P_1, P_2, P_3) = S$ and $Q_l \notin S$, the points P_1, P_2, P_3, Q_l cannot be coplanar for $l = 1, 2$. Therefore, we may assume by contradiction that P_i, P_j, Q_1, Q_2 are coplanar for some $i \neq j$. In this case, the lines $L(P_i, P_j) \subset S$, $L(Q_1, Q_2) = r$ intersect in $P = S \cap r$, hence P, P_i, P_j are collinear. But this contradicts the fact that P, P_i, P_j are in general position on S. We have thus proved that $\mathcal{R}$ is a projective frame.

Let now $P_i' = g(P_i), Q_j' = h(Q_j)$ for $i = 1, 2, 3, j = 1, 2$. The same argument as above shows that $P_1', P_2', P_3', Q_1', Q_2'$ are also in general position, so there exists a unique projectivity $f\colon \mathbb{P}^3(\mathbb{K}) \to \mathbb{P}^3(\mathbb{K})$ such that $f(P_i) = P_i', f(Q_j) = Q_j'$ for $i = 1, 2, 3, j = 1, 2$. Let us show that this projectivity coincides with g on S and with h on r. We have

$$f(S) = f(L(P_1, P_2, P_3)) = L(P_1', P_2', P_3') = S',$$
$$f(r) = f(L(Q_1, Q_2)) = L(Q_1', Q_2') = r',$$

so $f(P) = f(r \cap S) = r' \cap S' = P'$. Therefore, $f|_S$ and g coincide on the projective frame $\{P, P_1, P_2, P_3\}$ of S, so they coincide on the whole of S. In the same way one proves that $f|_r$ and h coincide on P, Q_1, Q_2, hence on r.

Finally, the fact that f is unique is now obvious: any projectivity satisfying the required properties must coincide with f on P_1, P_2, P_3, Q_1, Q_2, hence on the whole projective space $\mathbb{P}^3(\mathbb{K})$, since $\{P_1, P_2, P_3, Q_1, Q_2\}$ is a projective frame of $\mathbb{P}^3(\mathbb{K})$.

Note. Exercises 16 and 17 illustrate particular cases of a general fact regarding the extension of projective transformations defined on subspaces of a projective space. In the projective space $\mathbb{P}(V)$ let us consider the projective subspaces S_1, S_2, S_1', S_2', and let us fix projective isomorphisms $g_1\colon S_1 \to S_1', g_2\colon S_2 \to S_2'$. Suppose also that $L(S_1, S_2) = \mathbb{P}(V)$, $g_1(S_1 \cap S_2) = g_2(S_1 \cap S_2) = S_1' \cap S_2'$, and $g_1|_{S_1 \cap S_2} = g_2|_{S_1 \cap S_2}$. Then, there exists a projectivity $f\colon \mathbb{P}(V) \to \mathbb{P}(V)$ such that $f|_{S_i} = g_i$ for $i = 1, 2$. Moreover, if $S_1 \cap S_2 \neq \emptyset$, then such a projectivity is unique.

Exercise 18. Let r, s be distinct lines of $\mathbb{P}^2(\mathbb{K})$, let A, B be distinct points of $r \setminus s$, and take $C, D \in \mathbb{P}^2(\mathbb{K}) \setminus (r \cup s)$. Show that there exists a unique projectivity $f\colon \mathbb{P}^2(\mathbb{K}) \to \mathbb{P}^2(\mathbb{K})$ such that $f(A) = A, f(B) = B, f(s) = s, f(C) = D$.

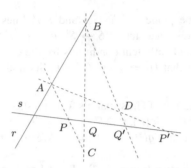

Fig. 2.3 The construction described in Exercise 18

Solution. Since $C, D \notin s$, each line $L(C, A)$, $L(C, B)$, $L(D, A)$, $L(D, B)$ meets s at exactly one point, so we can set $P = s \cap L(C, A)$, $P' = s \cap L(D, A)$, $Q = s \cap L(C, B)$, $Q' = s \cap L(D, B)$ (cf. Fig. 2.3). As $C, D \notin r$, the points P, P', Q, Q' lie on $s \setminus r$. Finally, we obviously have $P \neq Q$, $P' \neq Q'$. It follows that the quadruples $\{A, B, P, Q\}$ and $\{A, B, P', Q'\}$ define projective frames of $\mathbb{P}^2(\mathbb{K})$.

Let now f be the unique projectivity of $\mathbb{P}^2(\mathbb{K})$ such that $f(A) = A$, $f(B) = B$, $f(P) = P'$, $f(Q) = Q'$. Then we have $f(s) = f(L(P, Q)) = L(P', Q') = s$. Moreover, $f(C) = f(L(A, P) \cap L(B, Q)) = L(A, P') \cap L(B, Q') = D$, so f satisfies the required properties. Conversely, if g is a projectivity of $\mathbb{P}^2(\mathbb{K})$ satisfying the required properties, then $g(A) = A$, $g(B) = B$, $g(P) = g(L(A, C) \cap s) = L(A, D) \cap s = P'$ and $g(Q) = g(L(C, B) \cap s) = L(D, B) \cap s = Q'$, so $g = f$.

Exercise 19. In $\mathbb{P}^3(\mathbb{R})$, let r be the line of equations $x_0 - x_1 = x_2 - x_3 = 0$, let H, H' be the planes of equations $x_1 + x_2 = 0$, $x_1 - 2x_3 = 0$, respectively, and let $C = [1, 1, 0, 0]$. Compute the number of projectivities $f \colon \mathbb{P}^3(\mathbb{R}) \to \mathbb{P}^3(\mathbb{R})$ that satisfy the following conditions:

(i) $f(r) = r, f(H) = H', f(H') = H, f(C) = C$;
(ii) the fixed-point set of f contains a plane.

Solution. Let us first determine the incidence relations that hold among the subspaces described in the statement. It is immediate to check that $C \in r$ and $C \notin H \cup H'$. As a consequence, $r \cap H$ and $r \cap H'$ both consist of a single point: more precisely, an easy computation shows that, if $A = r \cap H$ and $B = r \cap H'$, then $A = [1, 1, -1, -1]$, $B = [2, 2, 1, 1]$. Moreover, since $H \neq H'$, the set $l = H \cap H'$ is a projective line, and the fact that $A \neq B$ readily implies that the lines r and l are skew.

Let now f be a projectivity satisfying the required conditions, and let S be a plane pointwise fixed by f. In order to understand the behaviour of f, we now determine the possible positions of S, showing first that S must contain l. We have $H \cap S = f(H \cap S) = H' \cap S$, so $H \cap S = H' \cap S$ is a projective subspace of $H \cap H' = l$. As $\dim(H \cap S) \geq 1$, we have $l = H \cap S = H' \cap S$, so in particular $l \subset S$, as desired.

Since l and r are skew, the intersection $r \cap S$ consists of one point. This point is fixed by f, so in order to determine S it is useful to study the fixed-point set of the restriction of f to r. Recall that by hypothesis $C \in r$, while by construction $A, B \in r$. We also have $f(A) = f(r \cap H) = r \cap H' = B$, and in the same way $f(B) = A$. Since $f(C) = C$, if $g: r \to r$ is the unique projectivity such that $g(A) = B$, $g(B) = A$, $g(C) = C$, then $f|_r = g$. Moreover, an easy computation shows that, if $r = \mathbb{P}(W)$, then g is induced by the unique linear isomorphism $\varphi: W \to W$ such that $\varphi(1, 1, -1, -1) = (2, 2, 1, 1)$ and $\varphi(2, 2, 1, 1) = (1, 1, -1, -1)$. Therefore, the fixed-point set of g (hence, of $f|_r$) is given by $\{C, D\}$, where $D = [1, 1, 2, 2]$. It follows that either $r \cap S = C$ or $r \cap S = D$.

Then, let $S_1 = L(l, C)$ and $S_2 = L(l, D)$. Since $C \notin l$, $D \notin l$ and $l \subset S$, if $C \in S$ then $S = S_1$, while if $D \in S$ then $S = S_2$. Moreover, since as observed above the lines r, s are skew, we have $L(S_1, r) = L(S_2, r) = \mathbb{P}^3(\mathbb{R})$.

Take $i \in \{1, 2\}$. As g fixes both $C = S_1 \cap r$ and $D = S_2 \cap r$, we may exploit Exercise 17 to show that there exists a unique projectivity $f_i : \mathbb{P}^3(\mathbb{R}) \to \mathbb{P}^3(\mathbb{R})$ such that $f_i|_{S_i} = \mathrm{Id}_{S_i}$ and $f_i|_r = g$. Moreover, what we have shown so far implies that, if f is a projectivity satisfying the conditions described in the statement, then f necessarily coincides either with f_1 or with f_2. We observe that $f_1 \neq f_2$, because otherwise the fixed-point set of f_1 would contain $S_1 \cup S_2$. Being the union of pairwise skew projective subspaces (cf. Sect. 1.2.5), the fixed-point set of f_1 would then coincide with the whole of $\mathbb{P}^3(\mathbb{R})$, contradicting the fact that $f_1(A) \neq A$.

In order to conclude it is now sufficient to show that both f_1 and f_2 satisfy the required conditions. By construction, for $i = 1, 2$ we have $f_i(C) = C, f_i(r) = r$, and the fixed-point set of f_i contains the plane S_i. Finally, we have $f_i(H) = f_i(L(l, A)) = L(l, B) = H'$ and $f_i(H') = f_i(L(l, B)) = L(l, A) = H$, as desired.

Exercise 20. Let $f: \mathbb{P}^1(\mathbb{K}) \to \mathbb{P}^1(\mathbb{K})$ be an injective function such that

$$\beta(P_1, P_2, P_3, P_4) = \beta(f(P_1), f(P_2), f(P_3), f(P_4))$$

for every quadruple P_1, P_2, P_3, P_4 of pairwise distinct points. Show that f is a projectivity.

Solution. Let P_1, P_2, P_3 be pairwise distinct points of $\mathbb{P}^1(\mathbb{K})$, and set $Q_i = f(P_i)$ for $i = 1, 2, 3$. Since f is injective, the sets $\{P_1, P_2, P_3\}$ and $\{Q_1, Q_2, Q_3\}$ are projective frames of $\mathbb{P}^1(\mathbb{K})$. Therefore, there exists a projectivity $g: \mathbb{P}^1(\mathbb{K}) \to \mathbb{P}^1(\mathbb{K})$ such that $g(P_i) = Q_i$ for $i = 1, 2, 3$. Moreover, since the cross-ratio is invariant under projectivities, for every $P \notin \{P_1, P_2, P_3\}$ we have

$$\beta(Q_1, Q_2, Q_3, g(P)) = \beta(g(P_1), g(P_2), g(P_3), g(P)) = \beta(P_1, P_2, P_3, P) = $$
$$= \beta(f(P_1), f(P_2), f(P_3), f(P)) = \beta(Q_1, Q_2, Q_3, f(P)).$$

As a consequence, $f(P) = g(P)$ for every $P \neq P_1, P_2, P_3$. On the other hand, for $i = 1, 2, 3$ we have $f(P_i) = g(P_i)$ by construction, so $f = g$, and f is a projectivity.

Exercise 21. (*Modulus of a quadruple of points*) Let

$$\mathcal{A} = \{P_1, P_2, P_3, P_4\}, \quad \mathcal{A}' = \{P'_1, P'_2, P'_3, P'_4\}$$

be quadruples of pairwise distinct points of $\mathbb{P}^1(\mathbb{C})$, and set

$$k = \beta(P_1, P_2, P_3, P_4), \quad k' = \beta(P_1', P_2', P_3', P_4').$$

(a) Show that the sets $\mathcal{A}, \mathcal{A}'$ are projectively equivalent if and only if

$$\frac{(k^2 - k + 1)^3}{k^2(k - 1)^2} = \frac{((k')^2 - k' + 1)^3}{(k')^2(k' - 1)^2}.$$

(b) Let G be the set of projectivities $f: \mathbb{P}^1(\mathbb{C}) \to \mathbb{P}^1(\mathbb{C})$ such that $f(\mathcal{A}) = \mathcal{A}$. For every $k \in \mathbb{C} \setminus \{0, 1\}$, compute the number $|G|$ of elements of G.

Solution. (a) As discussed in Sect. 1.5.2, the sets $\mathcal{A}, \mathcal{A}'$ are projectively equivalent if and only if k' belongs to the set

$$\Omega(k) = \left\{ k, \frac{1}{k}, 1 - k, \frac{1}{1 - k}, \frac{k - 1}{k}, \frac{k}{k - 1} \right\}.$$

(Observe that $k, k' \in \mathbb{C} \setminus \{0, 1\}$, since $P_i \neq P_j$ and $P_i' \neq P_j'$ for every $i \neq j$.)

Consider the rational function $j(t) = \dfrac{(t^2 - t + 1)^3}{t^2(t - 1)^2}$. In order to prove (a) it is sufficient to show that $k' \in \Omega(k)$ if and only if $j(k') = j(k)$.

Via a direct substitution, it is immediate to verify that $j(k) = j(k')$ if $k' \in \Omega(k)$.

Conversely, let us assume that $j(k) = j(k')$, and set $q(t) = (t^2 - t + 1)^3 - j(k)t^2$ $(t - 1)^2$. Of course we have $q(k') = 0$. Moreover, q is a polynomial of degree 6 admitting every element of $\Omega(k)$ as a root. Therefore, if $\Omega(k)$ contains 6 elements, then it coincides with the set of roots of q, and since $q(k') = 0$ we may deduce that $k' \in \Omega(k)$, as desired. If we set $\omega = \frac{1 + i\sqrt{3}}{2}$, then a direct computation shows that $|\Omega(k)| < 6$ only in the following cases:

- when $k \in \{\omega, \overline{\omega}\}$, and in this case $\Omega(k) = \{\omega, \overline{\omega}\}$;
- when $k \in \left\{-1, 2, \frac{1}{2}\right\}$, and in this case $\Omega(k) = \left\{-1, 2, \frac{1}{2}\right\}$.

Moreover, if $k \in \left\{-1, 2, \frac{1}{2}\right\}$ then $j(k) = \frac{27}{4}$ and $q(t) = (t + 1)^2(t - 2)^2 \left(t - \frac{1}{2}\right)^2$, while if $k \in \{\omega, \overline{\omega}\}$ then $j(k) = 0$ and $q(t) = (t - \omega)^3(t - \overline{\omega})^3$. In any case, $\Omega(k)$ coincides with the set of roots of q, and from the fact that $q(k') = 0$ we can deduce that $k' \in \Omega(k)$, as desired.

(b) Of course G is a group. If S_4 is the permutation group of $\{1, 2, 3, 4\}$, then for any given $f \in G$ there exists $\psi(f) \in S_4$ such that $f(P_i) = P_{\psi(f)(i)}$ for every $i = 1, 2, 3, 4$. Moreover, the map $\psi: G \to S_4$ thus defined is a group homomorphism. If $\psi(f) = \mathrm{Id}$, then f coincides with the identity on 3 distinct points of $\mathbb{P}^1(\mathbb{C})$, hence on the whole of $\mathbb{P}^1(\mathbb{C})$: therefore, the homomorphism ψ is injective, so we have $|G| = |\mathrm{Im}\,\psi| \leq |S_4| = 24$. Let us now investigate which permutations of the P_i are actually induced by a projectivity. To this aim, we will exploit some elementary facts about group actions on sets.

Let us consider the map $\eta \colon S_4 \times \Omega(k) \to \Omega(k)$ which is defined as follows: for every $h \in \Omega(k)$ and $\sigma \in S_4$, if $Q_1, Q_2, Q_3, Q_4 \in \mathbb{P}^1(\mathbb{C})$ are such that $\beta(Q_1, Q_2, Q_3, Q_4) = h$, then $\eta(\sigma, h) = \beta(Q_{\sigma(1)}, Q_{\sigma(2)}, Q_{\sigma(3)}, Q_{\sigma(4)})$. The properties of the cross-ratio described in Sect. 1.5.2 imply that $\eta(\sigma, h)$ does not depend on the choice of the Q_i; moreover, $\eta(\sigma, h) \in \Omega(k)$, so η is indeed well defined. Finally, it is immediate to check that $\eta(\sigma \circ \tau, h) = \eta(\sigma, \eta(\tau, h))$, so η defines an action of S_4 on $\Omega(k)$.

From the fundamental property of the cross-ratio (cf. Theorem 1.5.1) we deduce that $\sigma \in \mathrm{Im}\,\psi$ if and only if $\beta(P_1, P_2, P_3, P_4) = \beta(P_{\sigma(1)}, P_{\sigma(2)}, P_{\sigma(3)}, P_{\sigma(4)})$. In other words, $\mathrm{Im}\,\psi$ coincides with the stabilizer of k with respect to the action we have just introduced; moreover, this action is transitive, since every element of $\Omega(k)$ is the cross-ratio of an ordered quadruple obtained by permuting P_1, P_2, P_3, P_4. We thus have $|S_4| = |\mathrm{Stab}(k)|\,|\Omega(k)| = |\mathrm{Im}\,\psi|\,|\Omega(k)|$, so

$$|G| = |\mathrm{Im}\,\psi| = \frac{|S_4|}{|\Omega(k)|} = \frac{24}{|\Omega(k)|}.$$

Then, it follows from (a) that $|G| = 12$ if $k \in \{\omega, \overline{\omega}\}$, $|G| = 8$ if $k \in \left\{-1, 2, \frac{1}{2}\right\}$ and $|G| = 4$ otherwise.

Exercise 22. Let $f \colon \mathbb{P}^1(\mathbb{R}) \to \mathbb{P}^1(\mathbb{R})$ be the projectivity defined by

$$f([x_0, x_1]) = [-x_1, 2x_0 + 3x_1].$$

(a) Determine the fixed-point set of f.
(b) For $P = [2, 5] \in \mathbb{P}^1(\mathbb{R})$, compute the cross-ratio $\beta(A, B, P, f(P))$, where A and B are the fixed points of f.

Solution. The projectivity f is induced by the linear map $\varphi \colon \mathbb{R}^2 \to \mathbb{R}^2$ which is represented with respect to the canonical basis of $\mathbb{R}^2$ by the matrix $\begin{pmatrix} 0 & -1 \\ 2 & 3 \end{pmatrix}$. This matrix is diagonalizable, and it admits $(1, -1)$ and $(1, -2)$ as eigenvectors relative to the eingenvalues 1 and 2, respectively. It follows that $A = [1, -1]$ and $B = [1, -2]$ are the only fixed points of f.

We have seen in Sect. 1.5.4 that, since $A = [v]$ where v is an eigenvector of φ relative to the eigenvalue 1 and $B = [w]$ where w is an eigenvector of φ relative to the eigenvalue 2, the value $\beta(A, B, Q, f(Q))$ does not depend on $Q \in \mathbb{P}^1(\mathbb{R}) \setminus \{A, B\}$, and it is equal to $2/1 = 2$. Therefore, $\beta(A, B, P, f(P)) = 2$.

Exercise 23. (*Involutions of* $\mathbb{P}^1(\mathbb{K})$) Let f be a projectivity of $\mathbb{P}^1(\mathbb{K})$. Recall that f is an involution if $f^2 = \mathrm{Id}$, and that an involution f is non-trivial if $f \neq \mathrm{Id}$.

(a) If $M = \begin{pmatrix} a & b \\ c & d \end{pmatrix}$ is a matrix associated to f, show that f is a non-trivial involution if and only if $a + d = 0$.
(b) Show that f is a non-trivial involution if and only if there exist two distinct points Q_1, Q_2 that are switched by f, i.e., such that $f(Q_1) = Q_2$ and $f(Q_2) = Q_1$.

(c) Suppose that f is a non-trivial involution. Show that f has exactly either 0 or 2 fixed points, and that it has exactly 2 fixed points if $\mathbb{K} = \mathbb{C}$.

(d) Suppose that f fixes the points A, B. Show that f is a non-trivial involution if and only if, for any given point $P \in \mathbb{P}^1(\mathbb{K}) \setminus \{A, B\}$, the equality $\beta(A, B, P, f(P))$ $= -1$ holds (i.e., the characteristic of f is -1, cf. Sect. 1.5.4).

(e) Show that f is the composition of two involutions.

Solution. (a) First observe that, if $f \neq \mathrm{Id}$, then the minimal polynomial of M cannot have degree 1 (so it is equal to the characteristic polynomial of M). Moreover, we have $f^2 = \mathrm{Id}$ if and only if there exists $\lambda \in \mathbb{K}^*$ such that $M^2 = \lambda I$. It follows that f is a non-trivial involution if and only if the minimal polynomial and the characteristic polynomial of M coincide and are equal to $t^2 - \lambda$. Now the conclusion follows since the coefficient of t in the characteristic polynomial of M is equal to $-a - d$.

(b) If $f \neq \mathrm{Id}$ is an involution, it is sufficient to choose any point Q_1 not fixed by f and set $Q_2 = f(Q_1)$. Conversely, let Q_1, Q_2 be points that are exchanged by f, let P be any point of $\mathbb{P}^1(\mathbb{K}) \setminus \{Q_1, Q_2\}$ and set $P' = f(P)$. Then

$$\beta(Q_1, Q_2, P', P) = \beta(f(Q_1), f(Q_2), f(P'), f(P)) =$$
$$= \beta(Q_2, Q_1, f(P'), P') = \beta(Q_1, Q_2, P', f(P')),$$

where the first equality follows from the invariance of the cross-ratio under projectivities (cf. Sect. 1.5.1), while the second and the third one follow from the symmetries of the cross-ratio (cf. Sect. 1.5.2). Then $f^2(P) = f(P') = P$ and f is an involution.

(c) Point (a) implies that, if M is a matrix associated to f, then the characteristic polynomial of M is equal to $t^2 + \det M$, so either M does not admit any eigenvalue (when $-\det M$ is not a square in $\mathbb{K}$) or M admits two distinct eigenvalues relative to one-dimensional eigenspaces (when $-\det M$ is a square in $\mathbb{K}$). Therefore, f has exactly either 0 or 2 fixed points. Moreover, if $\mathbb{K}$ is algebraically closed, then M admits two distinct eigenvalues, so f has exactly two fixed points.

(d) As observed in Sect. 1.5.4, the cross-ratio $\beta(A, B, P, f(P))$ is independent of the choice of P. More precisely, in a projective frame $\mathcal{R}$ having A and B as fundamental points, f is represented by a matrix $N = \begin{pmatrix} \lambda & 0 \\ 0 & \mu \end{pmatrix}$, $\lambda, \mu \neq 0$. If $P \in \mathbb{P}^1(\mathbb{K}) \setminus \{A, B\}$ and $[P]_\mathcal{R} = [a, b]$, then $[f(P)]_\mathcal{R} = [\lambda a, \mu b]$, so $\beta(A, B, P, f(P)) = \frac{\mu}{\lambda}$. Therefore, $\beta(A, B, P, f(P)) = -1$ if and only if $\mu = -\lambda$, i.e., if and only if $N = \begin{pmatrix} \lambda & 0 \\ 0 & -\lambda \end{pmatrix}$. This condition is equivalent to the fact that N^2 is a multiple of the identity, i.e., to the fact that $f^2 = \mathrm{Id}$.

(e) If $f = \mathrm{Id}$ there is nothing to prove, so we suppose that there exists $A \in \mathbb{P}^1(\mathbb{K})$ such that $f(A) = A' \neq A$, and we set $A'' = f(A')$. If $A'' = A$, then f switches A and A', so it is an involution by point (b). Therefore, we can suppose that $A'' \neq A$, so that necessarily $A' \neq A''$, because otherwise we would have $f(A') = A' = f(A)$, which contradicts the fact that f is injective. As a consequence, being pairwise distinct, the points A, A', A'' define a projective frame of $\mathbb{P}^1(\mathbb{K})$, and by the Fundamental theorem of projective transformations there exists a projectivity $g \colon \mathbb{P}^1(\mathbb{K}) \to \mathbb{P}^1(\mathbb{K})$

such that $g(A) = A''$, $g(A') = A'$, $g(A'') = A$. Since g switches A and A'', g is an involution. Moreover, $f \circ g$ switches A' and A'', so it is an involution too. Therefore, $f = f \circ (g \circ g) = (f \circ g) \circ g$ is the composition of two involutions.

Exercise 24. Let A, B be distinct points of $\mathbb{P}^1(\mathbb{K})$. Show that there exists a unique non-trivial involution of $\mathbb{P}^1(\mathbb{K})$ having A and B as fixed points.

Solution (1). Take $P \in \mathbb{P}^1(\mathbb{K}) \setminus \{A, B\}$. If $f \colon \mathbb{P}^1(\mathbb{K}) \to \mathbb{P}^1(\mathbb{K})$ is a projectivity such that $f(A) = A$ and $f(B) = B$, then it follows from point (d) of Exercise 23 that f is a non-trivial involution if and only if $\beta(A, B, P, f(P)) = -1$, i.e., if and only if $f(P)$ is the unique point such that $A, B, P, f(P)$ is a harmonic quadruple. Since a projectivity of $\mathbb{P}^1(\mathbb{K})$ is uniquely determined by the values it takes on A, B, P, this concludes the proof.

Solution (2). If we fix a projective frame of $\mathbb{P}^1(\mathbb{K})$ having A and B as fundamental points, every projectivity $f \colon \mathbb{P}^1(\mathbb{K}) \to \mathbb{P}^1(\mathbb{K})$ such that $f(A) = A$ and $f(B) = B$ is represented in the induced coordinate system by a matrix $N = \begin{pmatrix} 1 & 0 \\ 0 & \lambda \end{pmatrix}$, $\lambda \in \mathbb{K}^*$. It is easily seen that N^2 is a multiple of the identity if and only if $\lambda = \pm 1$. Therefore, the unique non-trivial involution having A and B as fixed points is the projectivity represented by the matrix $\begin{pmatrix} 1 & 0 \\ 0 & -1 \end{pmatrix}$.

Exercise 25. Let $f \colon \mathbb{P}^1(\mathbb{K}) \to \mathbb{P}^1(\mathbb{K})$ be a projectivity, and let $A, B, C \in \mathbb{P}^1(\mathbb{K})$ be pairwise distinct points such that $f(A) = A, f(B) = C$. Show that A is the unique fixed point of f (i.e., f is parabolic, cf. Sect. 1.5.3) if and only if $\beta(A, C, B, f(C)) = -1$.

Solution. We endow $\mathbb{P}^1(\mathbb{K})$ with the homogeneous coordinates induced by the projective frame $\{A, B, C\}$. With respect to these coordinates, f is represented by a matrix M of the form $\begin{pmatrix} 1 & \lambda \\ 0 & \lambda \end{pmatrix}$ for some $\lambda \in \mathbb{K}^*$. Now, if $\lambda = 1$ the matrix M has exactly one eigenvalue, and this eigenvalue has geometric multiplicity one, so f is parabolic; otherwise, M has two distinct eigenvalues, and it is hyperbolic. So we need to show that $\beta(A, C, B, f(C)) = -1$ if and only if $\lambda = 1$. But the coordinates of $f(C)$ are $[1 + \lambda, \lambda]$, so

$$\beta(A, C, B, f(C)) = \beta([1, 0], [1, 1], [0, 1], [1 + \lambda, \lambda]) = -\lambda,$$

and this concludes the proof.

Exercise 26. Let A_1, A_2, A_3, A_4 be points of $\mathbb{P}^2(\mathbb{K})$ in general position, and set

$$P_1 = L(A_1, A_2) \cap L(A_3, A_4), \quad P_2 = L(A_2, A_3) \cap L(A_1, A_4), \quad r = L(P_1, P_2),$$
$$P_3 = L(A_2, A_4) \cap r, \qquad P_4 = L(A_1, A_3) \cap r.$$

Compute $\beta(P_1, P_2, P_3, P_4)$.

Solution (1). If we endow $\mathbb{P}^2(\mathbb{K})$ with homogeneous coordinates such that $A_1 = [1, 0, 0]$, $A_2 = [0, 1, 0]$, $A_3 = [0, 0, 1]$, $A_4 = [1, 1, 1]$, we easily obtain that $P_1 = [1, 1, 0]$ and $P_2 = [0, 1, 1]$. So $r = \{x_0 - x_1 + x_2 = 0\}$, hence $P_3 = [1, 2, 1]$ and $P_4 = [1, 0, -1]$. It follows that, if $r = \mathbb{P}(W)$, then a normalized basis of W associated to the projective frame $\{P_1, P_2, P_3\}$ is given by $(1, 1, 0)$, $(0, 1, 1)$. Since $(1, 0, -1) = (1, 1, 0) - (0, 1, 1)$, it follows that the required cross-ratio is equal to -1.

Solution (2). We set $t = L(A_1, A_3)$ and $Q = t \cap L(A_2, A_4)$ (cf. Fig. 2.4). Of course $A_2 \notin r \cup t$, $A_4 \notin r \cup t$, so the perspectivity $f : r \to t$ centred at A_2 and the perspectivity $g : t \to r$ centred at A_4 are well defined. By construction we have $f(P_1) = A_1$, $f(P_2) = A_3$, $f(P_3) = Q$, $f(P_4) = P_4$, so, being A_1, A_3, Q, P_4 pairwise distinct, the points P_1, P_2, P_3, P_4 are pairwise distinct too. Moreover, again by construction we have $g(A_1) = P_2$, $g(A_3) = P_1$, $g(Q) = P_3$, $g(P_4) = P_4$. Being the composition of two projectivities, the map $g \circ f : r \to r$ is a projectivity, so

$$\beta(P_1, P_2, P_3, P_4) = \beta(g(f(P_1)), g(f(P_2)), g(f(P_3)), g(f(P_4))) =$$
$$= \beta(P_2, P_1, P_3, P_4) = \frac{1}{\beta(P_1, P_2, P_3, P_4)},$$

where the first equality is due to the invariance of the cross-ratio under projectivities (cf. Sect. 1.5.1), while the last one is due to the symmetries of the cross-ratio (cf. Sect. 1.5.2). So $\beta(P_1, P_2, P_3, P_4)^2 = 1$, hence $\beta(P_1, P_2, P_3, P_4) = -1$ since the P_i are pairwise distinct, as observed above.

Note. (*Construction of the harmonic conjugate*) The previous exercise suggests a way to explicitly construct the *harmonic conjugate* of three pairwise distinct points P_1, P_2, P_3 of a projective line $r \subseteq \mathbb{P}^2(\mathbb{K})$, i.e., the point $P_4 \in r$ such that $\beta(P_1, P_2, P_3, P_4) = -1$. To this aim, let us consider a line $s \neq r$ passing through P and choose distinct points A_1, A_2 on $s \setminus \{P\}$. Let $A_4 = L(A_2, P_3) \cap L(A_1, P_2)$ and

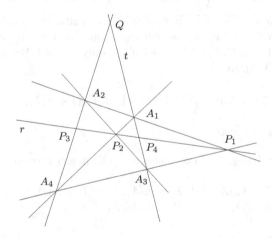

Fig. 2.4 Constructing a harmonic quadruple

$A_3 = L(P_1, A_4) \cap L(A_2, P_2)$, and let $P_4 = r \cap L(A_1, A_3)$ (observe that A_3, A_4, Q_4 are actually well defined). It is immediate to check that A_1, A_2, A_3, A_4 are in general position, and it is proven in Exercise 26 that $\beta(P_1, P_2, P_3, P_4) = -1$.

Exercise 27. Let P, Q, R, S be points of $\mathbb{P}^2(\mathbb{K})$ in general position, let $l_P = L(Q, R)$, $l_Q = L(P, R)$, $l_R = L(P, Q)$ and set

$$P' = L(P, S) \cap l_P, \quad Q' = L(Q, S) \cap l_Q, \quad R' = L(R, S) \cap l_R.$$

Also let $P'' \in l_P, Q'' \in l_Q, R'' \in l_R$ be the points that are uniquely determined by the following conditions:

$$\beta(Q, R, P', P'') = \beta(R, P, Q', Q'') = \beta(P, Q, R', R'') = -1.$$

Show that P'', Q'', R'' are collinear.

⚡ **Solution (1).** Let $T = L(Q', R') \cap l_P$, $W = L(T, S) \cap l_R$, $Z = L(T, S) \cap l_Q$. By applying Exercise 26 to the case when $A_1 = R, A_2 = R', A_3 = Q', A_4 = Q$, it is easy to verify that $\beta(S, T, W, Z) = -1$ (cf. Figs. 2.4 and 2.5). Thanks to the symmetries of the cross-ratio (cf. Sect. 1.5.2) we thus have $\beta(W, Z, S, T) = \beta(S, T, W, Z) = -1$. Moreover, the perspectivity $f: L(S, T) \to l_P$ centred at P maps W, Z, S, T to Q, R, P', T, respectively, so $\beta(Q, R, P', T) = -1$. Since by hypothesis $\beta(Q, R, P', P'') = -1$, we can conclude that $T = P''$.

Let now $g: l_Q \to l_R$ be the perspectivity centred at P''. By construction we have $g(P) = P$, $g(R) = Q$, and moreover $g(Q') = R'$ since $P'' = T$. Since the cross-ratio is invariant under projectivities, we then have

$$\beta(P, Q, R', g(Q'')) = \beta(g(P), g(R), g(Q'), g(Q'')) =$$

$$= \beta(P, R, Q', Q'') = \frac{1}{\beta(R, P, Q', Q'')} = -1,$$

so $g(Q'') = R''$. The definition of perspectivity implies now that $R'' = L(P'', Q'') \cap l_R$: in particular, the points P'', Q'', R'' are collinear.

Solution (2). By hypothesis the points P, Q, R, S define a projective frame of $\mathbb{P}^2(\mathbb{K})$, so we can choose homogeneous coordinates such that $P = [1, 0, 0]$, $Q = [0, 1, 0]$, $R = [0, 0, 1]$, $S = [1, 1, 1]$. We thus have $l_P = \{x_0 = 0\}$, $l_Q = \{x_1 = 0\}$, $l_R = \{x_2 = 0\}$, $L(P, S) = \{x_1 = x_2\}$, $L(Q, S) = \{x_0 = x_2\}$, $L(R, S) = \{x_0 = x_1\}$, hence $P' = [0, 1, 1]$, $Q' = [1, 0, 1]$, $R' = [1, 1, 0]$.

It is now easy to determine P'', Q'', R'': a normalized basis induced by the projective frame $\{Q, R, P'\}$ of l_P is given by $v_1 = (0, 1, 0)$, $v_2 = (0, 0, 1)$, so $P'' = [v_1 - v_2] = [0, 1, -1]$. In the same way one obtains $Q'' = [1, 0, -1], R'' = [1, -1, 0]$. It follows that P'', R'', Q'' all belong to the line of equation $x_0 + x_1 + x_2 = 0$.

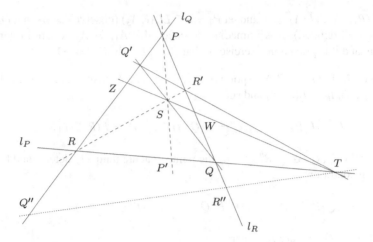

Fig. 2.5 The configuration described in Exercise 27

Exercise 28. Let $\mathbb{P}(V)$, $\mathbb{P}(W)$ be projective spaces over the field $\mathbb{K}$, let H, K be projective subspaces of $\mathbb{P}(V)$, and let $f\colon \mathbb{P}(V) \setminus H \to \mathbb{P}(W)$ be a degenerate projective transformation. Show that $f(K \setminus H)$ is a projective subspace of $\mathbb{P}(W)$ of dimension $\dim K - \dim(K \cap H) - 1$.

Solution. Let S, T be the linear subspaces of V such that $H = \mathbb{P}(S)$, $K = \mathbb{P}(T)$. The degenerate projective transformation f is induced by a linear map $\varphi\colon V \to W$ such that $\ker \varphi = S$, and it is immediate to check that $f(K \setminus H) = \mathbb{P}(\varphi(T))$. On the other hand, the dimension of the linear subspace $\varphi(T)$ is given by

$$\dim T - \dim(T \cap S) = (\dim K + 1) - (\dim(K \cap H) + 1) = \dim K - \dim(K \cap H),$$

so we finally have $\dim f(K \setminus H) = \dim K - \dim(K \cap H) - 1$.

Exercise 29. Let S, H be projective subspaces of the projective space $\mathbb{P}(V)$ such that $S \cap H = \emptyset$ and $L(S, H) = \mathbb{P}(V)$, and let $\pi_H \colon \mathbb{P}(V) \setminus H \to S$ be the projection onto S centred at H (cf. Sect. 1.2.7). Show that π_H is a degenerate projective transformation.

Solution (1). Let $n = \dim \mathbb{P}(V)$, $k = \dim S$, $h = \dim H$. An easy application of Grassmann's formula shows that $k + h = n - 1$. Moreover, it is easy to check that, if $P_0, \ldots, P_k$ are independent points of $S = \mathbb{P}(U)$ and $P_{k+1}, \ldots, P_{h+k+1}$ are independent points of H, then the set $\{P_0, \ldots, P_{h+k+1}\}$ is in general position in $\mathbb{P}(V)$, so it can be extended to a projective frame $\mathcal{R} = \{P_0, \ldots, P_{h+k+1}, Q\}$ of $\mathbb{P}(V)$.

Let us fix the system of homogeneous coordinates $x_0, \ldots, x_n$ induced by $\mathcal{R}$. The Cartesian equations of H and S are given by $x_0 = \cdots = x_k = 0$ and $x_{k+1} = \cdots = x_n = 0$, respectively (cf. Sects. 1.3.5 and 1.3.7). For any given $P = [y_0, \ldots y_n] \notin H$ it is easy to check that the subspace $L(H, P)$ is the set of points $[\lambda_0 y_0, \ldots, \lambda_0 y_k, \lambda_0 y_{k+1} + \lambda_1, \ldots, \lambda_0 y_n + \lambda_{h+1}]$, where $[\lambda_0, \ldots, \lambda_{h+1}] \in \mathbb{P}^{h+1}(\mathbb{K})$. So $L(H, P) \cap S$ is the point $[y_0, \ldots, y_k, 0, \ldots, 0]$ and π_H is the degenerate projective transformation induced by the linear map $\varphi \colon V \to U$ described in coordinates by the formula $(x_0, \ldots, x_n) \mapsto (x_0, \ldots, x_k, 0, \ldots, 0)$.

Solution (2). Let W, U be linear subspaces of V such that $H = \mathbb{P}(W)$ and $S = \mathbb{P}(U)$. It is easy to check that the conditions $S \cap H = \emptyset$, $L(S, H) = \mathbb{P}(V)$ imply that $W \cap U = \{0\}$, $W + U = V$, respectively. So $V = W \oplus U$, and the projection $p_U \colon V \to U$ mapping every $v \in V$ to the unique vector $p_U(v) \in U$ such that $v - p_U(v) \in W$ is well defined. The map p_U is linear, and we have $\ker p_U = W$. Therefore, for every $v \in V \setminus W$, the line spanned by the vector $p_U(v)$ coincides with the intersection of U with the subspace spanned by $W \cup \{v\}$. Hence for every $v \in V \setminus W$ we have $\pi_H([v]) = [p_U(v)]$, so π_H is the degenerate projective transformation induced by p_U.

Exercise 30. In $\mathbb{P}^3(\mathbb{R})$, let us consider the plane T_1 of equation $x_3 = 0$, the plane T_2 of equation $x_0 + 2x_1 - 3x_2 = 0$, and the point $Q = [0, 1, -1, 1]$, and let $f \colon T_1 \to T_2$ be the perspectivity centred at Q. Find Cartesian equations for the image through f of the line r obtained by intersecting T_1 with the plane $x_0 + x_1 = 0$.

Solution. By definition of perspectivity we have $f(r) = L(Q, r) \cap T_2$, so the Cartesian equations of $f(r)$ are given by the union of an equation of $L(Q, r)$ and an equation of T_2. Moreover, r is defined by the equations $x_0 + x_1 = x_3 = 0$, so the pencil of planes $\mathcal{F}_r$ centred at r has parametric equations

$$\lambda(x_0 + x_1) + \mu x_3 = 0, \qquad [\lambda, \mu] \in \mathbb{P}^1(\mathbb{R}).$$

By imposing that the generic plane of the pencil pass through Q one obtains $[\lambda, \mu] = [1, -1]$, so $L(Q, r)$ has equation $x_0 + x_1 - x_3 = 0$. Therefore, $f(r)$ is defined by the equations $x_0 + 2x_1 - 3x_2 = x_0 + x_1 - x_3 = 0$.

Exercise 31. Let $r, s \subset \mathbb{P}^2(\mathbb{K})$ be distinct lines, set $A = r \cap s$ and let $f \colon r \to s$ be a projective isomorphism. Prove that:

(a) f is a perspectivity if and only if $f(A) = A$.
(b) If $f(A) \neq A$, then there exist a line t of $\mathbb{P}^2(\mathbb{K})$ and two perspectivities $g \colon r \to t$, $h \colon t \to s$ such that $f = h \circ g$.
(c) Every projectivity $p \colon r \to r$ of r is the composition of at most three perspectivities.

Solution. (a) Every perspectivity between r and s fixes the point A (cf. Sect. 1.2.8). Conversely, if the isomorphism $f \colon r \to s$ fixes A, then we choose distinct points $P_1, P_2 \in r \setminus \{A\}$ and we set $Q_1 = f(P_1), Q_2 = f(P_2)$ (cf. Fig. 2.6). The lines $L(P_1, Q_1)$ and $L(P_2, Q_2)$ are distinct and meet at $O \notin r \cup s$. If $g \colon r \to s$ is the perspectivity centred at O, we have $g(A) = A$, $g(P_1) = Q_1$, $g(P_2) = Q_2$. The Fundamental theorem of projective transformations now implies that $f = g$, so f is a perspectivity.

(b) Choose pairwise distinct points $P_1, P_2, P_3 \in r \setminus \{A\}$ in such a way that the points $Q_1 = f(P_1)$, $Q_2 = f(P_2)$, $Q_3 = f(P_3)$ are distinct from A. Denote by M the point of intersection of the lines $L(P_1, Q_2)$ and $L(P_2, Q_1)$, by N the point of intersection of the lines $L(P_1, Q_3)$ and $L(P_3, Q_1)$, and by t the line $L(M, N)$ (cf. Fig. 2.6). It

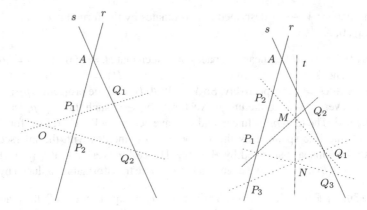

Fig. 2.6 Statements **a** on the *left* and **b** on the *right* of Exercise 31

is easy to check that the line t is distinct both from r and from s and does not contain the points P_1 and Q_1. Let us denote by $g : r \to t$ the perspectivity centred at Q_1 and by $h : t \to s$ the perspectivity centred at P_1. Since $(h \circ g)(P_i) = Q_i$ for $i = 1, 2, 3$, as in the previous point the Fundamental theorem of projective transformations implies that $f = h \circ g$.

When considering a projectivity $p : r \to r$, in order to prove (c) it is sufficient to apply (b) to the composition of p with any perspectivity $h : r \to t$ between r and any line $t \neq r$.

Exercise 32. (*Parametrization of a pencil of hyperplanes*) Let $\mathbb{P}(V)$ be a projective space of dimension n and let $H \subset \mathbb{P}(V)$ be a subspace of codimension 2. Denote by $\mathcal{F}_H$ the pencil of hyperplanes centred at H. If t is a line transverse to $\mathcal{F}_H$, i.e., if $t \subset \mathbb{P}(V)$ is a line disjoint from H, then denote by $f_t : t \to \mathcal{F}_H$ the map sending any point $P \in t$ to the hyperplane $L(P, H) \in \mathcal{F}_H$. Prove that f_t is a projective isomorphism (which is called the *parametrization of $\mathcal{F}_H$ via the transverse line t*).

Solution. Choose in $\mathbb{P}(V)$ a system of homogeneous coordinates such that H has equations $x_0 = x_1 = 0$ and the line t has equations $x_2 = \cdots = x_n = 0$; in this way x_0, x_1 is a system of homogeneous coordinates on t. In the dual system of homogeneous coordinates $a_0, \ldots, a_n$ on $\mathbb{P}(V)^*$, the pencil $\mathcal{F}_H$ has equations $a_2 = \cdots = a_n = 0$, so a_0, a_1 is a system of homogeneous coordinates on $\mathcal{F}_H$. With respect to these coordinates, the map $f_t : t \to \mathcal{F}_H$ is given by $[x_0, x_1] \mapsto [x_1, -x_0]$, so it is a projective isomorphism.

Note. One can use the fact that the parametrization f_t is a projective isomorphism in order to provide an alternative definition of the cross-ratio of four hyperplanes S_1, S_2, S_3, S_4 belonging to a pencil $\mathcal{F}_H$, without referring to the notion of dual projective space (cf. Sect. 1.5.1). If t is a transverse line, it is sufficient to set $\beta(S_1, S_2, S_3, S_4) = \beta(P_1, P_2, P_3, P_4)$, where $P_i = t \cap S_i$, $i = 1, \ldots, 4$: in fact, the invariance of the cross-ratio under projective isomorphisms and the fact that $f_t(P_i) = S_i$ ensure that the value $\beta(P_1, P_2, P_3, P_4)$ does not depend on the choice of t.

We also observe that one can prove that the cross-ratio $\beta(S_1 \cap t, S_2 \cap t, S_3 \cap t, S_4 \cap t)$ is independent of t without bringing the pencil $\mathcal{F}_H$ into play. To this aim it is sufficient to note that, if t' is any other transverse line, and $P_i' = t' \cap S_i$, $i = 1, 2, 3, 4$, then the points P_1', P_2', P_3', P_4' are the images of the points P_1, P_2, P_3, P_4 via the perspectivity between t and t' centred at H (cf. Sect. 1.2.8), so $\beta(P_1', P_2', P_3', P_4') = \beta(P_1, P_2, P_3, P_4)$.

When $n = 2$ we have thus proved that, if $\mathcal{F}_O$ is the pencil of lines of $\mathbb{P}^2(\mathbb{K})$ centred at O, then the parametrization of $\mathcal{F}_O$ obtained via the transverse line t is a projective isomorphism. Moreover, if the lines t_1, t_2 do not contain O, then the perspectivity between t_1 and t_2 centred at O is the composition $f_{t_2}^{-1} \circ f_{t_1}$. Since the composition of projective isomorphisms is a projective isomorphism, this argument provides an alternative proof of the fact that every perspectivity between two lines of the projective plane is a projective isomorphism.

Exercise 33. Let r and H be a line and a plane of $\mathbb{P}^3(\mathbb{K})$, respectively; suppose that $r \not\subseteq H$, and let $P = r \cap H$. Let $\mathcal{F}_r$ be the pencil of planes of $\mathbb{P}^3(\mathbb{K})$ centred at r, and let $\mathcal{F}_P$ be the pencil of lines of H centred at P. Prove that the map $\beta \colon \mathcal{F}_r \to \mathcal{F}_P$ defined by $\beta(K) = K \cap H$ is a well-defined projective isomorphism.

Solution. Let $s \subseteq H$ be a line such that $P \notin s$, and let $f_r \colon s \to \mathcal{F}_r$, $f_P \colon s \to \mathcal{F}_P$ be the maps defined by $f_r(Q) = L(r, Q)$, $f_P(Q) = L(P, Q)$ for every $Q \in s$. By construction, s does not contain P and is skew to r, so Exercise 32 ensures that f_r and f_P are well-defined projective isomorphisms. It is now immediate to check that the map β coincides with the composition $f_P \circ f_r^{-1}$, so β is a well-defined projective isomorphism.

Exercise 34. Let $r, s \subset \mathbb{P}^2(\mathbb{K})$ be distinct lines, let $A = r \cap s$, and let $f \colon r \to s$ be a projective isomorphism such that $f(A) = A$. Let also

$$W(f) = \{L(P_1, f(P_2)) \cap L(P_2, f(P_1)) \mid P_1, P_2 \in r, \ P_1 \neq P_2\}.$$

Prove that $W(f)$ is a projective line containing A.

Solution (1). We have seen in Exercise 31 that f is a perspectivity centred at $O \in \mathbb{P}^2(\mathbb{K}) \setminus (r \cup s)$. Let $l = L(A, O)$, and let $\mathcal{F}_A$ be the pencil of lines of $\mathbb{P}^2(\mathbb{K})$ centred at A: by construction $r, s, l \in \mathcal{F}_A$. In order to show that $W(f)$ is contained in a line passing through A it is sufficient to show that, as M varies in $W(f) \setminus \{A\}$, the cross-ratio $\beta(s, r, l, L(A, M))$ does not depend on M: namely, if this is the case and $k = \beta(s, r, l, L(A, M))$, then we necessarily have $W(f) \subseteq t$, where t is the unique line of $\mathcal{F}_A$ such that $\beta(s, r, l, t) = k$.

So let P_1, P_2 be distinct points of r, and let

$$M = L(P_1, f(P_2)) \cap L(P_2, f(P_1)) = L(P_1, s \cap L(O, P_2)) \cap L(P_2, s \cap L(O, P_1)).$$

Of course, if $P_1 = A$ or $P_2 = A$ then $M = A$, so we can suppose that P_1, P_2 are distinct from A. Then it is easy to check that $M \neq A$. We now show that $\beta(s, r, l, L(A, M)) = -1$. As already observed, this implies that $W(f)$ is contained in the line $t \in \mathcal{F}_A$ such that $\beta(s, r, l, t) = -1$.

Let $w = L(O, M)$, and observe that $A \notin w$, so w is transverse to the pencil $\mathcal{F}_A$. If $N = s \cap w$ and $Z = r \cap w$, then by the Note following Exercise 32 we have

$$\beta(s, r, l, L(A, M)) = \beta(s \cap w, r \cap w, l \cap w, L(A, M) \cap w) = \beta(N, Z, O, M).$$

On the other hand it is easy to check that, if we apply the construction described in the statement of Exercise 26 to the case when $A_1 = P_1$, $A_2 = f(P_1)$, $A_3 = P_2$, $A_4 = f(P_2)$, then we get $\beta(O, M, N, Z) = -1$ (cf. Figs. 2.4 and 2.7). Thanks to the symmetries of the cross-ratio (cf. Sect. 1.5.2) we then have $\beta(N, Z, O, M) = -1$, so $\beta(s, r, l, L(A, M)) = -1$, as desired. We have thus proved that the inclusion $W(f) \subseteq t$ holds.

Let us check that also the converse inclusion holds. For every $P \in r \setminus \{A\}$ we have $L(P, f(A)) \cap L(f(P), A) = A$, so $A \in W(f)$. So let $R \in t \setminus \{A\}$ and let v be any line passing through R and distinct both from t and from $L(O, R)$. Let $P_1 = v \cap r$, $P_2 = f^{-1}(v \cap s)$. Since $R \neq A$, $v \neq t$ and $v \neq L(O, R)$, the points P_1, P_2 are distinct and they are also distinct from A. Moreover $L(P_1, f(P_2)) = v$, so what has been proved above implies that $L(P_1, f(P_2)) \cap L(P_2, f(P_1)) \in v \cap t = \{R\}$, and $R \in W(f)$. We have thus shown that $t \subseteq W(f)$, hence $W(f) = t$, as desired.

Solution (2). Let us fix a point $P \in r \setminus \{A\}$, take $P_0 \in r \setminus \{A, P\}$ and let $M = L(P, f(P_0)) \cap L(P_0, f(P))$. We will show that, if $t = L(A, M)$, then $W(f) = t$.

We first prove the inclusion $W(f) \subseteq t$. Observe that, if $g: r \to t$ is the perspectivity centred at $f(P)$ and $h: t \to s$ is the perspectivity centred at P, then f and $h \circ g$ coincide on A, P, P_0, so $f = h \circ g$ (cf. Fig. 2.8). We easily deduce that for every $P_1 \in r \setminus \{P\}$ we have $L(P_1, f(P)) \cap L(P, f(P_1)) = g(P_1) \in t$.

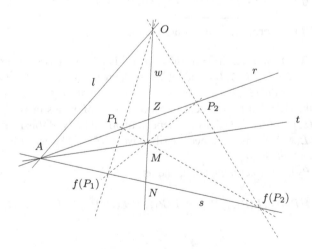

Fig. 2.7 The configurations described in Solution (1) of Exercise 34

Let us now prove that indeed $L(P_1, f(P_2)) \cap L(P_2, f(P_1)) \in t$ for every $P_1, P_2 \in r$, $P_1 \neq P_2$. This obviously holds if $P_1 = A$ or $P_2 = A$, and also holds if $P_1 = P$ or $P_2 = P$ thanks to the previous considerations. Therefore, we may assume $P_1, P_2 \in r \setminus \{A, P\}$. We consider the hexagon with vertices

$$P_1, \quad Q_2 = f(P_2), \quad P, \quad Q_1 = f(P_1), \quad P_2, \quad Q = f(P).$$

By Pappus' Theorem (cf. Exercise 13 and Figs. 2.2, 2.8) the points

$$L(P_1, Q_2) \cap L(P_2, Q_1), \quad L(P, Q_2) \cap L(P_2, Q), \quad L(P, Q_1) \cap L(P_1, Q)$$

are collinear. We have observed above that the second and the third of these points are distinct and lie on the line t, so also the point $L(P_1, Q_2) \cap L(P_2, Q_1) = L(P_1, f(P_2)) \cap L(P_2, f(P_1))$ lies on t. We have thus proved that $W(f) \subseteq t$.

We now come to the opposite inclusion. We observe that $A \in W(f)$ because for every $P \in r \setminus \{A\}$ we have $L(P, f(A)) \cap L(A, f(P)) = A$. So let $Q \in t \setminus \{A\}$. We first show that f cannot be a perspectivity centred at Q. In fact, it is immediate to check that, if P_1, P_2 are distinct points of $r \setminus \{A\}$, then the points $P_1, P_2, f(P_1), f(P_2)$ are in general position, so the points $A = L(P_1, P_2) \cap L(f(P_1), f(P_2))$, $B = L(P_1, f(P_2)) \cap L(P_2, f(P_1))$, $L(P_1, f(P_1)) \cap L(P_2, f(P_2))$ are not collinear (cf. Exercise 6). But we have proved above that $A \in t$, $B \in t$, so if f were a perspectivity centred at Q, then we would have $L(P_1, f(P_1)) \cap L(P_2, f(P_2)) = Q \in t$, and the points A, B, Q would be collinear, a contradiction. So there exists a line $v \neq t$ containing Q and such that the points $R_1 = v \cap r$ and $R_2 = f^{-1}(v \cap s)$ are distinct. We thus have $L(R_1, f(R_2)) = v$, hence $L(R_1, f(R_2)) \cap L(R_2, f(R_1)) \in t$ by the considerations above. Therefore $L(R_1, f(R_2)) \cap L(R_2, f(R_1)) = t \cap v = Q$, so $Q \in W(f)$. We have thus proved that $t \subseteq W(f)$.

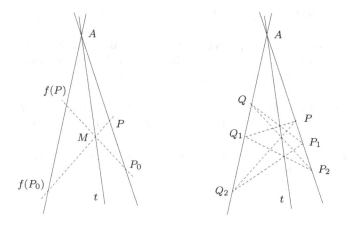

Fig. 2.8 Solution (2) of Exercise 34: on the *left*, f is described as the composition of two perspectivities; on the *right*, the inclusion $W(f) \subseteq t$ as a consequence of Pappus' Theorem

Solution (3). By Exercise 31, f is a perspectivity centred at $O \in \mathbb{P}^2(\mathbb{K}) \setminus (r \cup s)$. Take points $B \in r$, $C \in s$ such that A, B, C, O are in general position, and endow $\mathbb{P}^2(\mathbb{K})$ with the system of homogeneous coordinates induced by the projective frame $\{A, B, C, O\}$. Then we have $r = \{x_2 = 0\}$, $s = \{x_1 = 0\}$. Let now P, P' be distinct points in $r \setminus \{A\}$. We have $P = [a, 1, 0]$, $P' = [a', 1, 0]$ for some $a, a' \in \mathbb{K}$. Using that

$$f(P) = L(O, P) \cap s = L([1, 1, 1], [a, 1, 0]) \cap \{x_1 = 0\},$$

one easily proves that $f(P) = [1 - a, 0, 1]$. In a similar way one gets $f(P') = [1 - a', 0, 1]$. We thus have $L(P, f(P')) = \{x_0 - ax_1 + (a' - 1)x_2 = 0\}$ and $L(P', f(P)) = \{x_0 - a'x_1 + (a - 1)x_2 = 0\}$, so $L(P, f(P')) \cap L(P', f(P)) = [1 - a - a', -1, 1]$. It follows that, if P, P' are distinct points in $r \setminus \{A\}$, then the point $L(P, f(P')) \cap L(P', f(P))$ lies on $t \setminus \{A\}$, where t is the line (through A) of equation $x_1 + x_2 = 0$, and every point of $t \setminus \{A\}$ arises in this way. On the other hand, for every $P \in r \setminus \{A\}$ one has $L(P, f(A)) \cap L(A, f(P)) = A$, so $W(f) = t$, as desired.

Exercise 35. Let $r, s \subset \mathbb{P}^2(\mathbb{K})$ be distinct lines, set $A = r \cap s$, and let $f : r \to s$ be a projective isomorphism such that $f(A) \neq A$. Let also $B = f^{-1}(A) \in r$ and set

$$W(f) = \{L(P, f(P')) \cap L(P', f(P)) \mid P, P' \in r, \ P \neq P', \ \{P, P'\} \neq \{A, B\}\}.$$

Prove that $W(f)$ is a projective line.

Solution (1). Let us fix a point $P_1 \in r \setminus \{A, B\}$. If P_2, P_3 are distinct points of $r \setminus \{A, B, P_1\}$ and we set $M = L(P_1, f(P_2)) \cap L(P_2, f(P_1))$, $N = L(P_1, f(P_3)) \cap L(P_3, f(P_1))$, $t = L(M, N)$ (cf. Fig. 2.9), then it is easy to check that the points M, N and the line t are well defined and that $f = h \circ g$, where $g : r \to t$ is the perspectivity centred at $f(P_1)$ and $h : t \to s$ is the perspectivity centred at P_1. It readily follows that

$$L(f(P_1), P) \cap L(P_1, f(P)) = g(P) \in t \qquad \forall P \in r \setminus \{P_1\}. \tag{2.3}$$

In order to prove that $W(f) \subseteq t$, let us first show that t does not depend on the choice of P_1. If $C = f(A) \in s$, since $f(A) \neq A$ the points A, B, C are in general position. Together with our previous considerations, this implies that

$$\begin{aligned}
B &= L(P_1, A) \cap L(f(P_1), B) = L(P_1, f(B)) \cap L(f(P_1), B) \in t \\
C &= L(f(P_1), A) \cap L(P_1, C) = L(f(P_1), A) \cap L(P_1, f(A)) \in t.
\end{aligned} \tag{2.4}$$

Therefore, the line t contains both B and C, so that $t = L(B, C)$. In particular, t does not depend on the choice of P_1.

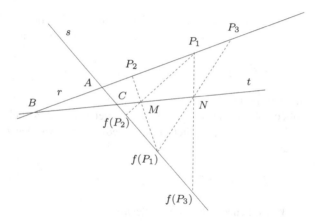

Fig. 2.9 The construction of the set $W(f)$ described in Solution (1) of Exercise 35

Combining (2.3) with the fact that t is independent of P_1 we deduce that

$$L(P, f(P')) \cap L(P', f(P)) \in t \quad \forall P \in r \setminus \{A, B\}, \ P' \in r \setminus \{P\}.$$

Now the inclusion $W(f) \subseteq t$ readily follows from the fact that the expression $L(P, f(P')) \cap L(P', f(P))$ is symmetric in P, P'.

Let us now check that $t \subseteq W(f)$. We first observe that, by (2.4), the points $B = t \cap r, C = t \cap s$ belong to $W(f)$. Let then $Q \in t \setminus \{B, C\}$. If $f(P) = L(P, Q) \cap s$ for every $P \in r \setminus \{A, B\}$, then f is the perspectivity centred at Q, so $f(A) = A$, against the hypothesis. Therefore, there exists $P_1 \in r \setminus \{A, B\}$ such that $f(P_1) \neq L(P_1, Q) \cap s$. Let $P_2 = f^{-1}(L(P_1, Q) \cap s)$. By construction $P_2 \neq P_1$. Let now $Q' = L(P_1, f(P_2)) \cap L(P_2, f(P_1))$. By construction $Q' \in L(P_1, f(P_2)) = L(P_1, Q)$, and our previous considerations imply that $Q' \in W(f) \subseteq t$. It follows that $Q' = t \cap L(P_1, Q) = Q$, so $Q \in W(f)$. We have thus shown that $t \subseteq W(f)$, as desired.

Solution (2). Let us set $C = f(A) \in s$. The points A, B, C are not collinear, so we may choose a system of homogeneous coordinates x_0, x_1, x_2 such that $A = [1, 0, 0]$, $B = [0, 1, 0], C = [0, 0, 1]$. We then have $r = \{x_2 = 0\}, s = \{x_1 = 0\}$, and we can endow r and s with the systems of homogeneous coordinates induced by x_0, x_1, x_2. Since $f(B) = A$ and $f(A) = C$, with respect to these coordinates the isomorphism f is represented by the matrix $\begin{pmatrix} 0 & \lambda \\ 1 & 0 \end{pmatrix}$, for some $\lambda \in \mathbb{K}^*$.

If $P = [a, b, 0], P' = [a', b', 0]$ are distinct points of r such that $\{P, P'\} \neq \{A, B\}$, we thus have $f(P) = [\lambda b, 0, a], f(P') = [\lambda b', 0, a']$; so $L(P, f(P'))$ and $L(P', f(P))$ have equations $ba'x_0 - aa'x_1 - \lambda bb'x_2 = 0$ and $ab'x_0 - aa'x_1 - \lambda bb'x_2 = 0$, respectively. Since $P \neq P'$ we have $ab' - ba' \neq 0$, so the lines $L(P, f(P'))$ and $L(P', f(P))$ meet at $[0, \lambda bb', -aa']$ (observe that we cannot have $\lambda bb' = aa' = 0$ because $\{P, P'\} \neq \{A, B\}$). If t is the line of equation $x_0 = 0$, we can conclude that $W(f) \subseteq t$.

On the other hand, take $P = [\lambda, 1, 0]$ and $P' = [a', b', 0]$ with $a' \neq \lambda b'$. Then $P \neq P'$ and $\{P, P'\} \neq \{A, B\}$, and the previous computation shows that

$$L(P, f(P')) \cap L(P', f(P)) = [0, b', -a'].$$

This proves that every point of t, except possibly $[0, 1, -\lambda]$, belongs to $W(f)$. On the other hand, of course we can choose distinct elements $a, a' \in \mathbb{K}$ such that $aa' = \lambda^2$. If $P = [a, 1, 0]$, $P' = [a', 1, 0]$, then it is immediate to check that $P \neq P'$ and $\{P, P'\} \neq \{A, B\}$, and

$$L(P, f(P')) \cap L(P', f(P)) = [0, 1, -\lambda].$$

We thus have $t \subseteq W(f)$, hence $W(f) = t$, as desired.

Note. In Solution (1), after proving (2.3) one can conclude that $W(f) = t$ by exploiting Pappus' Theorem in the same spirit as in Solution (2) of Exercise 34.

Exercise 36. (a) Let A, A', C, C' be points of $\mathbb{P}^1(\mathbb{K})$ such that $A \notin \{C, C'\}$ and $A' \notin \{C, C'\}$. Prove that there exists a unique non-trivial involution $f : \mathbb{P}^1(\mathbb{K}) \to \mathbb{P}^1(\mathbb{K})$ such that $f(A) = A', f(C) = C'$.

(b) Let A, B, C, A', B', C' be points of $\mathbb{P}^1(\mathbb{K})$ such that each quadruple A, B, C, C' and A', B', C', C contains only pairwise distinct points. Prove that there exists a unique involution $f : \mathbb{P}^1(\mathbb{K}) \to \mathbb{P}^1(\mathbb{K})$ such that $f(A) = A'$, $f(B) = B'$, $f(C) = C'$ if and only if

$$\beta(A, B, C, C') = \beta(A', B', C', C).$$

(c) Let $r \subseteq \mathbb{P}^2(\mathbb{K})$ be a projective line and let P_1, P_2, P_3, P_4 be points of $\mathbb{P}^2(\mathbb{K}) \setminus r$ in general position. For every $i \neq j$ let $s_{ij} = L(P_i, P_j)$, and set

$$A = r \cap s_{12}, \quad B = r \cap s_{13}, \quad C = r \cap s_{14}$$
$$A' = r \cap s_{34}, \quad B' = r \cap s_{24}, \quad C' = r \cap s_{23}$$

(cf. Fig. 2.10). Prove that there exists a unique involution f of r such that $f(A) = A', f(B) = B', f(C) = C'$.

Solution. (a) The case when $A = A'$ and $C = C'$ has already been settled in Exercise 24.

Let us now suppose $A = A'$, $C \neq C'$ (the case when $A \neq A'$, $C = C'$ being similar). If f satisfies the conditions of the statement, then necessarily $f(A) = A' = A$, $f(C) = C', f(C') = f(f(C)) = C$. Moreover, by the Fundamental theorem of projective transformations, there exists a unique projectivity mapping A, C, C' to A, C', C, respectively. By point (b) of Exercise 23, this projectivity is an (obviously non-trivial) involution, and this concludes the proof in the case when $A = A', C \neq C'$.

We finally suppose $A \neq A'$, $C \neq C'$. Our hypothesis implies that the triples A, A', C and A, A', C' define two projective frames of $\mathbb{P}^1(\mathbb{K})$, so there exists a unique

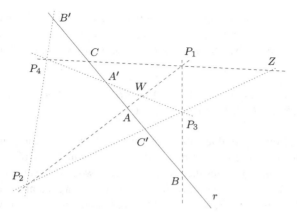

Fig. 2.10 Exercise 36, point (c): the case when $Z \notin r$

projectivity $f: \mathbb{P}^1(\mathbb{K}) \to \mathbb{P}^1(\mathbb{K})$ such that $f(A) = A', f(A') = A, f(C) = C'$. Moreover, by point (b) of Exercise 23, the projectivity f is a non-trivial involution. On the other hand, any projectivity satisfying the required conditions necessarily maps A, A', C to A', A, C', respectively, so it must coincide with f.

(b) If an involution f with the required properties exists, then $f(C') = f(f(C)) = C$, so

$$\beta(A, B, C, C') = \beta(f(A), f(B), f(C), f(C')) = \beta(A', B', C', C)$$

thanks to the invariance of the cross-ratio with respect to projective transformations.

We then suppose that $\beta(A, B, C, C') = \beta(A', B', C', C)$, and we show that the required involution exists (the fact that such involution is unique readily follows from the Fundamental theorem of projective transformations). By point (a), there exists a non-trivial involution $f: \mathbb{P}^1(\mathbb{K}) \to \mathbb{P}^1(\mathbb{K})$ such that $f(A) = A', f(C) = C'$. Since $f(C') = C$, we have $\beta(A', B', C', C) = \beta(A, B, C, C') = \beta(f(A), f(B), f(C), f(C')) = \beta(A', f(B), C', C)$, where the first equality follows from the hypothesis, while the second one is due to the invariance of the cross-ratio with respect to projective transformations. Since the points A', B', C', C are pairwise distinct, the condition $\beta(A', B', C', C) = \beta(A', f(B), C', C)$ implies that $f(B) = B'$, as desired.

(c) We first observe that the points A, B, C are pairwise distinct, because otherwise the line r would contain P_1, against the hypothesis. In a similar way one proves that the points A', B', C' are also pairwise distinct, because otherwise the line r would pass through P_2, P_3 or P_4, against the hypothesis.

Let us set $W = s_{12} \cap s_{34}, T = s_{13} \cap s_{24}, Z = s_{23} \cap s_{14}$. Since P_1, P_2, P_3, P_4 are in general position, the points W, T, Z are not collinear (cf. Exercise 6).

We first consider the case when r does not contain Z. Since $C = r \cap s_{14}$ and $C' = r \cap s_{23}$, we have that $C \neq C'$. Also observe that $C' \neq A$ (since $P_2 \notin r$), and $C' \neq B$ (since $P_3 \notin r$). In a similar way, since $P_4 \notin r$ we have $C \neq A'$ and $C \neq B'$. Therefore, the quadruples A, B, C, C' and A', B', C', C contain only pairwise distinct points, so by point (b) we are left to show that $\beta(A, B, C, C') = \beta(A', B', C', C)$.

Let $g: r \to s_{23}$ be the perspectivity centred at P_1 and $h: s_{23} \to r$ the perspectivity centred at P_4. By construction, g maps the points A, B, C, C' to the points P_2, P_3, Z, C', respectively, and h maps P_2, P_3, Z, C' to B', A', C, C', respectively. We thus have

$$\beta(A, B, C, C') = \beta(h(g(A)), h(g(B)), h(g(C)), h(g(C'))) =$$
$$= \beta(B', A', C, C') = \beta(A', B', C', C),$$

where the first equality is due to the invariance of the cross-ratio under projective isomorphisms, while the last one is due to the symmetries of the cross-ratio (cf. Sect. 1.5.2). This concludes the proof in the case when r does not contain Z.

If $Z \in r$ but $T \notin r$, then we have $B \neq B'$; a similar argument as above shows that the quadruples A, C, B, B' and A', C', B', B contain only pairwise distinct points. By point (b), in order to conclude it is sufficient to show that $\beta(A, C, B, B') = \beta(A', C', B', B)$. This equality can be proved as above by considering first the perspectivity between r and $L(P_2, P_4)$ centred at P_1 and then the perspectivity between $L(P_2, P_4)$ and r centred at P_3.

Finally, if r passes both through Z and through T, then necessarily $W \notin r$, because otherwise W, T, Z would be collinear. Then, a suitable variation of the previous argument allows one to conclude the proof also in this case.

Note. A *quadrilateral* of $\mathbb{P}^2(\mathbb{K})$ is an unordered set of 4 points in general position (called *vertices*) of the projective plane. A *pair of opposite sides* of a quadrilateral Q is a pair of lines whose union contains the vertices of Q (observe that any quadrilateral has exactly three pairs of opposite sides).

Let r be a line that does not contain any vertex of Q, and observe that the union of each pair of opposite sides of Q meets r at exactly two points. Point (c) of Exercise 36 can be reformulated as follows: there exists an involution of r that switches the points of each pair determined on r by a pair of opposite sides of Q.

Exercise 37. Let $\mathbb{P}(V)$ be a projective space of dimension n and let S_1, S_2 be distinct hyperplanes of $\mathbb{P}(V)$. Show that a projective isomorphism $f: S_1 \to S_2$ is a perspectivity if and only if $f(A) = A$ for every $A \in S_1 \cap S_2$.

Solution (1). It is sufficient to show that, if $f: S_1 \to S_2$ is a projective isomorphism that is the identity on $S_1 \cap S_2$, then f is a perspectivity. Let us choose points $P_1, P_2 \in S_1 \setminus (S_1 \cap S_2)$, set $Q_1 = f(P_1)$, $Q_2 = f(P_2)$ and denote by A the point at which the line $L(P_1, P_2)$ meets the subspace $S_1 \cap S_2$. Since f sends collinear points to collinear points and $f(A) = A$ by hypothesis, the point A is collinear with Q_1 and Q_2. So the subspace $L(P_1, P_2, Q_1, Q_2)$ is a plane, and the lines $L(P_1, Q_1)$ and $L(P_2, Q_2)$ meet at a point O. It is immediate to check that $O \notin S_1 \cup S_2$; therefore, we can consider the perspectivity $g: S_1 \to S_2$ centred at O. The projective isomorphisms f and g coincide on P_1, P_2 and on the whole of $S_1 \cap S_2$. Therefore, the Fundamental theorem of projective transformations ensures that $f = g$.

Solution (2). We now give an analytic proof of the statement, by carrying out some computations in a suitably chosen system of homogeneous coordinates $x_0, \ldots, x_n$

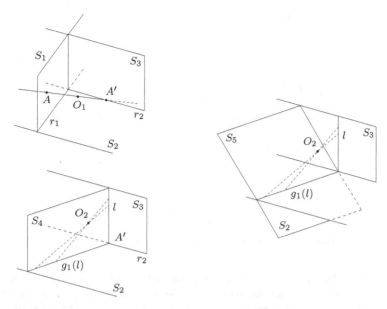

Fig. 2.11 On the *left*: on the *top*, the construction of g_1; on the *bottom*, the choice of l and the construction of S_4. On the *right*: the choice of S_5 and the conclusion of the proof

such that S_1 has equation $x_n = 0$ and S_2 has equation $x_{n-1} = 0$. Using the fact that f is the identity on the subspace of equations $x_{n-1} = x_n = 0$, it is not difficult to show that the map f is described by the formula $[x_0, \dots, x_{n-1}, 0] \mapsto [x_0 + a_0 x_{n-1}, \dots, x_{n-2} + a_{n-2} x_{n-1}, 0, a_n x_{n-1}]$, where $a_0, \dots a_{n-2} \in \mathbb{K}$ and $a_n \in \mathbb{K}^*$. Then, one can check directly that f is the perspectivity centred at $O = [a_0, \dots, a_{n-2}, -1, a_n]$.

Note. Exercise 37 extends point (a) of Exercise 31 (that provides a characterization of perspectivities between lines in a projective plane) to the case of any dimension.

Exercise 38. Let S_1, S_2 be distinct planes of $\mathbb{P}^3(\mathbb{K})$. Prove that any projective isomorphism $f : S_1 \rightarrow S_2$ is the composition of at most three perspectivities.

Solution. Let $r_1 = S_1 \cap S_2$ and take a point $A \in S_1 \setminus r_1$ such that the point $A' = f(A)$ does not belong to r_1 (cf. Fig. 2.11). Let $S_3 \neq S_2$ be a plane passing through A' and not containing A, and choose a point $O_1 \in L(A, A') \setminus \{A, A'\}$. If $\pi_1 : S_3 \rightarrow S_1$ is the perspectivity centred at O_1, then $g_1 = f \circ \pi_1 : S_3 \rightarrow S_2$ is a projective isomorphism. Moreover, $A' \in r_2 = S_3 \cap S_2$ and $g_1(A') = A'$.

Let $l \neq r_2$ be a line contained in S_3 and passing through A'. Since $g_1(A') = A'$, the map g_1 transforms l into a line passing through A'. Let now S_4 be the plane containing l and $g_1(l)$, so that $l = S_3 \cap S_4$. Then Exercise 31 implies that $g_1|_l$ is a perspectivity centred at a point $O_2 \in S_4$ such that $O_2 \notin S_3$.

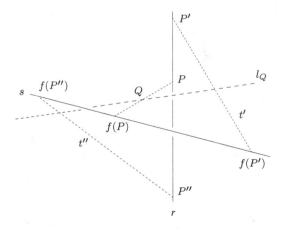

Fig. 2.12 The construction described in the solution of Exercise 39

Let now S_5 be a plane containing $g_1(l)$ and distinct both from S_2 and from S_4, and let us consider the perspectivity $\pi_2 \colon S_5 \to S_3$ centred at O_2. Then $g_1 \circ \pi_2 \colon S_5 \to S_2$ is a projective isomorphism that fixes the line $g_1(l) = S_5 \cap S_2$ pointwise. Now Exercise 37 implies that there exists a perspectivity π_3 such that $g_1 \circ \pi_2 = \pi_3$, so that $f \circ \pi_1 \circ \pi_2 = \pi_3$. This implies the conclusion, since the inverse of a perspectivity is again a perspectivity.

Exercise 39. Let $r, s \subset \mathbb{P}^3(\mathbb{K})$ be skew lines, and let $f \colon r \to s$ be a projective isomorphism. Prove that there exist infinitely many lines l such that f coincides with the perspectivity centred at l.

Solution. Take $P \in r$, and let $t = L(P, f(P))$. We will show that, for every $Q \in t \setminus \{P, f(P)\}$, there exists a line l_Q with the following properties: l_Q passes through Q, l_Q is skew both to r and to s, and f coincides with the perspectivity centred at l_Q. When Q varies in $t \setminus \{P, f(P)\}$, the lines l_Q are pairwise distinct, and this implies the conclusion.

Let us fix $Q \in t \setminus \{P, f(P)\}$, and choose $P', P'' \in r \setminus \{P\}$, such that $P' \neq P''$. We set $t' = L(P', f(P'))$, $t'' = L(P'', f(P''))$ (cf. Fig. 2.12). If t', t'' were not skew, then $P', P'', f(P'), f(P'')$ would be coplanar, so r and s would also be coplanar, a contradiction. In the same way one proves that t is skew both to t' and to t'', so $Q \notin t' \cup t''$. Therefore, Exercise 8 implies that there exists a unique line l_Q that contains Q and meets both t' and t''. Observe that no plane S can contain $l_Q \cup r$ (or $l_Q \cup s$), because otherwise each of the lines t', t'' would contain at least two points of S, thus t' and t'' would be coplanar, a contradiction. So l_Q is skew both to r and to s. By construction, the perspectivity from r to s centred at l_Q maps P to $f(P)$, P' to $f(P')$ and P'' to $f(P'')$. By the Fundamental theorem of projective transformations, we deduce that this perspectivity coincides with f, whence the conclusion.

Note. An alternative solution of Exercise 39 can be obtained by following the strategy described in the Note following Exercise 178.

Exercise 40. Let P_1 and P_2 be distinct points of $\mathbb{P}^2(\mathbb{K})$ and let $\mathcal{F}_1$, $\mathcal{F}_2$ be the pencils of lines centred at P_1 and P_2, respectively. Let $f\colon \mathcal{F}_1 \to \mathcal{F}_2$ be any function. Show that the following facts are equivalent:

(i) f is a projective isomorphism such that $f(L(P_1, P_2)) = L(P_1, P_2)$,
(ii) there exists a line r disjoint from $\{P_1, P_2\}$ and such that $f(s) = L(s \cap r, P_2)$ for every $s \in \mathcal{F}_1$.

Solution. (i) $\Rightarrow$ (ii). Via the duality correspondence (cf. Sect. 1.4.2), the pencils $\mathcal{F}_1$, $\mathcal{F}_2$ correspond to lines l_1, l_2 of the dual projective plane $\mathbb{P}^2(\mathbb{K})^*$. Moreover, the projective isomorphism $f\colon \mathcal{F}_1 \to \mathcal{F}_2$ induces the dual projective isomorphism $f_*\colon l_1 \to l_2$. The condition $f(L(P_1, P_2)) = L(P_1, P_2)$ translates into the fact that f_* fixes the point at which l_1 and l_2 meet, and this implies in turn that f_* is a perspectivity (cf. Exercise 31). Let $R \in \mathbb{P}^2(\mathbb{K})^*$ be the centre of this perspectivity, and let $r \subset \mathbb{P}^2(\mathbb{K})$ be the line corresponding to R via duality. Since $R \notin l_1 \cup l_2$ we have that $P_1 \notin r$ and $P_2 \notin r$, and the fact that $f_*(Q) = L(R, Q) \cap l_2$ for every $Q \in l_1$ implies that $f(s) = L(P_2, r \cap s)$ for every $s \in \mathcal{F}_1$, as desired.

(ii) $\Rightarrow$ (i). For $i = 1, 2$, let $g_i\colon \mathcal{F}_i \to r$ be the map defined by $g_i(s) = s \cap r$. We have shown in Exercise 32 and in the Note following it that the map g_i is a well-defined projective isomorphism. It follows that $f = g_2^{-1} \circ g_1$ is a projective isomorphism. The fact that $f(L(P_1, P_2)) = L(P_1, P_2)$ readily follows from the definition of f.

Exercise 41. In $\mathbb{P}^2(\mathbb{R})$ consider the point $P = [1, 2, 1]$ and the lines

$$l_1 = \{x_0 + x_1 = 0\}, \quad m_1 = \{x_0 + 3x_2 = 0\},$$
$$l_2 = \{x_0 - x_1 = 0\}, \quad m_2 = \{x_2 = 0\},$$
$$l_3 = \{x_0 + 2x_1 = 0\}, \quad m_3 = \{3x_0 + x_2 = 0\}.$$

Determine the points $Q \in \mathbb{P}^2(\mathbb{R})$ for which there exists a projectivity $f\colon \mathbb{P}^2(\mathbb{R}) \to \mathbb{P}^2(\mathbb{R})$ such that $f(P) = Q$ and $f(l_i) = m_i$, $i = 1, 2, 3$.

Solution. First observe that the lines l_1, l_2, l_3 (m_1, m_2, m_3, respectively) belong to the pencil $\mathcal{F}_O$ centred at $O = [0, 0, 1]$ (to the pencil $\mathcal{F}_{O'}$ centred at $O' = [0, 1, 0]$, respectively). Let $r = L(O, P) = \{2x_0 - x_1 = 0\}$.

Suppose now that f is a projectivity such that $f(l_i) = m_i$ for every $i = 1, 2, 3$. Then $f(O) = O'$, and the dual projectivity associated to f induces a projective isomorphism between $\mathcal{F}_O$ and $\mathcal{F}_{O'}$. This isomorphism maps l_i to m_i for $i = 1, 2, 3$. Therefore, if $r' = f(r)$, then $\beta(l_1, l_2, l_3, r) = \beta(f(l_1), f(l_2), f(l_3), f(r)) = \beta(m_1, m_2, m_3, r')$ (cf. Sect. 1.5.1 for the definition and the basic properties of the cross-ratio of concurrent lines). We now fix on $\mathcal{F}_O$ (on $\mathcal{F}_{O'}$, respectively) a projective frame with respect to which the line of equation $ax_0 + bx_1 = 0$ (the line of equation $ax_0 + bx_2 = 0$, respectively) has homogeneous coordinates equal to $[a, b]$. With this choice, the lines $l_1, l_2, l_3, r \in \mathcal{F}_O$ have coordinates $[1, 1], [1, -1], [1, 2], [2, -1]$, respectively, so $\beta(l_1, l_2, l_3, r) = -9$. On the other hand, the lines $m_1, m_2, m_3 \in \mathcal{F}_{O'}$ have coordinates $[1, 3], [0, 1], [3, 1]$, respectively, so if $[a_0, b_0]$ are the coordinates of r' in $\mathcal{F}_{O'}$, then we have

$$-9 = \beta([1, 3], [0, 1], [3, 1], [a_0, b_0]) = \frac{1 \cdot b_0 - 3 \cdot a_0}{1 \cdot 1 - 3 \cdot 3} \cdot \frac{3 \cdot 1 - 0 \cdot 3}{a_0 \cdot 1 - b_0 \cdot 0},$$

hence $[a_0, b_0] = [1, 27]$ and $r' = \{x_0 + 27x_2 = 0\}$. We deduce that $f(P) \in f(r) = r'$. Moreover, since f is injective, we have $f(P) \neq f(O)$, so $f(P) \in r' \setminus \{O'\}$.

Let now $Q \in r' \setminus \{O'\}$, and let us show that there exists a projectivity $f : \mathbb{P}^2(\mathbb{R}) \to \mathbb{P}^2(\mathbb{R})$ such that $f(l_i) = m_i$ for $i = 1, 2, 3$, and $f(P) = Q$. Let $A_1 \in l_1 \setminus \{O\}$ and $A_2 \in l_2 \setminus (\{O\} \cup (l_2 \cap L(A_1, P))$. By construction, the points O, A_1, A_2, P define a projective frame of $\mathbb{P}^2(\mathbb{R})$. In a similar way, if $B_1 \in m_1 \setminus \{O\}$ and $B_2 \in m_2 \setminus (\{O'\} \cup L(B_1, Q))$, then the points O', B_1, B_2, Q provide a projective frame of $\mathbb{P}^2(\mathbb{R})$. Let $f : \mathbb{P}^2(\mathbb{R}) \to \mathbb{P}^2(\mathbb{R})$ be the unique projectivity such that $f(O) = O'$, $f(A_1) = B_1$, $f(A_2) = B_2$, $f(P) = Q$. Since f transforms lines into lines, we obviously have $f(l_1) = m_1, f(l_2) = m_2$, so we are left to show that $f(l_3) = m_3$.

The dual projectivity f_* associated to f maps $\mathcal{F}_O$ to $\mathcal{F}_{O'}$ and satisfies $f_*(l_1) = m_1$, $f_*(l_2) = m_2, f_*(r) = r'$, so thanks to the invariance of the cross-ratio under projective isomorphisms we get

$$\beta(m_1, m_2, m_3, r') = \beta(l_1, l_2, l_3, r) =$$
$$= \beta(f_*(l_1), f_*(l_2), f_*(l_3), f_*(r)) = \beta(m_1, m_2, f_*(l_3), r')$$

and $f(l_3) = m_3$, as desired.

Exercise 42. In $\mathbb{P}^2(\mathbb{R})$ consider the lines l_1, l_2, l_3 having equations $x_2 = 0$, $x_2 - x_1 = 0, x_2 - 2x_1 = 0$, respectively, and the line l_4 having equation $\alpha(x_0 - x_1) + x_2 - 4x_1 = 0$, where $\alpha \in \mathbb{R}$. Consider also the lines m_1, m_2, m_3, m_4 having equations $x_1 = 0, x_1 - x_0 = 0, x_0 = 0, x_1 - \gamma x_0 = 0$, respectively, where $\gamma \in \mathbb{R}$.

Find the values of α and γ for which there exists a projectivity f of $\mathbb{P}^2(\mathbb{R})$ such that $f(l_i) = m_i$ for $i = 1, \ldots, 4$, and that transforms the line $x_0 = 0$ into the line $x_2 = 0$.

Choose a pair of such values, and describe explicitly a projectivity satisfying the required properties.

Solution (1). It is easy to check that the lines l_1, l_2, l_3 all intersect at $R = [1, 0, 0]$, while the lines m_1, m_2, m_3, m_4 all intersect at $S = [0, 0, 1]$. Therefore, a necessary condition for the existence of f is that also the line l_4 passes through R, i.e., that $\alpha = 0$. Moreover, if f exists, then the restriction of f to the line $r = \{x_0 = 0\}$ is a projective isomorphism between r and the line $s = \{x_2 = 0\}$. We observe that r intersects the lines $l_i, i = 1, \ldots, 4$ at the points $P_1 = [0, 1, 0]$, $P_2 = [0, 1, 1]$, $P_3 = [0, 1, 2]$, $P_4 = [0, 1, 4]$, respectively, while s intersects the lines $m_i, i = 1, \ldots, 4$, at the points $Q_1 = [1, 0, 0]$, $Q_2 = [1, 1, 0]$, $Q_3 = [0, 1, 0]$, $Q_4 = [1, \gamma, 0]$, respectively. Therefore, if f esists, then the restriction $f|_r : r \to s$ is a projective isomorphism such that $f(P_i) = Q_i$ for $i = 1, \ldots, 4$, and the cross-ratio $\beta(P_1, P_2, P_3, P_4)$ must coincide with the cross-ratio $\beta(Q_1, Q_2, Q_3, Q_4)$. An easy computation now shows that $\beta(P_1, P_2, P_3, P_4) = \frac{2}{3}$ and $\beta(Q_1, Q_2, Q_3, Q_4) = \frac{\gamma}{\gamma - 1}$. Therefore, by imposing $\frac{\gamma}{\gamma - 1} = \frac{2}{3}$, we obtain a second necessary condition for the existence of f, namely $\gamma = -2$.

We now prove that the conditions $\alpha = 0$ and $\gamma = -2$ are also sufficient for the existence of a projectivity f with the required properties. In fact, if $\gamma = -2$ then there exists a (unique) projective isomorphism $g: r \to s$ such that $g(P_i) = Q_i$ for $i = 1, \ldots, 4$. Let us endow r, s with the systems of homogeneous coordinates $[x_1, x_2]$, $[x_0, x_1]$, respectively, and let B be a 2×2 invertible matrix representing g with respect to these coordinate systems. Then the matrix $A = \begin{pmatrix} 0 & B \\ 0 & \\ \hline 1 & 0\ 0 \end{pmatrix}$ represents (with respect to the standard homogeneous coordinates of $\mathbb{P}^2(\mathbb{R})$) a projectivity f satisfying the required properties.

Let us now construct g so that $g([1, 0]) = [1, 0]$, $g([1, 1]) = [1, 1]$, $g([1, 2]) = [0, 1]$. In this way $g(P_i) = Q_i$ for $i = 1, \ldots, 3$, so also $g(P_4) = Q_4$, due to the condition imposed on the cross-ratios. An easy computation shows that g is induced e.g. by the matrix $B = \begin{pmatrix} 2 & -1 \\ 0 & 1 \end{pmatrix}$. Therefore, the projectivity $\mathbb{P}^2(\mathbb{R})$ induced by the matrix $A = \begin{pmatrix} 0 & 2 & -1 \\ 0 & 0 & 1 \\ 1 & 0 & 0 \end{pmatrix}$ satisfies the required properties.

We observe that, once the necessary conditions $\alpha = 0$ and $\gamma = -2$ have been established, one could also directly construct f as follows. Let U be a point in $l_3 \setminus \{R, P_3\}$, e.g. $U = [1, 1, 2]$, and let V be a point in $m_3 \setminus \{S, Q_3\}$, e.g. $V = [0, 1, 1]$. Since $\{R, P_1, P_2, U\}$ and $\{S, Q_1, Q_2, V\}$ are both projective frames of $\mathbb{P}^2(\mathbb{R})$, there exists a unique projectivity f of $\mathbb{P}^2(\mathbb{R})$ that maps R, P_1, P_2, U to S, Q_1, Q_2, V, respectively. As $r = L(P_1, P_2)$ and $l_3 = L(R, U)$, it easily follows that $f(r) = L(Q_1, Q_2) = s$ and $f(l_3) = L(S, V) = m_3$. Since $P_3 = r \cap l_3$ and $Q_3 = s \cap m_3$, one can conclude that $f(P_3) = Q_3$. Finally, the fact that $f(P_4) = Q_4$ follows from the condition imposed on the cross-ratios.

Solution (2). If we consider a line of $\mathbb{P}^2(\mathbb{R})$ as a point of $\mathbb{P}^2(\mathbb{R})^*$, then the lines described in the statement of the exercise correspond to the points

$$L_1 = [0, 0, 1], \ L_2 = [0, -1, 1], \ L_3 = [0, -2, 1], \ L_4 = [\alpha, -\alpha - 4, 1], \ R = [1, 0, 0]$$

$$M_1 = [0, 1, 0], \ M_2 = [-1, 1, 0], \ M_3 = [1, 0, 0], \ M_4 = [-\gamma, 1, 0], \ S = [0, 0, 1],$$

respectively. Therefore, we are looking for a projectivity g of $\mathbb{P}^2(\mathbb{R})^*$ such that $g(L_i) = M_i$ for $i = 1, \ldots, 4$, and $g(R) = S$. Since M_1, M_2, M_3, M_4 are collinear (as they correspond to lines in a pencil), the points L_1, L_2, L_3, L_4 must be collinear too, and this occurs if and only if $\alpha = 0$. Moreover, a necessary (and sufficient) condition for the existence of a projectivity g such that $g(L_i) = M_i$ for $i = 1, \ldots, 4$ is that the cross-ratio $\beta(L_1, L_2, L_3, L_4)$ coincides with the cross-ratio $\beta(M_1, M_2, M_3, M_4)$; it can be checked that this occurs if and only if $\gamma = -2$.

An easy computation (similar to the one carried out in solution (1)) shows that the projectivity g associated e.g. to the matrix $H = \begin{pmatrix} 0 & 1 & 0 \\ 0 & 1 & 2 \\ 1 & 0 & 0 \end{pmatrix}$ is such that $g(L_i) = M_i$ for $i = 1, \dots, 4$ and $g(R) = S$.

We now come back to $\mathbb{P}^2(\mathbb{R})$ via duality. Thus, if we consider the points of $\mathbb{P}^2(\mathbb{R})^*$ as lines of $\mathbb{P}^2(\mathbb{R})$, then the matrix ${}^t H^{-1} = \begin{pmatrix} 0 & 1 & -\frac{1}{2} \\ 0 & 0 & \frac{1}{2} \\ 1 & 0 & 0 \end{pmatrix}$ induces a projectivity that satisfies all the required properties.

Exercise 43. (a) Let r, s be distinct lines of $\mathbb{P}^2(\mathbb{K})$ passing through the point A. Let B, C, D be pairwise distinct points of $r \setminus \{A\}$, and let B', C', D' be pairwise distinct points of $s \setminus \{A\}$. Prove that the lines $L(B, B')$, $L(C, C')$, $L(D, D')$ all intersect if and only if $\beta(A, B, C, D) = \beta(A, B', C', D')$.

(b) Let A, B be distinct points of a line r of $\mathbb{P}^2(\mathbb{K})$. Let r_1, r_2, r_3 be pairwise distinct lines passing through A and distinct from r, and let s_1, s_2, s_3 be pairwise distinct lines passing through B and distinct from r. Prove that the points $r_1 \cap s_1, r_2 \cap s_2, r_3 \cap s_3$ are collinear if and only if $\beta(r, r_1, r_2, r_3) = \beta(r, s_1, s_2, s_3)$.

(c) Let A, B be pairwise distinct points of $\mathbb{P}^2(\mathbb{K})$ and let $\mathcal{F}_A$ ($\mathcal{F}_B$, respectively) be the pencil of lines centred at A (at B, respectively). Let $f \colon \mathcal{F}_A \to \mathcal{F}_B$ be a projective isomorphism such that $f(L(A, B)) = L(A, B)$. Prove that the set

$$Q = \bigcup_{s \in \mathcal{F}_A} s \cap f(s)$$

is the union of two distinct lines.

Solution. (a) We denote by O the point at which the distinct lines $L(B, B')$ and $L(C, C')$ intersect; it is immediate to check that $O \notin r \cup s$. Let $\varphi_O \colon r \to s$ be the perspectivity centred at O, which maps the points A, B, C to the points A, B', C', respectively. Since the cross-ratio is invariant under projective isomorphisms, we have $\beta(A, B, C, D) = \beta(A, B', C', \varphi_O(D))$. Therefore, $\beta(A, B, C, D) = \beta(A, B', C', D')$ if and only if $\varphi_O(D) = D'$, i.e., if and only if $L(D, D')$ passes through O.

(b) Via duality, the lines of the pencil centred at A (at B, respectively) form a line of the dual projective plane. Using this, it is not difficult to show that the statement of (b) is just the dual statement of (a), so the conclusion follows from the Duality principle (cf. Theorem 1.4.1).

(c) By hypothesis $f(L(A, B)) = L(A, B)$, so $L(A, B) \subseteq Q$. Let r_1, r_2 be distinct lines passing through A and distinct from $L(A, B)$. Observe that $r_i \neq f(r_i)$ for $i = 1, 2$, and $r_1 \cap f(r_1), r_2 \cap f(r_2)$ are distinct points. If t is the line joining the points $r_1 \cap f(r_1)$ and $r_2 \cap f(r_2)$, then point (b) implies that, for every $r \in \mathcal{F}_A \setminus \{L(A, B)\}$, the point $r \cap f(r)$ belongs to t, and so $Q \subseteq L(A, B) \cup t$. On the other hand, for every $P \in t$, if $l = L(A, P)$ then $f(l)$ is a line passing through B and such that $f(l) \cap l \in t$, so $f(l) \cap l = P$. This proves that every point of t belongs to Q, thus $Q = L(A, B) \cup t$.

Note. It is possible to prove point (c) via an argument based on the Duality principle. In fact, the statement of Exercise 40 implies that, if $f: \mathcal{F}_A \to \mathcal{F}_B$ is a projective isomorphism such that $f(L(A, B)) = L(A, B)$, then there exists a line $l \subseteq \mathbb{P}^2(\mathbb{C})$ such that $A \notin l$, $B \notin l$, and $f(s) = L(B, l \cap s)$ for every $s \in \mathcal{F}_A$. This proves that, if $s \in \mathcal{F}_A$ is distinct from $r = L(A, B)$, then $s \cap f(s) = l \cap s$, so in particular $s \cap f(s) \subset l$. It is now easy to show (as described above) that indeed $\mathcal{Q} = l \cup r$.

Exercise 44. (*Invariant sets of projectivities of the projective plane*) Let f be a projectivity of $\mathbb{P}^2(\mathbb{K})$, where $\mathbb{K} = \mathbb{C}$ or $\mathbb{K} = \mathbb{R}$. Find the possible configurations of fixed points, invariant lines, axes (an axis is a pointwise fixed line) and centres (a centre is a fixed point such that every line passing through it is invariant) of f.

Solution. If $\mathbb{K} = \mathbb{C}$, in a suitable system of homogeneous coordinates the projectivity f is represented by one of the following Jordan matrices (see also Sect. 1.5.3):

(a) $A = \begin{pmatrix} 1 & 0 & 0 \\ 0 & \lambda & 0 \\ 0 & 0 & \mu \end{pmatrix}$ with $\lambda, \mu \in \mathbb{K} \setminus \{0, 1\}$, $\lambda \neq \mu$;

(b) $A = \begin{pmatrix} 1 & 0 & 0 \\ 0 & 1 & 0 \\ 0 & 0 & \lambda \end{pmatrix}$ with $\lambda \in \mathbb{K} \setminus \{0, 1\}$;

(c) $A = \begin{pmatrix} 1 & 0 & 0 \\ 0 & 1 & 0 \\ 0 & 0 & 1 \end{pmatrix}$;

(d) $A = \begin{pmatrix} 1 & 1 & 0 \\ 0 & 1 & 0 \\ 0 & 0 & \lambda \end{pmatrix}$ with $\lambda \in \mathbb{K} \setminus \{0, 1\}$;

(e) $A = \begin{pmatrix} 1 & 1 & 0 \\ 0 & 1 & 0 \\ 0 & 0 & 1 \end{pmatrix}$;

(f) $A = \begin{pmatrix} 1 & 1 & 0 \\ 0 & 1 & 1 \\ 0 & 0 & 1 \end{pmatrix}$.

Let now $\mathbb{K} = \mathbb{R}$. If a matrix (hence, every matrix) representing f has only real eigenvalues, then in a suitable system of homogeneous coordinates of $\mathbb{P}^2(\mathbb{R})$ the projectivity f is represented by one of the matrices listed above. Otherwise, every matrix representing f has one real and two conjugate non-real eigenvalues. In this case, there exists a system of homogeneous coordinates of $\mathbb{P}^2(\mathbb{R})$ with respect to which f is represented by the matrix

(g) $A = \begin{pmatrix} a & -b & 0 \\ b & a & 0 \\ 0 & 0 & 1 \end{pmatrix}$ with $a \in \mathbb{R}$, $b \in \mathbb{R}^*$.

Let us now analyze the cases listed above, focusing on the fixed-point set and on the existence of invariant lines, axes and centres.

Recall that $P = [X]$ is a fixed point of f if and only if X is an eigenvector of A, and a line r of equation ${}^t\!CX = 0$ is invariant under f if and only if C is an eigenvector of the matrix ${}^t\!A$ (cf. Sect. 1.4.5).

We set $P_0 = [1, 0, 0]$, $P_1 = [0, 1, 0]$, $P_2 = [0, 0, 1]$ and we denote by r_i the line of equation $x_i = 0$ for $i = 0, 1, 2$.

Case (a). The eigenspaces of the matrix A are the lines generated by $(1, 0, 0)$, $(0, 1, 0)$, $(0, 0, 1)$, respectively, so f has three fixed points, namely the points P_0, P_1, P_2. Moreover, f admits three invariant lines, namely the lines r_0, r_1, r_2. We observe that $f|_{r_i}$ is a hyperbolic projectivity of r_i for $i = 0, 1, 2$.

Case (b). The eigenvectors of the matrix A are of the form $(0, 0, c)$ or $(a, b, 0)$, so P_2 is a fixed point and the line r_2 consists of fixed points, i.e., it is an axis. Any line passing through P_2 has an equation of the form $ax_0 + bx_1 = 0$; since $(a, b, 0)$ is an eigenvector of ${}^t\!A = A$, every line passing through P_2 is invariant, so P_2 is a center. Summarizing, we have a centre P_2 and an axis r_2 such that $P_2 \notin r_2$.

Case (c). In this case f is the identity, so every point is a centre and every line is an axis.

Case (d). The same argument as above shows that f has two fixed points, namely the points P_0 and P_2, and two invariant lines, namely the lines r_1 and r_2. Therefore, there are no centres and no axes. We observe that $f|_{r_1}$ is a hyperbolic projectivity and $f|_{r_2}$ is a parabolic projectivity.

Case (e). The eigenvectors of A are of the form $(a, 0, b)$, so the line r_1 is an axis. Moreover, the eigenvectors of ${}^t\!A$ are of the form $(0, a, b)$, so any line of equation $ax_1 + bx_2 = 0$ is invariant; these lines are exactly the elements of the pencil centred at P_0, so P_0 is a center. In this case, the centre P_0 belongs to the axis r_1.

Case (f). It is readily seen that there exists a unique fixed point P_0 and, by looking at ${}^t\!A$, a unique invariant line r_2. We observe that the fixed point P_0 belongs to the invariant line r_2, and $f|_{r_2}$ is a parabolic projectivity.

Case (g). Recall that this case occurs only when $\mathbb{K} = \mathbb{R}$ and A has one real and two conjugate non-real eigenvalues. Therefore, P_2 is the unique fixed point and r_2 is the unique invariant line; moreover, the invariant line does not contain the fixed point, and the restriction $f|_{r_2}$ is an elliptic projectivity.

Note. The solution of Exercise 44 shows that projectivities of $\mathbb{P}^2(\mathbb{K})$ can be partitioned into 6 (when $\mathbb{K} = \mathbb{C}$) or 7 (when $\mathbb{K} = \mathbb{R}$) disjoint families, which can be characterized according to the number of fixed points, invariant lines, centres and axes, and to the incidence relations that hold among these objects. These families are classified by the Jordan form (when $\mathbb{K} = \mathbb{C}$) or by the "real Jordan form" (when $\mathbb{K} = \mathbb{R}$) of the associated matrices.

Exercise 45. Let $P = [1, 1, 1]$ and $l = \{x_0 + x_1 - 2x_2 = 0\}$ be a point and a line of $\mathbb{P}^2(\mathbb{R})$, respectively. Find an explicit formula for a projectivity $f \colon \mathbb{P}^2(\mathbb{R}) \to \mathbb{P}^2(\mathbb{R})$ such that $f(l) = l$ and that the fixed-point set of f coincides with $\{P\}$.

Solution. If g is the projectivity of $\mathbb{P}^2(\mathbb{R})$ induced by the matrix $B = \begin{pmatrix} 1 & 1 & 0 \\ 0 & 1 & 1 \\ 0 & 0 & 1 \end{pmatrix}$,

then $[1, 0, 0]$ is the unique fixed point of g, and it is contained in the line $\{x_2 = 0\}$, which is invariant under g (cf. Exercise 44). Therefore, if $h\colon \mathbb{P}^2(\mathbb{R}) \to \mathbb{P}^2(\mathbb{R})$ is a projectivity such that $h([1, 0, 0]) = P$, $h(\{x_2 = 0\}) = l$, then the map $h \circ g \circ h^{-1}$ gives the required projectivity.

Since $l = \mathbb{P}(W)$, where $W \subseteq \mathbb{R}^3$ is the linear subspace generated by $(1, 1, 1)$ and $(2, 0, 1)$, such a projectivity h is induced for example by the invertible matrix $A = \begin{pmatrix} 1 & 2 & 1 \\ 1 & 0 & 0 \\ 1 & 1 & 0 \end{pmatrix}$. As $A^{-1} = \begin{pmatrix} 0 & 1 & 0 \\ 0 & -1 & 1 \\ 1 & 1 & -2 \end{pmatrix}$, we have $ABA^{-1} = \begin{pmatrix} 3 & 1 & -3 \\ 0 & 0 & 1 \\ 1 & 0 & 0 \end{pmatrix}$, so we can set $f([x_0, x_1, x_2]) = [3x_0 + x_1 - 3x_2, x_2, x_0]$.

Exercise 46. In $\mathbb{P}^2(\mathbb{R})$ consider the points

$$P_1 = [1, 0, 0], \quad P_2 = [0, 1, 0], \quad P_3 = [0, 0, 1], \quad P_4 = [1, 1, 1],$$

$$Q_1 = [1, -1, -1], \quad Q_2 = [1, 3, 1], \quad Q_3 = [1, 1, -1], \quad Q_4 = [1, 1, 1].$$

(a) Provide an explicit formula for a projectivity f of $\mathbb{P}^2(\mathbb{R})$ such that $f(P_i) = Q_i$ for $i = 1, 2, 3, 4$. Is such a projectivity unique?
(b) Find all the lines of $\mathbb{P}^2(\mathbb{R})$ that are invariant under f.

Solution. (a) The points P_1, P_2, P_3, P_4 are in general position, so they define a projective frame with associated normalized basis $\{(1, 0, 0), (0, 1, 0), (0, 0, 1)\}$. In a similar way, an easy computation shows that Q_1, Q_2, Q_3, Q_4 define a projective frame with associated normalized basis $\{(1, -1, -1), (1, 3, 1), (-1, -1, 1)\}$. By the Fundamental theorem of projective transformations, there exists a unique projectivity f satisfying the required properties, and such an f is induced by the linear map defined by the matrix $B = \begin{pmatrix} 1 & 1 & -1 \\ -1 & 3 & -1 \\ -1 & 1 & 1 \end{pmatrix}$.

(b) An easy computation shows that the characteristic polynomial of B is $(2 - t)^2(1 - t)$. The eigenspace of B relative to the eigenvalue 2 is 2-dimensional and it is described by the equation $x_0 - x_1 + x_2 = 0$, where (x_0, x_1, x_2) are the standard coordinates of $\mathbb{R}^3$. The eigenspace of B relative to the eigenvalue 1 is 1-dimensional, and it is generated by the vector $v = (1, 1, 1)$. Now it follows from Exercise 44 that, if r is the line of $\mathbb{P}^2(\mathbb{R})$ of equation $x_0 - x_1 + x_2 = 0$ and $P = [1, 1, 1] = [v] \in \mathbb{P}^2(\mathbb{R})$, then r is pointwise fixed by f, and a line of $\mathbb{P}^2(\mathbb{R})$ distinct from r is f-invariant if and only if it passes through P.

Exercise 47. In $\mathbb{P}^2(\mathbb{R})$ consider the points

$$P_1 = [1, 0, 0], \quad P_2 = [0, -1, 1], \quad P_3 = [0, 0, -1], \quad P_4 = [1, -1, 2],$$

$$Q_1 = [3, 1, -1], \quad Q_2 = [-1, -3, 3], \quad Q_3 = [-1, 1, 3], \quad Q_4 = [1, -1, 5].$$

(a) Construct a projectivity f of $\mathbb{P}^2(\mathbb{R})$ such that $f(P_i) = Q_i$ for $i = 1, 2, 3, 4$. Is such a projectivity unique?

(b) Prove that there exist a point P and a line r such that $P \notin r$, $f(P) = P$ and r is pointwise fixed by f.

(c) Let s be a line passing through P and set $Q = s \cap r$. Prove that the cross-ratio $\beta(P, Q, R, f(R))$ is independent of s and of R, when s varies in the pencil of lines centred at P and R varies in $s \setminus \{P, Q\}$.

Solution. It is easy to check that the points P_1, P_2, P_3, P_4 define a projective frame of $\mathbb{P}^2(\mathbb{R})$ with associated normalized basis $v_1 = (1, 0, 0)$, $v_2 = (0, -1, 1)$, $v_3 = (0, 0, 1)$. Moreover, the points Q_1, Q_2, Q_3, Q_4 define a projective frame of $\mathbb{P}^2(\mathbb{R})$ with associated normalized basis $w_1 = (3, 1, -1)$, $w_2 = (-1, -3, 3)$, $w_3 = (-1, 1, 3)$. Therefore, by the Fundamental theorem of projective transformations, there exists a unique projectivity f of $\mathbb{P}^2(\mathbb{R})$ such that $f(P_i) = Q_i$ for $i = 1, 2, 3, 4$, and this projectivity is induced e.g. by the linear map $\varphi \colon \mathbb{R}^3 \to \mathbb{R}^3$ such that $\varphi(v_i) = w_i$ for $i = 1, 2, 3$. Since $(0, 1, 0) = -v_2 + v_3$ and $-w_2 + w_3 = (0, 4, 0)$, the map φ is represented, with respect to the canonical basis of $\mathbb{R}^3$, by the matrix $A = \begin{pmatrix} 3 & 0 & -1 \\ 1 & 4 & 1 \\ -1 & 0 & 3 \end{pmatrix}$.

The characteristic polynomial of A is equal to $(4 - t)^2(2 - t)$. It is immediate to check that $\dim \ker(A - 4I) = 2$, so φ admits a 2-dimensional eigenspace V_4 relative to the eigenvalue 4, and a 1-dimensional eigenspace V_2 relative to the eigenvalue 2. It follows that $P = \mathbb{P}(V_2)$ and $r = \mathbb{P}(V_4)$ are the point and the line requested in (b). An easy computation shows that $P = [1, -1, 1]$ and that r has equation $x_0 + x_2 = 0$.

Let us now take $Q \in r$. We have seen in Sect. 1.5.4 that, since $P = [v]$, where v is an eigenvector of φ relative to the eigenvalue 2, and $Q = [w]$, where w is an eigenvector of φ relative to the eigenvalue 4, for every $R \in L(P, Q) \setminus \{P, Q\}$ we have $\beta(P, Q, R, f(R)) = 4/2 = 2$.

Exercise 48. Let r, s be distinct lines of $\mathbb{P}^2(\mathbb{R})$ and set $R = r \cap s$. Let A, B, C be pairwise distinct points of $r \setminus \{R\}$, and let $g \colon r \to r$ be the unique projectivity such that $g(A) = A$, $g(R) = R$ and $g(B) = C$. For every $P \in \mathbb{P}^2(\mathbb{R}) \setminus r$, set $t(P) = L(B, P) \cap s$, and $h(P) = L(C, t(P)) \cap L(A, P)$.

(a) Prove that there exists a unique projectivity $f \colon \mathbb{P}^2(\mathbb{R}) \to \mathbb{P}^2(\mathbb{R})$ such that $f|_{\mathbb{P}^2(\mathbb{R}) \setminus r} = h$ and $f|_r = g$.

(b) Find the fixed points of f.

(c) Show that f is an involution if and only if $\beta(A, R, B, C) = -1$ (Fig. 2.13).

Solution. If M, N are distinct points of $s \setminus \{R\}$, then the points A, B, M, N define a projective frame of $\mathbb{P}^2(\mathbb{R})$. Let us fix the system of homogeneous coordinates induced by this frame. Then $A = [1, 0, 0]$, $B = [0, 1, 0]$, $r = \{x_2 = 0\}$, $s = \{x_0 = x_1\}$, $R = [1, 1, 0]$, and $C = [1, \beta, 0]$ with $\beta \neq 0, 1$.

(a) Take $P \in \mathbb{P}^2(\mathbb{R}) \setminus r$. Then $P = [a, b, c]$ with $c \neq 0$. The line $L(B, P)$ has equation $cx_0 - ax_2 = 0$, so $t(P) = L(B, P) \cap s = [a, a, c]$. Therefore, the line $L(C, t(P))$ has equation $-c\beta x_0 + cx_1 + a(\beta - 1)x_2 = 0$. Since $L(A, P)$ has equation $cx_1 - bx_2 = 0$, we thus get $h(P) = [a(\beta - 1) + b, \beta b, \beta c]$. It follows that, on the set $\mathbb{P}^2(\mathbb{R}) \setminus r$, the map h coincides with the projectivity $f \colon \mathbb{P}^2(\mathbb{R}) \to \mathbb{P}^2(\mathbb{R})$ represented

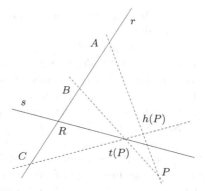

Fig. 2.13 The construction described in Exercise 48

(with respect to the chosen coordinates) by the invertible matrix $\begin{pmatrix} \beta - 1 & 1 & 0 \\ 0 & \beta & 0 \\ 0 & 0 & \beta \end{pmatrix}$.

Therefore, in order to prove (a) we are left to show that $f|_r = g$. However, since $f([x_0, x_1, x_2]) = [(\beta - 1)x_0 + x_1, \beta x_1, \beta x_2]$ for every $[x_0, x_1, x_2] \in \mathbb{P}^2(\mathbb{R})$, we have $f(A) = f([1, 0, 0]) = [1, 0, 0] = A, f(R) = f([1, 1, 0]) = [1, 1, 0] = R$ and $f(B) = f([0, 1, 0]) = [1, \beta, 0] = C$, so $f|_r$ and g coincide on three pairwise distinct points of r, so that $f|_r = g$.

(b) If $P \in s \setminus \{R\}$, then $t(P) = P$, and $f(P) = h(P) = P$. Moreover, we have proved in point (a) that $f(R) = R$ and $f(A) = A$. Therefore, every point of $s \cup \{A\}$ is fixed by f. Now, if $f(M) = M$ for some $M \notin s \cup \{A\}$, then f is the identity (cf. Sect. 1.2.5), and this contradicts the fact that $f(B) \neq B$. We have thus shown that the fixed-point set of f is equal to $s \cup \{A\}$.

(c) Since $f(A) = A$, $f(R) = R$ and $f(B) = C$, Exercise 23 implies that $f^2|_r = (f|_r)^2 = \mathrm{Id}_r$ if and only if $\beta(A, R, B, C) = -1$. Therefore, if f is an involution, then $\beta(A, R, B, C) = -1$. Conversely, let us suppose that $\beta(A, R, B, C) = -1$. Then $f^2|_r = \mathrm{Id}_r$. But $f|_s = \mathrm{Id}_s$, so $f^2|_{r\cup s} = \mathrm{Id}_{r\cup s}$. Since the fixed-point set of a projectivity is given by the union of pairwise skew projective subspaces (cf. Sect. 1.2.5), we deduce that f^2 coincides with the identity on $L(r, s) = \mathbb{P}^2(\mathbb{R})$, and this concludes the proof.

Exercise 49. Let $f : \mathbb{P}^2(\mathbb{K}) \to \mathbb{P}^2(\mathbb{K})$ be a non-trivial involution.

(a) Show that the fixed-point set of f is given by $l \cup \{P\}$, where l, P are a line and a point of $\mathbb{P}^2(\mathbb{K})$, respectively, and $P \notin l$.

(b) Show that there exists an affine chart $h : \mathbb{P}^2(\mathbb{K}) \setminus l \to \mathbb{K}^2$ such that $h(f(h^{-1}(v))) = -v$ for every $v \in \mathbb{K}^2$.

Solution. (a) We first observe that the minimal polynomial of a linear endomorphism g of $\mathbb{K}^3$ cannot be irreducible of degree 2. In fact, if this were the case, then g would not have any eigenvalue. However, the characteristic polynomial of g would be divisible by the minimal polynomial of g. Therefore, having degree 3, the characteristic polynomial of g would be divisible by a linear factor, and this would imply that g admits an eigenvalue, a contradiction.

Let now $\varphi \colon \mathbb{K}^3 \to \mathbb{K}^3$ be a linear map inducing f, and let m be the minimal polynomial of φ. As $f \neq \mathrm{Id}$, we have that $\deg m \geq 2$. Let us show that m is the product of two non-proportional linear factors. By hypothesis there exists $\lambda \in \mathbb{K}^*$ such that $\varphi^2 = \lambda \, \mathrm{Id}_{\mathbb{K}^3}$. If λ is not a square in $\mathbb{K}$ then m, being a factor of $t^2 - \lambda$, is irreducible of degree 2, and this contradicts what we have proved above. So let $\alpha \in \mathbb{K}$ be a square root of λ. Since m divides $t^2 - \lambda = (t - \alpha)(t + \alpha)$ and $\deg m \geq 2$, we have $m = (t - \alpha)(t + \alpha)$. Also observe that, since $\alpha \neq 0$ and $\mathbb{K} \subseteq \mathbb{C}$, we have $\alpha \neq -\alpha$, so m is indeed the product of two non-proportional linear factors.

Therefore, $\mathbb{K}^3$ decomposes as the direct sum of two eigenspaces W_1, W_2 of φ having dimensions 1 and 2, respectively. If $P = \mathbb{P}(W_1)$, $l = \mathbb{P}(W_2)$, then $P \notin l$, and the fixed-point set of f coincides with $\{P\} \cup l$.

(b) After replacing φ by $\alpha^{-1}\varphi$ or by $-\alpha^{-1}\varphi$, we may suppose that $\varphi|_{W_1} = \mathrm{Id}_{W_1}$ and $\varphi|_{W_2} = -\mathrm{Id}_{W_2}$. Now, if $v_1 \in W_1 \setminus \{0\}$ and $\{v_2, v_3\}$ is a basis of W_2, then $\{v_1, v_2, v_3\}$ is a basis of $\mathbb{K}^3$. It is not difficult to show that, with respect to the homogeneous coordinates of $\mathbb{P}^2(\mathbb{K})$ induced by $\{v_1, v_2, v_3\}$, the map f is described by the formula $f([x_0, x_1, x_2]) = [x_0, -x_1, -x_2]$. Moreover, with respect to these coordinates we have $l = \mathbb{P}(\{x_0 = 0\})$. After setting $h \colon \mathbb{P}^2(\mathbb{K}) \setminus l \to \mathbb{K}^2$, $h[1, x, y] = (x, y)$, we finally have $h(f(h^{-1}(v))) = -v$ for every $v \in \mathbb{K}^2$.

Note. If $\mathbb{K} = \mathbb{C}$ or $\mathbb{K} = \mathbb{R}$, Exercise 49 admits an easier solution that makes use of the Jordan form (or of the "real Jordan form") (cf. Exercise 44).

Exercise 50. Let $f \colon \mathbb{P}^2(\mathbb{Q}) \to \mathbb{P}^2(\mathbb{Q})$ be a projectivity such that $f^4 = \mathrm{Id}$, $f^2 \neq \mathrm{Id}$. Compute the number of fixed points of f.

Solution (1). Let $\varphi \colon \mathbb{Q}^3 \to \mathbb{Q}^3$ be a linear map inducing f, and let $m, p \in \mathbb{Q}[t]$ be the minimal and the characteristic polynomial of φ, respectively. Since the fixed-point set of f coincides with the projection on $\mathbb{P}^2(\mathbb{Q})$ of the set of eigenvectors of φ, we analyze the number and the dimensions of the eigenspaces of φ. To this aim, we study the factorization of m and p.

Since $f^4 = \mathrm{Id}_{\mathbb{P}^2(\mathbb{Q})}$, there exists $\lambda \in \mathbb{Q}^*$ such that $\varphi^4 = \lambda \, \mathrm{Id}_{\mathbb{Q}^3}$, so m divides $t^4 - \lambda$ in $\mathbb{Q}[t]$. Let us prove that λ is positive. If λ were negative, then $t^4 - \lambda$ would not have any rational root, so φ would not have any eigenvalue. As a consequence, p would not have any rational root so it would be irreducible, being of degree 3. By Hamilton-Cayley Theorem, we would have $m = p$, so m would have degree 3. Being divided by m, the polynomial $t^4 - \lambda$ would be divided by a linear factor, thus admitting a rational root. So λ would be positive, against our hypothesis. We have thus shown that λ is positive.

Let us prove that p is the product of a linear factor and an irreducible factor of degree 2. This implies that φ has exactly one eigenspace, and that this eigenspace is one-dimensional, so that f necessarily has exactly one fixed point. So let $\alpha \in \mathbb{R}^*$ be the positive fourth root of λ. If p were irreducible over $\mathbb{Q}$, then Hamilton-Cayley Theorem would imply that $m = p$, so the polynomial $t^4 - \lambda = (t - \alpha)(t + \alpha)(t - i\alpha)(t + i\alpha)$ would be divided by a factor of degree 3 and irreducible over

Q. By analyzing the cases $\alpha \in \mathbb{Q}$, $\alpha \in \mathbb{R} \setminus \mathbb{Q}$, it can be easily shown that this is impossible, so p is reducible, and it is divided by a linear factor. As a consequence, φ admits an eigenvalue, so α is rational. We now observe that $t^2 + \alpha^2$ must divide m, because otherwise m would divide $t^2 - \alpha^2$, and we would have $f^2 = \text{Id}$. Therefore, we may conclude that either $p = m = (t - \alpha)(t^2 + \alpha^2)$ or $p = m = (t + \alpha)(t^2 + \alpha^2)$, as desired.

Solution (2). Set $g = f^2$. We have seen in Exercise 49 that the fixed-point set of g is given by $\{P\} \cup l$, where P is a point and l is a line not containing P. Now, if a point Q is fixed by g, then $g(f(Q)) = f^3(Q) = f(g(Q)) = f(Q)$, and $f(Q)$ is also fixed by g. It follows that $f(\{P\} \cup l) \subseteq \{P\} \cup l$, and $f(P) = P, f(l) = l$, since f is a projectivity. So P is a fixed point of f.

Suppose now that $Q \neq P$ is another fixed point of f, and set $s = L(P, Q)$. Of course we have $f(s) = s$ and, if s is endowed with homogeneous coordinates such that $P = [1, 0]$, $Q = [0, 1]$, then the map $f|_s$ can be represented by a matrix of the form $\begin{pmatrix} \alpha & 0 \\ 0 & \beta \end{pmatrix}$, $\alpha, \beta \in \mathbb{Q}^*$. From $f^4 = \text{Id}$ it follows that $\alpha^4 = \beta^4$, so $\alpha^2 = \beta^2$, and $g|_s = f^2|_s = \text{Id}_s$, which contradicts the fact that the fixed-point set of g is contained in $\{P\} \cup l$, while the line s passes through P, so it contains points in $\mathbb{P}^2(\mathbb{K}) \setminus (\{P\} \cup l)$. This proves that P is the unique fixed point of f.

Note. Let $\mathcal{H}$ be the set of projectivities that satisfy the conditions described in the statement. The solutions above show that any $f \in \mathcal{H}$ has exactly one fixed point, but they do not exclude the possibility that $\mathcal{H} = \emptyset$ (and, if this were the case, then any answer to the question "How many fixed points has f?" would be correct!). However, it is easy to check that the map $f \colon \mathbb{P}^2(\mathbb{Q}) \to \mathbb{P}^2(\mathbb{Q})$ defined by $f([x_0, x_1, x_2]) = [x_1, -x_0, x_2]$ belongs to the set $\mathcal{H}$.

Note. It is easy to check that the solutions above may be adapted to deal with the case when the field $\mathbb{Q}$ is replaced by the field $\mathbb{R}$. Therefore, the statement of Exercise 50 is still true if we replace $\mathbb{P}^2(\mathbb{Q})$ by $\mathbb{P}^2(\mathbb{R})$.

On the contrary, every projectivity f of $\mathbb{P}^2(\mathbb{C})$ such that $f^4 = \text{Id}$ is induced by a diagonalizable linear map, so it admits at least three fixed points. Moreover, in this case the conditions $f^4 = \text{Id}, f^2 \neq \text{Id}$ are not sufficient to determine the number of fixed points of f: if $g, h \colon \mathbb{P}^2(\mathbb{C}) \to \mathbb{P}^2(\mathbb{C})$ are the projectivities induced by the matrices $\begin{pmatrix} 1 & 0 & 0 \\ 0 & 1 & 0 \\ 0 & 0 & i \end{pmatrix}$, $\begin{pmatrix} 1 & 0 & 0 \\ 0 & -1 & 0 \\ 0 & 0 & i \end{pmatrix}$, respectively, then $g^4 = h^4 = \text{Id}, g^2 \neq \text{Id}, h^2 \neq \text{Id}$, but g and h do not have the same number of fixed points.

Exercise 51. Let P_1, P_2, P_3 be points of $\mathbb{P}^2(\mathbb{K})$ in general position; let r be a line such that $P_i \notin r$ for $i = 1, 2, 3$.

(a) Prove that there exists a unique projectivity f of $\mathbb{P}^2(\mathbb{K})$ such that

$$f(P_1) = P_1, \quad f(P_2) = P_3, \quad f(P_3) = P_2, \quad f(r) = r.$$

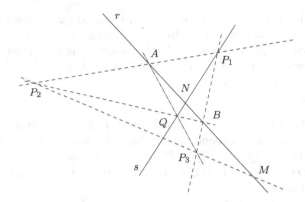

Fig. 2.14 The construction described in the solution of Exercise 51

(b) Prove that the fixed-point set of f is the union of a point $M \in r$ and a line $s \subseteq \mathbb{P}^2(\mathbb{K})$ such that $M \notin s$.

Solution. (a) Set $A = L(P_1, P_2) \cap r$, $B = L(P_1, P_3) \cap r$ (cf. Fig. 2.14). It is immediate to check that A, B, P_2, P_3 define a projective frame of $\mathbb{P}^2(\mathbb{K})$.

If f is a projectivity satisfying the required properties, then $f(L(P_1, P_2)) = L(f(P_1), f(P_2)) = L(P_1, P_3)$, so $f(A) = f(r \cap L(P_1, P_2)) = r \cap L(P_1, P_3) = B$. In a similar way one shows that $f(B) = A$ so, as by hypothesis $f(P_2) = P_3$ and $f(P_3) = P_2$, the Fundamental theorem of projective transformations implies that the required projectivity is unique, if it exists.

So let $f: \mathbb{P}^2(\mathbb{K}) \to \mathbb{P}^2(\mathbb{K})$ be the unique projectivity such that $f(P_2) = P_3$, $f(P_3) = P_2, f(A) = B, f(B) = A$. We have

$$f(P_1) = f(L(A, P_2) \cap L(B, P_3)) = L(B, P_3) \cap L(A, P_2) = P_1$$

and

$$f(r) = f(L(A, B)) = L(f(A), f(B)) = L(B, A) = r,$$

so f satisfies the properties described in the statement.

(b) If $M = L(P_2, P_3) \cap r$ then

$$f(M) = f(L(P_2, P_3)) \cap f(r) = L(P_2, P_3) \cap r = M.$$

Moreover $f(L(A, P_3)) = L(B, P_2)$ and $f(L(B, P_2)) = L(A, P_3)$, so also the point $Q = L(A, P_3) \cap L(B, P_2)$ is fixed by f. As $L(A, P_3) \cap r = A$, $L(B, P_2) \cap r = B$, we also have $Q \notin r$, and $Q \neq P_1$ because otherwise B would lie on $L(P_1, P_2)$ and P_1, P_2, P_3 would be collinear. So if $s = L(Q, P_1)$, then the point $N = s \cap r$ is well defined. Since Q and P_1 are fixed by f, we have $f(s) = s$, so $f(N) = f(s \cap r) = s \cap r = N$. The restriction of f to s fixes the three distinct points P_1, Q, N, so it coincides with the identity of s.

Since P_2, P_3, A, B are in general position, the points $M = L(P_2, P_3) \cap L(A, B)$, $P_1 = L(P_3, B) \cap L(P_2, A)$, $Q = L(A, P_3) \cap L(B, P_2)$ are not collinear (cf. Exercise 6), so $M \notin s = L(P_1, Q)$. Therefore, the fixed-point set of f contains the line s and the point $M \in r$, which does not lie on s. Now, f cannot have other fixed points, because otherwise f would coincide with the identity on a projective frame of $\mathbb{P}^2(\mathbb{K})$, so it would be the identity, and this would contradict the fact that $f(P_2) = P_3 \neq P_2$.

Exercise 52. Let P_1, P_2, P_3, P_4 be points of $\mathbb{P}^1(\mathbb{K})$ such that $\beta(P_1, P_2, P_3, P_4) = -1$ and let $(\mathbb{P}^1(\mathbb{K}) \setminus \{P_4\}, g)$ be any affine chart. Show that $g(P_3)$ is the midpoint of the segment with endpoints $g(P_1), g(P_2)$, i.e., show that, if $\alpha_i = g(P_i)$ for $i = 1, 2, 3$, then $\alpha_3 = \frac{\alpha_1 + \alpha_2}{2}$.

Solution (1). By definition of affine chart (cf. Sect. 1.3.8) we may endow $\mathbb{P}^1(\mathbb{K})$ with a system of homogeneous coordinates such that $P_1 = [1, \alpha_1]$, $P_2 = [1, \alpha_2]$, $P_3 = [1, \alpha_3]$, $P_4 = [0, 1]$. Then it follows from Sect. 1.5.1 that

$$-1 = \frac{(1 - 0)(\alpha_2 - \alpha_3)}{(\alpha_3 - \alpha_1)(0 - 1)} = \frac{\alpha_2 - \alpha_3}{\alpha_1 - \alpha_3},$$

hence $\alpha_3 - \alpha_1 = \alpha_2 - \alpha_3$ and $\alpha_3 = \frac{\alpha_1 + \alpha_2}{2}$, as desired.

Solution (2). Thanks to the symmetries of the cross-ratio (cf. Sect. 1.5.2) we have

$$\beta(P_2, P_1, P_3, P_4) = \beta(P_1, P_2, P_3, P_4)^{-1} = (-1)^{-1} = -1 = \beta(P_1, P_2, P_3, P_4),$$

so there exists a projectivity $f: \mathbb{P}^1(\mathbb{K}) \to \mathbb{P}^1(\mathbb{K})$ such that $f(P_1) = P_2, f(P_2) = P_1$, $f(P_3) = P_3, f(P_4) = P_4$. As $f(P_4) = P_4$, the map $h: \mathbb{K} \to \mathbb{K}$ defined by $h = g \circ f \circ g^{-1}$ is a well-defined affinity. Therefore, there exist $\lambda \in \mathbb{K}^*$, $\mu \in \mathbb{K}$ such that $h(x) = \lambda x + \mu$ for every $x \in \mathbb{K}$. Moreover, since $f(P_1) = P_2$ and $f(P_2) = P_1$, we have $h(\alpha_1) = \alpha_2$, $h(\alpha_2) = \alpha_1$, so $\lambda \alpha_1 + \mu = \alpha_2$ and $\lambda \alpha_2 + \mu = \alpha_1$. It follows that $\lambda = -1$ and $\mu = \alpha_1 + \alpha_2$, so $h(\alpha_3) = -\alpha_3 + \alpha_1 + \alpha_2$. Moreover, from $f(P_3) = P_3$ we deduce that $h(\alpha_3) = \alpha_3$, so $-\alpha_3 + \alpha_1 + \alpha_2 = \alpha_3$, as desired.

Chapter 3
Exercises on Curves and Hypersurfaces

Affine and projective hypersurfaces. Conjugation and
complexification. Affine charts and projective closure.
Singularities, tangent space and multiplicity. Intersection of
curves in the projective plane. Plane cubics. Linear systems of
plane curves.

Abstract Solved problems on (real and complex) affine and projective hypersurfaces, tangent space and singularities, plane cubics, linear systems of plane curves.

Assumption: Throughout all the chapter, the symbol $\mathbb{K}$ will denote $\mathbb{R}$ or $\mathbb{C}$.

Exercise 53. Let $\mathcal{I}$, $\mathcal{J}$ be hypersurfaces of $\mathbb{P}^n(\mathbb{K})$, $n \geq 2$, and let P be a point of $\mathbb{P}^n(\mathbb{K})$. In addition, denote by $\overline{C_P(\mathcal{I})}$, $\overline{C_P(\mathcal{J})}$ and $\overline{C_P(\mathcal{I} + \mathcal{J})}$ the projective tangent cones at P to $\mathcal{I}$, $\mathcal{J}$ and $\mathcal{I} + \mathcal{J}$, respectively. Show that

$$m_P(\mathcal{I} + \mathcal{J}) = m_P(\mathcal{I}) + m_P(\mathcal{J}), \quad \overline{C_P(\mathcal{I} + \mathcal{J})} = \overline{C_P(\mathcal{I})} + \overline{C_P(\mathcal{J})}.$$

Solution. It is possible to choose homogeneous coordinates $x_0, \dots, x_n$ of $\mathbb{P}^n(\mathbb{K})$ such that $P = [1, 0, \dots, 0]$. Let $y_i = \frac{x_i}{x_0}$, $i = 1, \dots, n$, be the usual affine coordinates of U_0, and let $f(y_1, \dots, y_n) = 0$, $f'(y_1, \dots, y_n) = 0$ be the equations of the affine parts $\mathcal{I} \cap U_0$, $\mathcal{J} \cap U_0$ of $\mathcal{I}$, $\mathcal{J}$, respectively. If $d = m_P(\mathcal{I})$ and $d' = m_P(\mathcal{J})$, then one has

$$f(y_1, \dots, y_n) = f_d(y_1, \dots, y_n) + g(y_1, \dots, y_n),$$
$$f'(y_1, \dots, y_n) = f'_{d'}(y_1, \dots, y_n) + g'(y_1, \dots, y_n),$$

where f_d and $f'_{d'}$ are homogeneous of degrees d and d' respectively, g is a sum of monomials of degree greater than d and g' is a sum of monomials of degree greater than d'. Note that $f_d = 0$ and $f'_{d'} = 0$ are the equations of the affine tangent cones $C_P(\mathcal{I})$ and $C_P(\mathcal{J})$, respectively.

Now the affine part $(\mathcal{I} + \mathcal{J}) \cap U_0$ of $\mathcal{I} + \mathcal{J}$ has equation

$$f_d(y_1, \dots, y_n) f'_{d'}(y_1, \dots, y_n) + h(y_1, \dots, y_n) = 0,$$

where $f_d f'_{d'}$ is homogeneous of degree $d + d'$, and h is a sum of monomials of degree greater than $d + d'$. It follows immediately that $m_P(\mathcal{I} + \mathcal{J}) = d + d' = m_P(\mathcal{I}) + m_P(\mathcal{J})$.

Moreover, from the equation of $(\mathcal{I} + \mathcal{J}) \cap U_0$ written above one deduces that the affine tangent cone $C_P(\mathcal{I} + \mathcal{J})$ has equation $f_d f'_{d'} = 0$, so that $C_P(\mathcal{I} + \mathcal{J}) = C_P(\mathcal{I}) + C_P(\mathcal{J})$. The claim follows by passing to the projective closures.

Note. It is also possible to prove the equality concerning the multiplicities of P for $\mathcal{I}$, $\mathcal{J}$ and $\mathcal{I} + \mathcal{J}$ without resorting to a particular choice of coordinates. First of all we observe that for every line $r \subseteq \mathbb{P}^n(\mathbb{K})$ one has

$$I(\mathcal{I} + \mathcal{J}, r, P) = I(\mathcal{I}, r, P) + I(\mathcal{J}, r, P) \geq m_P(\mathcal{I}) + m_P(\mathcal{J}),$$

so that $m_P(\mathcal{I} + \mathcal{J}) \geq m_P(\mathcal{I}) + m_P(\mathcal{J})$.

Moreover, since the sum of the projective tangent cones to $\mathcal{I}$ and to $\mathcal{J}$ at P is a hypersurface and the complement of the support of a hypersurface is non-empty (cf. Sect. 1.7.2), there exists a point $Q \in \mathbb{P}^n(\mathbb{K})$ that belongs neither to the projective tangent cone to $\mathcal{I}$ at P nor the projective tangent cone to $\mathcal{J}$ at P. If $r = L(P, Q)$, then one has $I(\mathcal{I}, r, P) = m_P(\mathcal{I})$, $I(\mathcal{J}, r, P) = m_P(\mathcal{J})$. Hence

$$m_P(\mathcal{I} + \mathcal{J}) \leq I(\mathcal{I} + \mathcal{J}, r, P) = I(\mathcal{I}, r, P) + I(\mathcal{J}, r, P) = m_P(\mathcal{I}) + m_P(\mathcal{J}).$$

The argument that we have just described also shows that the support of the projective tangent cone to $\mathcal{I} + \mathcal{J}$ at P coincides with the union of the supports of the projective tangent cones to $\mathcal{I}$ and to $\mathcal{J}$ at P.

Exercise 54. Let $n \geq 2$ and let $\mathcal{I}$ be a hypersurface of $\mathbb{P}^n(\mathbb{K})$. Let $\mathcal{I} = m_1 \mathcal{I}_1 + \cdots + m_k \mathcal{I}_k$ be the decomposition of $\mathcal{I}$ into irreducible components. Show that the point P is singular for $\mathcal{I}$ if and only if at least one of the following conditions holds:

  (i)   there exists j such that $P \in \mathcal{I}_j$ and $m_j \geq 2$;
 (ii)   there exists j such that $P \in \text{Sing}(\mathcal{I}_j)$;
(iii)   there exist $j \neq s$ such that $P \in \mathcal{I}_j \cap \mathcal{I}_s$.

Solution. The claim is an immediate consequence of Exercise 53. However we give an alternative proof.

Let $F = 0$ be an equation of $\mathcal{I}$ and let $F = c F_1^{m_1} \cdot \ldots \cdot F_k^{m_k}$ be the factorization of F, where $c \in \mathbb{K}^*$, the F_i are mutually coprime irreducible homogeneous polynomials and $m_i > 0$ for every $i = 1, \ldots, k$. By Leibniz's rule

$$\nabla F(P) = c \sum_{i=1}^{n} m_i F_1^{m_1}(P) \cdot \ldots \cdot F_i^{m_i - 1}(P) \cdot \ldots \cdot F_k^{m_k}(P) \nabla F_i(P). \qquad (3.1)$$

If P belongs to at least two distinct components of $\mathcal{I}$, or it belongs to a component $\mathcal{I}_j$ of multiplicity $m_j > 1$, then all the terms of the right hand side of (3.1) vanish and therefore P is singular for $\mathcal{I}$. This case corresponds to conditions (i) and (iii).

If P belongs to only one component $\mathcal{I}_j$ of $\mathcal{I}$ and $\mathcal{I}_j$ is a component of $\mathcal{I}$ of multiplicity 1, by Eq. (3.1) one has $\nabla F(P) = c' \nabla F_j(P)$, where $c' = c \prod_{i \neq j} F_i^{m_i}(P)$ is a non-zero scalar. Hence in this case $\nabla F(P) = 0$ if and only if $\nabla F_j(P) = 0$, that is if $P \in \mathrm{Sing}(\mathcal{I}_j)$.

Exercise 55. Let $n \geq 2$. Show that:

(a) Two hypersurfaces $\mathcal{I}$ and $\mathcal{J}$ of $\mathbb{P}^n(\mathbb{C})$ have a non-empty intersection.
(b) If $\mathcal{I}$ is a smooth hypersurface of $\mathbb{P}^n(\mathbb{C})$, then $\mathcal{I}$ is irreducible.

Solution. (a) If $\mathcal{I}$ is defined by the equation $F(x_0, \ldots, x_n) = 0$ and $\mathcal{J}$ is defined by the equation $G(x_0, \ldots, x_n) = 0$, we set:

$$F_1(x_0, x_1, x_2) = F(x_0, x_1, x_2, 0, \ldots, 0), \quad G_1(x_0, x_1, x_2) = G(x_0, x_1, x_2, 0, \ldots, 0).$$

By Bézout's Theorem there exists a point $[a, b, c] \in \mathbb{P}^2(\mathbb{C})$ such that $F_1(a, b, c) = G_1(a, b, c) = 0$. Thus the point $P = [a, b, c, 0, \ldots, 0] \in \mathbb{P}^n(\mathbb{C})$ belongs to both $\mathcal{I}$ and $\mathcal{J}$, as required.

(b) Assume by contradiction that $\mathcal{I}$ is reducible and that consequently there exist hypersurfaces $\mathcal{I}_1$ and $\mathcal{I}_2$ of $\mathbb{P}^n(\mathbb{C})$ such that $\mathcal{I} = \mathcal{I}_1 + \mathcal{I}_2$. By (a) there exists $P \in \mathbb{P}^n(\mathbb{C})$ such that $P \in \mathcal{I}_1$ and $P \in \mathcal{I}_2$. By Exercise 54 the point P is singular for $\mathcal{I}$, contradicting the assumption that $\mathcal{I}$ is smooth.

Exercise 56. Show that, if $\mathcal{C}$ is a reduced curve of $\mathbb{P}^2(\mathbb{K})$, then $\mathrm{Sing}(\mathcal{C})$ is a finite set.

Solution. Let $F(x_0, x_1, x_2) = 0$ be an equation of $\mathcal{C}$ and let $d = \deg F$. If $d = 1$, then the curve $\mathcal{C}$ is a line and therefore it has no singular points. So we can assume $d \geq 2$.

Since $\mathcal{C}$ is reduced, we may write $\mathcal{C} = \mathcal{C}_1 + \cdots + \mathcal{C}_m$, where the $\mathcal{C}_i$ are distinct irreducible curves. By Exercise 54 one has $\mathrm{Sing}(\mathcal{C}) = \bigcup_i \mathrm{Sing}(\mathcal{C}_i) \cup \bigcup_{i \neq j}(\mathcal{C}_i \cap \mathcal{C}_j)$. By Bézout's Theorem the set $\mathcal{C}_i \cap \mathcal{C}_j$ is finite for every $i \neq j$, since $\mathcal{C}_i$ and $\mathcal{C}_j$ are irreducible.

Hence to complete the proof it suffices to show that, if $\mathcal{C}$ is irreducible, then $\mathrm{Sing}(\mathcal{C})$ is finite. Since $F \neq 0$, one has $F_{x_j} \neq 0$ for at least one index $j \in \{0, 1, 2\}$. The curve $\mathcal{D}$ defined by the equation $F_{x_j} = 0$ has degree $d - 1$ and one has $\mathrm{Sing}(\mathcal{C}) \subseteq \mathcal{C} \cap \mathcal{D}$. Since $\mathcal{C}$ is irreducible and has degree strictly greater than $\mathcal{D}$, the curves $\mathcal{C}$ and $\mathcal{D}$ have no common components. By Bézout's Theorem $\mathcal{C} \cap \mathcal{D}$ is a finite set, and therefore $\mathrm{Sing}(\mathcal{C})$ is also finite.

Exercise 57. Let $\mathcal{I}$ be a hypersurface of $\mathbb{P}^n(\mathbb{K})$ whose support contains a hyperplane $H \subseteq \mathbb{P}^n(\mathbb{K})$. Show that H is an irreducible component of $\mathcal{I}$. In particular, if $\mathcal{I}$ has degree greater than 1, then $\mathcal{I}$ is reducible.

Solution. It is possible to choose homogeneous coordinates such that H is defined by the equation $x_0 = 0$. Let $F(x_0, \ldots, x_n) = 0$ be an equation of $\mathcal{I}$. There exists a homogeneous (possibly zero) polynomial $F_1(x_0, x_1, \ldots, x_n)$ and a homogeneous (possibly zero) polynomial $F_2(x_1, \ldots, x_n)$ such that $F(x_0, \ldots, x_n) = x_0 F_1(x_0, \ldots, x_n) + F_2(x_1, \ldots, x_n)$. The fact that $H \subseteq \mathcal{I}$ implies that for every $(a_1, \ldots, a_n) \in \mathbb{K}^n$ we have $F(0, a_1, \ldots, a_n) = 0$, so that $F_2(a_1, \ldots, a_n) = 0$. Then $F_2 = 0$ by the Identity principle of polynomials. So x_0 divides F, namely H is an irreducible component of $\mathcal{I}$.

Exercise 58. Let $\mathcal{I}$ be a hypersurface of $\mathbb{P}^n(\mathbb{K})$ and let $P \in \mathcal{I}$ be a point. Assume that $H \subseteq \mathbb{P}^n(\mathbb{K})$ is a hyperplane passing through P and not contained in $\mathcal{I}$. Prove that:

(a) $m_P(\mathcal{I} \cap H) \geq m_P(\mathcal{I})$; in particular, if $\mathcal{I}$ is singular at P, then $\mathcal{I} \cap H$ is singular at P for any H.
(b) The hypersurface $\mathcal{I} \cap H$ of H is singular at P if and only if H is contained in the tangent space $T_P(\mathcal{I})$.
(c) There exists a hyperplane H of $\mathbb{P}^n(\mathbb{K})$ such that $\mathcal{I}$ and $\mathcal{I} \cap H$ have the same multiplicities at P.

Solution (1). (a) If r is a line of H passing through P, then $I(\mathcal{I}, r, P) = I(\mathcal{I} \cap H, r, P)$, so the result immediately follows from the definition of multiplicity of a hypersurface at a point.

(b) By definition, the hypersurface $\mathcal{I} \cap H$ is singular at P if and only if $I(\mathcal{I}, r, P) = I(\mathcal{I} \cap H, r, P) \geq 2$ for any line r contained in H and passing through P. As the tangent space $T_P(\mathcal{I})$ is the union of all lines that are tangent to $\mathcal{I}$ at P, the previous condition is satisfied if and only H is contained in $T_P(\mathcal{I})$.

(c) Let r be a line through P such that the multiplicity of intersection $I(\mathcal{I}, r, P)$ is minimal, and is therefore equal to the multiplicity of $\mathcal{I}$ at P. Then any hyperplane H containing r has the required property.

Solution (2). (a) Let $m = m_P(\mathcal{I}) \geq 1$. We can choose homogeneous coordinates such that $P = [1, 0, \ldots, 0]$ and $H = \{x_n = 0\}$. In the affine coordinates $y_i = \frac{x_i}{x_0}$, $i = 1, \ldots, n$, the affine part $\mathcal{I} \cap U_0$ is defined by the equation $f(y_1, \ldots, y_n) = 0$, where

$$f(y_1, \ldots, y_n) = f_m(y_1, \ldots, y_n) + h(y_1, \ldots, y_n),$$

f_m is a non-zero homogeneous polynomial of degree m and h contains only monomials of degree $\geq m + 1$. The affine coordinates mentioned above induce affine coordinates $(y_1, \ldots, y_{n-1})$ on the affine part $H \cap U_0$ of H. In these coordinates the affine part of $\mathcal{I} \cap H$ is defined by the equation $g(y_1, \ldots, y_{n-1}) = 0$, where

$$g(y_1, \ldots, y_{n-1}) = f_m(y_1, \ldots, y_{n-1}, 0) + h(y_1, \ldots, y_{n-1}, 0).$$

From this we easily deduce that $m_P(\mathcal{I} \cap H) \geq m = m_P(\mathcal{I})$, as desired.

(b) The hypersurface $\mathcal{I} \cap H$ is singular at P if and only if $g(y_1, \ldots, y_{n-1})$ does not contain any linear term. This latter fact occurs if either $m > 1$ or $m = 1$ and the hyperplane $f_1(y_1, \ldots, y_n) = 0$ contains (and consequently coincides with) $H \cap U_0$. In the first case $T_P(\mathcal{I}) = \mathbb{P}^n(\mathbb{K})$ and so evidently $H \subseteq T_P(\mathcal{I})$. In the second case $T_P(\mathcal{I})$ is the projective closure of the affine hyperplane $f_1(y_1, \ldots, y_n) = 0$, which coincides with $H \cap U_0$; therefore, also in this case we get that $H \subseteq T_P(\mathcal{I})$.

(c) Since f_m is non-zero, by the Identity principle of polynomials there exists a point $Q = [1, v_1, \ldots, v_n] \in U_0$ such that $f_m(v_1, \ldots, v_n) \neq 0$. By applying the considerations made in (a) to a hyperplane $H \subseteq \mathbb{P}^n(\mathbb{K})$ containing the line $L(P, Q)$, we see that $\mathcal{I} \cap H$ has multiplicity m at P.

Exercise 59. Given a hypersurface $\mathcal{I}$ of $\mathbb{P}^n(\mathbb{C})$, show that there exists a hypersurface $\mathcal{J}$ of $\mathbb{P}^n(\mathbb{R})$ such that $\mathcal{I}$ is the complexification $\mathcal{J}_\mathbb{C}$ of $\mathcal{J}$ if and only if $\sigma(\mathcal{I}) = \mathcal{I}$.

Solution. If $\mathcal{I}$ is the complexification of a real hypersurface, then $\mathcal{I}$ can be defined by means of an equation $F = 0$ with $F \in \mathbb{R}[x_0, \ldots, x_n]$. As $\sigma(F) = F$, also the conjugate hypersurface $\sigma(\mathcal{I})$ is defined by $F = 0$, so it coincides with $\mathcal{I}$.

Conversely, assume that $\mathcal{I}$ is a hypersurface such that $\sigma(\mathcal{I}) = \mathcal{I}$ and $\mathcal{I} = [F]$ with $F \in \mathbb{C}[x_0, \ldots, x_n]$. The conjugate hypersurface $\sigma(\mathcal{I})$ is defined by $\sigma(F) = 0$, so there exists $\lambda \in \mathbb{C}^*$ such that $\sigma(F) = \lambda F$. Let us write $F(x) = A(x) + iB(x)$, with $A, B \in \mathbb{R}[x_0, \ldots, x_n]$. Since $F \neq 0$, up to replacing F by iF we may assume $A \neq 0$, so that A defines a real hypersurface $\mathcal{J}$. We have

$$A(x) = \frac{1}{2}(F(x) + \sigma(F)(x)) = \frac{1 + \lambda}{2} F(x),$$

so $\lambda \neq -1$ and the polynomials F and A define the same hypersurface of $\mathbb{P}^n(\mathbb{C})$, i.e. $\mathcal{I} = \mathcal{J}_\mathbb{C}$.

Exercise 60. Let r be a line of $\mathbb{P}^2(\mathbb{C})$. Prove that:

(a) If $r = \sigma(r)$, then r contains infinitely many real points.
(b) If $r \neq \sigma(r)$, then the point $P = r \cap \sigma(r)$ is the only real point of r.

Solution. (a) If $r = \sigma(r)$, by Exercise 59 the line r is the complexification of a real line s, so it contains infinitely many real points.

(b) If r and $\sigma(r)$ are distinct lines, the point $P = r \cap \sigma(r)$ is a real point, because $\sigma(P)$ belongs both to $\sigma(r)$ and to $\sigma(\sigma(r)) = r$ and hence it coincides with P. In addition, if $Q \in r$ is a real point, then $Q = \sigma(Q) \in r \cap \sigma(r) = \{P\}$, so P is the only real point of r.

Exercise 61. Let $\mathcal{I}$ be an irreducible hypersurface of $\mathbb{P}^n(\mathbb{R})$ of odd degree. Show that the complexified hypersurface $\mathcal{I}_\mathbb{C}$ is irreducible too.

Solution. Let $\mathcal{J}$ be an irreducible component of $\mathcal{I}_\mathbb{C}$. If $\sigma(\mathcal{J}) \neq \mathcal{J}$, the hypersurfaces $\mathcal{J}$ and $\sigma(\mathcal{J})$ are distinct irreducible components of $\mathcal{I}_\mathbb{C}$ having the same degree. Since by hypothesis $\deg \mathcal{I}_\mathbb{C} = \deg \mathcal{I}$ is odd, then there exists at least one component $\mathcal{J}$ of $\mathcal{I}_\mathbb{C}$ such that $\sigma(\mathcal{J}) = \mathcal{J}$. By Exercise 59 there exists a hypersurface $\mathcal{K}$ of $\mathbb{P}^n(\mathbb{R})$ such that $\mathcal{J} = \mathcal{K}_\mathbb{C}$. Then $\mathcal{K}$ is a component of $\mathcal{I}$ and consequently $\mathcal{K} = \mathcal{I}$, as $\mathcal{I}$ is irreducible. It follows that $\mathcal{I}_\mathbb{C} = \mathcal{K}_\mathbb{C} = \mathcal{J}$ is irreducible, as required.

Exercise 62. If $\mathcal{C}$ denotes a cubic of $\mathbb{P}^2(\mathbb{C})$, show that:

(a) If $\mathcal{C}$ is reducible and $P \in \mathcal{C}$ is a singular point, then $\mathcal{C}$ contains a line passing through P.

(b) If C has only one singular point P, and P is either a node or an ordinary cusp, then C is irreducible.

(c) If C has an inflection point P, then C is reducible if and only if C contains the tangent line at P.

(d) If C has finitely many singular points and finitely many inflection points, then C is irreducible.

Solution. (a) If C is reducible, then $C = Q + l$, where l is a line and Q is a conic (which may be reducible too). Of course, if $P \in l$ the result is trivial, so we may suppose that $P \in Q \setminus l$. Since P is singular for C, then P must be singular for Q too (cf. Exercise 54). In particular Q must be singular, so it is the union of two (possibly coincident) lines, because every singular conic is reducible. It follows that in any case P lies on a line contained in C.

(b) Assume by contradiction that C is reducible. As seen in (a), we have $C = Q + l$, where l is a line passing through P and Q is a (possibly reducible) conic. Furthermore P, being a double point, is a smooth point for Q; the hypothesis that P is the only singular point of C readily implies that $l \cap Q = \{P\}$. As a consequence l is the tangent line to Q at P, which easily implies that l is the only principal tangent to C at P (cf. Exercise 53). This yields a contradiction if P is a node of C. In addition $I(C, l, P) = \infty$, which leads to a contradiction if P is an ordinary cusp. Therefore C is irreducible.

(c) Evidently C is reducible if it contains the tangent line at P. Conversely, assume that C is reducible, say $C = Q + l$, where l is a line and Q is a (possibly reducible) conic. As the inflection point P is a non-singular point, then either $P \in l \setminus Q$ or $P \in Q \setminus l$. In the first case the tangent τ_P to C at P coincides with l, so τ_P is contained C. If instead $P \in Q \setminus l$, then $I(Q, \tau_P, P) = I(C, \tau_P, P) \geq 3$; since Q has degree 2, by Bézout's Theorem necessarily $\tau_P \subseteq Q \subseteq C$.

(d) Assume by contradiction that C is reducible, and therefore $C = l + Q$, where l is a line and Q is a conic (which may be reducible, too). If $l \cap Q$ is a finite set, all the infinitely many points of $l \setminus Q$ are inflection points of C, a contradiction. Hence l and Q have infinitely many points in common, so Bézout's Theorem implies that $l \subseteq Q$. Then all points of l are singular for C, which again contradicts the hypotheses. It follows that C is irreducible.

Exercise 63. Let $F(x_0, x_1, x_2) = 0, G(x_0, x_1, x_2) = 0$ be the equations of two reduced curves of $\mathbb{P}^2(\mathbb{C})$. Prove that F and G define the same curve if and only if $V(F) = V(G)$.

Solution. First of all recall that the polynomials F and G define the same curve if and only if one of them is a scalar multiple of the other (and therefore they have in particular the same degree d), that is if and only if $[F] = [G]$ in $\mathbb{P}(V)$, where $V = \mathbb{C}[x_0, x_1, x_2]_d$ is the vector space consisting of all homogeneous polynomials of degree d and of the zero polynomial. Hence, if $[F] = [G]$ then one obviously has $V(F) = V(G)$ (and this is the reason why the support of a curve is well defined!).

Assume now that $V(F) = V(G)$. In addition, let $F = F_1 F_2 \cdot \ldots \cdot F_k$ be the decomposition of F into irreducible factors; note that the F_i are pairwise coprime, because $[F]$ is reduced. Fix $i \in \{1, \ldots, k\}$. By Sect. 1.7.1 the polynomial F_i is homogeneous. Moreover, from the fact that $V(F) \subseteq V(G)$ one deduces that $V(F_i)$ is contained in $V(G)$. Since $V(F_i)$ consists of infinitely many points, Bézout's Theorem implies that the curves defined by F_i and G have a common irreducible component, i.e. that F_i, being irreducible, divides G. Since this is true for every i and the F_i are pairwise coprime, one can conclude that F divides G. On the other hand, by symmetry G divides F too, and this obviously implies that each of F and G is a scalar multiple of the other, as requested.

Note. Let $F(x_0, x_1, x_2) = 0$, $G(x_0, x_1, x_2) = 0$ be the equations of two curves of $\mathbb{P}^2(\mathbb{C})$ (possibly not reduced and possibly not of the same degree) such that $V(F) \subseteq V(G)$. The solution of Exercise 63 shows that every irreducible factor of F divides G, so that every irreducible component of F is actually an irreducible component of G.

Exercise 64. If C is an irreducible curve of $\mathbb{P}^2(\mathbb{R})$ of odd degree, prove that:

(a) The support of C contains infinitely many points.
(b) If $\mathcal{D}$ is a curve of odd degree, then $C \cap \mathcal{D} \neq \emptyset$.

Solution. (a) Denote by d the degree of C and let r be a line of $\mathbb{P}^2(\mathbb{R})$ that is not a component of C. The points of $C \cap r$ correspond to the roots of a non-zero homogeneous polynomial g in two variables of degree d (cf. Sect. 1.7.6). As by hypothesis d is odd, by Theorem 1.7.2 the polynomial g has at least one real root, so $C \cap r$ is not empty. If we choose $P \in \mathbb{P}^2(\mathbb{R}) \setminus C$, when r varies in the pencil of lines centred at P we find infinitely many points of C.

(b) Of course, it is sufficient to consider the case when $\mathcal{D}$ is irreducible and distinct from C. Denote by m the degree of $\mathcal{D}$. Consider a system of homogeneous coordinates x_0, x_1, x_2 of $\mathbb{P}^2(\mathbb{R})$ such that the point of coordinates $[0, 0, 1]$ belongs neither to $C \cup \mathcal{D}$ nor to the union of the lines joining two points of $C \cap \mathcal{D}$. Let $F(x_0, x_1, x_2) = 0$ be an equation of C and $G(x_0, x_1, x_2) = 0$ an equation of $\mathcal{D}$ and let $R(x_0, x_1) = \mathrm{Ris}(F, G, x_2)$. In the chosen coordinates, the polynomial $R(x_0, x_1)$ is non-zero and homogeneous of degree dm; in addition, the irreducible factors of $R(x_0, x_1)$ of degree 1 are in one-to-one correspondence with the points of $C \cap \mathcal{D}$ (cf. Sect. 1.9.3). As dm is odd, by Theorem 1.7.2 the polynomial $R(x_0, x_1)$ has at least one irreducible factor of degree 1, which corresponds to a point $Q \in C \cap \mathcal{D}$.

Exercise 65. Assume that C is an irreducible curve of degree d of $\mathbb{P}^2(\mathbb{R})$ that contains a non-singular point P. Prove that:

(a) If $\mathcal{D}$ is a curve of degree m such that md is even and $I(C, \mathcal{D}, P) = 1$, then there exists $Q \in C \cap \mathcal{D}$ such that $Q \neq P$.
(b) The support of C contains infinitely many points.

Solution. (a) Since $\mathcal{C}$ is irreducible by hypothesis and $I(\mathcal{C}, \mathcal{D}, P) = 1$, then $\mathcal{C}$ is not a component of $\mathcal{D}$, so by the real Bézout's Theorem the set $\mathcal{C} \cap \mathcal{D}$ is finite.

Assume that x_0, x_1, x_2 is a system of homogeneous coordinates satisfying the properties listed in the solution of Exercise 64(b) and such that $P = [1, 0, 0]$; so the points of $\mathcal{C} \cap \mathcal{D}$ are in one-to-one correspondence with the linear factors of the polynomial $R(x_0, x_1) = \mathrm{Ris}(F, G, x_2)$, which is non-zero and homogeneous of degree dm. Recall that, by definition, the multiplicity of a factor of degree 1 in the factorization of $R(x_0, x_1)$ coincides with the multiplicity of intersection between $\mathcal{C}$ and $\mathcal{D}$ at the corresponding point (cf. Sect. 1.9.3). As $I(\mathcal{C}, \mathcal{D}, P) = 1$, then $R(x_0, x_1) = x_1 S(x_0, x_1)$, where $S(x_0, x_1)$ is a homogeneous polynomial of degree $dm - 1$ which is not divisible by x_1. Since $dm - 1$ is odd, by Theorem 1.7.2 the polynomial $S(x_0, x_1)$ has at least one irreducible factor of degree 1, which corresponds to a point $Q \in \mathcal{C} \cap \mathcal{D}$ such that $Q \neq P$.

(b) Since a line of $\mathbb{P}^2(\mathbb{R})$ contains infinitely many points, we can consider only the case $d > 1$. Let $\mathcal{D}$ be a conic of $\mathbb{P}^2(\mathbb{R})$ passing through P, such that $\mathcal{D}$ is non-singular at P and the lines $T_P(\mathcal{D})$ and $T_P(\mathcal{C})$ are distinct. By Exercise 109 we have $I(\mathcal{C}, \mathcal{D}, P) = 1$, so by part (a) there exists a point $Q \in \mathcal{C} \cap \mathcal{D}$ such that $Q \neq P$.

Let $P_1, P_2, P_3 \in \mathbb{P}^2(\mathbb{R}) \backslash \mathcal{C}$ be points such that P, P_1, P_2, P_3 are in general position. The conics passing through P, P_1, P_2, P_3 form a pencil $\mathcal{F} = \{\mathcal{D}_{[\lambda,\mu]} \mid [\lambda, \mu] \in \mathbb{P}^1(\mathbb{R})\}$ having $\{P, P_1, P_2, P_3\}$ as base-point set. Since imposing that a conic pass through a point and that it is tangent at that point to a given line are linear conditions (cf. Sect. 1.9.6), the set of conics of $\mathcal{F}$ tangent to $\mathcal{C}$ at P (i.e. to $T_P(\mathcal{C})$ at P, cf. Exercise 109) either coincides with $\mathcal{F}$ or contains at most one conic. The reducible conics $L(P, P_1) + L(P_2, P_3)$ and $L(P, P_2) + L(P_1, P_3)$ belong to $\mathcal{F}$ and, as P, P_1, P_2, P_3 are in general position, they are not both tangent to $T_P(\mathcal{C})$ in P. Therefore there exists at most one $[\lambda_0, \mu_0] \in \mathbb{P}^1(\mathbb{R})$ such that $\mathcal{D}_{[\lambda_0,\mu_0]}$ is tangent to $\mathcal{C}$ at P. By part (a), for every $[\lambda, \mu] \neq [\lambda_0, \mu_0]$ there exists a point $Q_{[\lambda,\mu]} \in \mathcal{C} \cap \mathcal{D}_{[\lambda,\mu]}$ such that $Q_{[\lambda,\mu]} \neq P$. Since two distinct conics of the pencil meet exactly at P, P_1, P_2, P_3 and P_1, P_2, P_3 do not lie on $\mathcal{C}$, when $[\lambda, \mu]$ varies in $\mathbb{P}^1(\mathbb{R}) \backslash \{[\lambda_0, \mu_0]\}$ the points $Q_{[\lambda,\mu]}$ are all distinct, so they form an infinite set.

Exercise 66. Analyse the singularities of the projective cubic $\mathcal{C}$ of $\mathbb{P}^2(\mathbb{C})$ defined by the equation

$$F(x_0, x_1, x_2) = x_0^3 + 2x_0^2 x_2 + x_0 x_2^2 + x_1^2 x_2 = 0.$$

Solution. The singular points of $\mathcal{C}$ are given by the solutions of the system

$$\begin{cases} F_{x_0} = 3x_0^2 + 4x_0 x_2 + x_2^2 = 0 \\ F_{x_1} = 2x_1 x_2 = 0 \\ F_{x_2} = 2x_0^2 + 2x_0 x_2 + x_1^2 = 0 \end{cases}.$$

From $F_{x_1} = 0$ one deduces $x_2 = 0$ or $x_1 = 0$. It is immediate to check that if $x_2 = 0$ then one has $x_0 = x_1 = 0$. If instead $x_1 = 0$, then from $F_{x_2} = 0$ one deduces $x_0 = 0$ or $x_0 = -x_2$. In addition, if $x_0 = x_1 = 0$ then from $F_{x_0} = 0$ one deduces $x_2 = 0$, while

$x_0 = -x_2$ automatically implies $F_{x_0} = 0$. It follows that the only singular point of $\mathcal{C}$ is given by $[1, 0, -1]$.

Considering on U_0 the coordinates $u = \frac{x_1}{x_0}$, $v = \frac{x_2}{x_0}$, the affine part $\mathcal{C} \cap U_0$ of $\mathcal{C}$ has equation $f(u, v) = 1 + 2v + v^2 + u^2 v = 0$. Furthermore, if $\tau \colon \mathbb{C}^2 \to \mathbb{C}^2$ is the translation defined by $\tau(u, v) = (u, v + 1)$, then the curve $\tau(\mathcal{C} \cap U_0)$ has equation $f(\tau^{-1}(u, v)) = v^2 + u^2 v - u^2 = (v - u)(v + u) + u^2 v = 0$. Hence $(0, 0)$ is an ordinary double point of $\tau(\mathcal{C} \cap U_0)$. It follows that $[1, 0, -1]$ is an ordinary double point of $\mathcal{C}$.

Exercise 67. Consider the projective curve $\mathcal{C}$ of $\mathbb{P}^2(\mathbb{C})$ of equation

$$F(x_0, x_1, x_2) = x_0 x_2^2 - x_1^3 + x_0 x_1^2 + 5x_0^2 x_1 - 5x_0^3 = 0$$

and the point $Q = [0, 1, 0]$. Check that $\mathcal{C}$ is non-singular and determine the points $P \in \mathcal{C}$ such that the tangent to $\mathcal{C}$ at P passes through the point Q.

Solution. Assume that (x_0, x_1, x_2) is a solution of the system

$$\begin{cases} F_{x_0} = x_2^2 + x_1^2 + 10x_0 x_1 - 15x_0^2 = 0 \\ F_{x_1} = (x_0 + x_1)(5x_0 - 3x_1) = 0 \\ F_{x_2} = 2x_0 x_2 = 0 \end{cases}.$$

From $F_{x_2} = 0$ one deduces that $x_0 = 0$ or $x_2 = 0$. In the former case from $F_{x_1} = 0$ one deduces $x_1 = 0$, so that $F_{x_0} = 0$ allows one to conclude that $x_2 = 0$, too. In the latter case from $F_{x_1} = 0$ one deduces $x_0 = -x_1$ or $x_1 = \frac{5}{3}x_0$. Together with the condition $x_2 = 0$, either of these equalities, when plugged in $F_{x_0} = 0$, implies $x_0 = x_1 = 0$. In either case we have shown $x_0 = x_1 = x_2 = 0$, so that $\mathcal{C}$ is non-singular.

The tangent to $\mathcal{C}$ at the point $[y_0, y_1, y_2]$ has equation $F_{x_0}(y_0, y_1, y_2)x_0 + F_{x_1}(y_0, y_1, y_2)x_1 + F_{x_2}(y_0, y_1, y_2)x_2 = 0$, and therefore it contains Q if and only if $F_{x_1}(y_0, y_1, y_2) = 0$. It follows that the points of $\mathcal{C}$ whose tangent contains Q are precisely those determined by the solutions of the system

$$\begin{cases} F(y_0, y_1, y_2) = y_0 y_2^2 - y_1^3 + y_0 y_1^2 + 5y_0^2 y_1 - 5y_0^3 = 0 \\ F_{x_1}(y_0, y_1, y_2) = (y_0 + y_1)(5y_0 - 3y_1) = 0 \end{cases}$$

which is satisfied by the points of coordinates $[0, 0, 1], [1, -1, 2\sqrt{2}], [1, -1, -2\sqrt{2}], [3\sqrt{3}, 5\sqrt{3}, 2i\sqrt{10}], [3\sqrt{3}, 5\sqrt{3}, -2i\sqrt{10}]$.

Note. We are going to show in Exercise 81 that, if $\mathcal{C}$ is a smooth curve of $\mathbb{P}^2(\mathbb{C})$ of degree greater than 1 and $Q \in \mathbb{P}^2(\mathbb{C})$, then the set of lines tangent to $\mathcal{C}$ and passing through Q is always finite and non-empty.

Moreover, in the solution of Exercise 67 we observed that the points of the cubic $\mathcal{C}$ whose tangent τ_P contains Q are the intersection points of $\mathcal{C}$ with the conic $\mathcal{Q}$ defined by the equation $F_{x_1} = 0$. By Bézout's Theorem the number of these points, counted with multiplicity, is equal to 6. Actually, it is not hard to check that the point

$P = [0, 0, 1]$ is an inflection point and that $I(\mathcal{C}, \mathcal{Q}, P) = 2$, while the remaining 4 points are not inflection points and $\mathcal{C}$ and $\mathcal{Q}$ intersect at these points with multiplicity 1.

Exercise 68. Denote by $\mathcal{C}$ the curve of $\mathbb{C}^2$ of equation $f(x, y) = xy^2 - y^4 + x^3 - 2x^2y = 0$. Determine:

(a) The improper points and the asymptotes of $\mathcal{C}$.
(b) The singular points of $\mathcal{C}$, their multiplicities and their principal tangents, identifying which of them are ordinary singular points.
(c) The equation of the tangent to $\mathcal{C}$ at the point $P = (4, -4)$.

Solution. (a) If we identify $\mathbb{C}^2$ with the affine chart U_0 of $\mathbb{P}^2(\mathbb{C})$ by means of the map $j_0 \colon \mathbb{C}^2 \to U_0$ defined by $j_0(x_1, x_2) = [1, x_1, x_2]$, then the projective closure $\overline{\mathcal{C}}$ of $\mathcal{C}$ has equation

$$F(x_0, x_1, x_2) = x_0 x_1 x_2^2 - x_2^4 + x_0 x_1^3 - 2x_0 x_1^2 x_2 = 0.$$

By computing the intersection between $\overline{\mathcal{C}}$ and the line $x_0 = 0$, it turns out that the only improper point is $P = [0, 1, 0]$.

With respect to the affine coordinates $u = \frac{x_0}{x_1}, v = \frac{x_2}{x_1}$ of the affine chart U_1, the point P has coordinates $(0, 0)$ and the affine part $\overline{\mathcal{C}} \cap U_1$ has equation $uv^2 - v^4 + u - 2uv = 0$. So P is a simple point of $\overline{\mathcal{C}}$ and the tangent to $\overline{\mathcal{C}} \cap U_1$ at P has equation $u = 0$. Therefore the tangent to $\overline{\mathcal{C}}$ at P is the line $x_0 = 0$; consequently $\mathcal{C}$ has no asymptotes.

(b) Recall that the singular points of $\mathcal{C}$ are the proper points that are singular for $\overline{\mathcal{C}}$. In order to determine the singular points of $\overline{\mathcal{C}}$, it suffices to solve the system

$$\begin{cases} F_{x_0} = x_1 x_2^2 + x_1^3 - 2x_1^2 x_2 = x_1(x_2 - x_1)^2 = 0 \\ F_{x_1} = x_0 x_2^2 + 3x_0 x_1^2 - 4x_0 x_1 x_2 = 0 \\ F_{x_2} = 2x_0 x_1 x_2 - 4x_2^3 - 2x_0 x_1^2 = 0 \end{cases}$$

which has as unique solution the point $Q = [1, 0, 0]$, corresponding to $(0, 0) \in \mathbb{C}^2$. Inspecting the equation of $\mathcal{C}$ we realize that $(0, 0)$ is a triple point; since the homogeneous part of degree 3 of $f(x, y)$ is $xy^2 + x^3 - 2x^2y = x(x - y)^2$, we see that the principal tangents to $\mathcal{C}$ at the origin are the lines $x = 0$ and $x - y = 0$ (the latter one with multiplicity 2). So the origin is a non-ordinary singular point.

(c) As $F_{x_0}(1, 4, -4) = 256$, $F_{x_1}(1, 4, -4) = 128$, $F_{x_2}(1, 4, -4) = 192$, the equation of the projective tangent to $\overline{\mathcal{C}}$ at $[1, 4, -4]$ is $4x_0 + 2x_1 + 3x_2 = 0$. It follows that the tangent to $\mathcal{C}$ at P has equation $2x + 3y + 4 = 0$.

Exercise 69. Consider the curve $\mathcal{C}$ of $\mathbb{C}^2$ of equation $f(x, y) = x - xy^2 + 1 = 0$.

(a) Determine the singular points and the asymptotes of $\mathcal{C}$.
(b) Determine the inflection points of the projective closure of $\mathcal{C}$, check that they are collinear and compute the equation of the line containing them.

Solution. (a) If we identify $\mathbb{C}^2$ with the affine chart U_0 of $\mathbb{P}^2(\mathbb{C})$ by means of the map $j_0 \colon \mathbb{C}^2 \to U_0$ defined by $j_0(x_1, x_2) = [1, x_1, x_2]$, the projective closure $\overline{C}$ of C has equation $F(x_0, x_1, x_2) = x_0^2 x_1 - x_1 x_2^2 + x_0^3 = 0$.

In order to determine the singular points of C it is enough to observe that the only solution of the system

$$\begin{cases} F_{x_0} = x_0(2x_1 + 3x_0) = 0 \\ F_{x_1} = (x_0 - x_2)(x_0 + x_2) = 0 \\ F_{x_2} = -2x_1 x_2 = 0 \end{cases}$$

is the homogeneous triple $[x_0, x_1, x_2] = [0, 1, 0]$. It follows that $P = [0, 1, 0]$ is the only singular point of $\overline{C}$, and that C has no singular point.

By intersecting $\overline{C}$ with the line at infinity $x_0 = 0$, we easily see that the improper points of C are P and $Q = [0, 0, 1]$ and that Q is a smooth point. As $F_{x_0}(0, 0, 1) = F_{x_2}(0, 0, 1) = 0$, the (principal) tangent to $\overline{C}$ at Q has equation $x_1 = 0$. So $x = 0$ is the equation of an asymptote of C.

In order to determine the principal tangents to $\overline{C}$ at P, we can use the affine coordinates $u = \frac{x_0}{x_1}$, $v = \frac{x_2}{x_1}$ defined on the affine chart U_1. In those coordinates $P = (0, 0)$ and the affine part $\overline{C} \cap U_1$ has equation $u^2 - v^2 + u^3 = (u - v)(u + v) + u^3 = 0$. So P is a double point, and the principal tangents to $\overline{C} \cap U_1$ at P have equations $u + v = 0$, $u - v = 0$. Therefore the principal tangents to $\overline{C}$ at P have equations $x_0 + x_2 = 0$, $x_0 - x_2 = 0$, and C has as asymptotes the lines $y = 1$ and $y = -1$, in addition to the line $x = 0$ found above.

(b) Since

$$H_F(x_0, x_1, x_2) = \det \begin{pmatrix} 2x_1 + 6x_0 & 2x_0 & 0 \\ 2x_0 & 0 & -2x_2 \\ 0 & -2x_2 & -2x_1 \end{pmatrix} = 8(x_0^2 x_1 - 3x_0 x_2^2 - x_1 x_2^2),$$

the inflection points of $\overline{C}$ are the simple points of $\overline{C}$ whose coordinates are solutions of the system

$$\begin{cases} x_0^2 x_1 - x_1 x_2^2 + x_0^3 = 0 \\ x_0^2 x_1 - 3x_0 x_2^2 - x_1 x_2^2 = 0 \end{cases}.$$

If we subtract the second equation from the first one, easy computations show that the inflection points of $\overline{C}$ are the points $[0, 0, 1]$, $[-12, 9, 4i\sqrt{3}]$, $[-12, 9, -4i\sqrt{3}]$. These points lie on the line of equation $3x_0 + 4x_1 = 0$.

Note. As shown in Exercise 62, a cubic having only one node and no other singular point is irreducible. Therefore, the fact that the inflection points of the projective closure of C are collinear follows from a general result that we will show in Exercise 88.

Exercise 70. If C is the curve of $\mathbb{R}^2$ of equation $f(x, y) = x^3 - xy^2 + x^2 y - y^3 + xy^4 = 0$, determine:

(a) The singular points of the projective closure $\overline{C}$ of C, computing, for each of these points, the multiplicity, the principal tangents and the multiplicity of intersection of $\overline{C}$ with each principal tangent.
(b) The points at infinity and the asymptotes of C.
(c) All the lines of the pencil with centre $[1, 0, 0]$ that are tangent to $\overline{C}$ at two distinct points.

Solution. (a) The projective closure $\overline{C}$ of C has equation

$$F(x_0, x_1, x_2) = x_0^2(x_1^3 - x_1 x_2^2 + x_1^2 x_2 - x_2^3) + x_1 x_2^4 = 0;$$

note that $F = x_0^2(x_1 - x_2)(x_1 + x_2)^2 + x_1 x_2^4$.

The singular points of $\overline{C}$ are characterized by the property that their homogeneous coordinates annihilate the gradient of F, i.e. they are the solutions of the system

$$\begin{cases} F_{x_0} = 2x_0(x_1 - x_2)(x_1 + x_2)^2 = 0 \\ F_{x_1} = x_0^2(x_1 + x_2)^2 + 2x_0^2(x_1 - x_2)(x_1 + x_2) + x_2^4 = 0 \\ F_{x_2} = -x_0^2(x_1 + x_2)^2 + 2x_0^2(x_1 - x_2)(x_1 + x_2) + 4x_1 x_2^3 = 0 \end{cases}.$$

Hence the singular points are $A = [0, 1, 0]$ and $B = [1, 0, 0]$.

We see from the equation of C that B is a triple point with the line τ_1 of equation $x_1 - x_2 = 0$ and the line τ_2 of equation $x_1 + x_2 = 0$ (counted twice) as principal tangents, so that B is a non-ordinary triple point.

In the affine chart U_0 a parametrization of $\tau_1 \cap U_0$ is given by the map $\gamma \colon \mathbb{R} \to \mathbb{R}^2$, $\gamma(t) = (t, t)$. Since $\gamma(0) = B$ and the polynomial $f(\gamma(t)) = t^5$ admits 0 as a root of multiplicity 5, one has $I(C, \tau_1, B) = 5$. Similarly, we have $I(C, \tau_2, B) = 5$.

In order to study the point $A = [0, 1, 0]$, we work in the chart U_1 with affine coordinates $u = \frac{x_0}{x_1}, v = \frac{x_2}{x_1}$. In these coordinates $A = (0, 0)$ and $\overline{C} \cap U_1$ has equation $u^2(1 - v)(1 + v)^2 + v^4 = 0$. Hence A is a non-ordinary double point (cusp) of $\overline{C} \cap U_1$, and the principal tangent to $\overline{C}$ at A is given by the projective closure τ_3 of the affine line $u = 0$, namely by the line $\tau_3 = \{x_0 = 0\}$. In addition $I(\overline{C}, \tau_3, A) = 4$, and therefore A is a non-ordinary cusp.

(b) Intersecting $\overline{C}$ with the line $x_0 = 0$ we find that the points $A = [0, 1, 0]$ and $P = [0, 0, 1]$ are the points at infinity of C. By what we have seen in (a), the point A does not give rise to an asymptote. We have also seen that P is a simple point of C; since $\nabla F(0, 0, 1) = (0, 1, 0)$, the tangent to C at P is the line of equation $x_1 = 0$. So we find that the affine line of equation $x = 0$ is an asymptote of C.

(c) The lines of the pencil with centre $B = [1, 0, 0]$ have equation $a x_1 + b x_2 = 0$ as $[a, b]$ varies in $\mathbb{P}^1(\mathbb{R})$.

The line of the pencil corresponding to $b = 0$, i.e. the line $x_1 = 0$, meets the requirement, since it is tangent to $\overline{C}$ both at B and at P.

So let us assume $b \neq 0$, for instance $b = -1$, and see for which values of $a \in \mathbb{R}$ the line $x_2 = a x_1$ is tangent to $\overline{C}$ at two distinct points. Substituting $x_2 = a x_1$ in the equation of the curve we get the equation

$$x_1^3((1-a)(1+a)^2 x_0^2 + a^4 x_1^2) = 0$$

which shows that for every value of a the line intersects C at $B = [1, 0, 0]$ with multiplicity at least 3, as it is obvious since B is a triple point. Hence we have to see for which values of a the remaining intersection points coincide but are different from B.

The equation $(1-a)(1+a)^2 x_0^2 + a^4 x_1^2 = 0$ has two coincident solutions only if either $a = 0$ or $(1-a)(1+a) = 0$.

The value $a = 0$ is acceptable, since the corresponding line $x_2 = 0$ is tangent to $\overline{C}$ at B (triple point) and at A (cusp).

On the other hand neither of the values $a = 1$ and $a = -1$ is acceptable since the corresponding lines $x_2 = x_1$ and $x_2 = -x_1$ (that we denoted by τ_1 and τ_2) intersect $\overline{C}$ only at B with multiplicity 5.

Exercise 71. Consider the curve C of $\mathbb{P}^2(\mathbb{C})$ of equation

$$F(x_0, x_1, x_2) = x_0^2 x_1^2 - x_0 x_1 x_2^2 - 3x_1^4 - x_0^2 x_2^2 - 2x_0 x_1^3 = 0;$$

compute its singular points, their multiplicities and their principal tangents.
Furthermore, say whether C is reducible.

Solution. The singular points of C are given by the solutions of the system

$$\begin{cases} F_{x_0} = 2x_0 x_1^2 - x_1 x_2^2 - 2x_0 x_2^2 - 2x_1^3 = 0 \\ F_{x_1} = 2x_0^2 x_1 - x_0 x_2^2 - 12x_1^3 - 6x_0 x_1^2 = 0. \\ F_{x_2} = -2x_0 x_2 (x_0 + x_1) = 0 \end{cases}$$

An easy computation shows that the singular points of C are $P = [0, 0, 1]$, $Q = [1, 0, 0]$, $R = [1, -1, 2]$ and $S = [1, -1, -2]$.

Now observe that P, R and S lie on the line r of equation $x_0 + x_1 = 0$. Hence one has $I(C, r, P) + I(C, r, R) + I(C, r, S) \geq 2 + 2 + 2 = 6 > 4$, so that by Bézout's Theorem the line r is an irreducible component of C. In fact, if $G(x_0, x_1, x_2) = x_0 x_1^2 - 3x_1^3 - x_0 x_2^2$, then one has $F(x_0, x_1, x_2) = (x_0 + x_1)G(x_0, x_1, x_2)$, and therefore C is reducible.

Taking the coordinates $u = \frac{x_1}{x_0}$, $v = \frac{x_2}{x_0}$ on U_0, the equation of the affine part $C \cap U_0$ of C becomes $f(u, v) = u^2 - uv^2 - 3u^4 - v^2 - 2v^3 = 0$. Since in these coordinates Q corresponds to the origin, and the homogeneous part of smallest degree of f is $(u + v)(u - v)$, Q is an ordinary double point of C with principal tangents of equation $x_1 + x_2 = 0$ and $x_1 - x_2 = 0$, respectively.

In order to compute the multiplicity and the principal tangents of the singular points of C that lie on r, we now exploit the information on the factorization of F that we have obtained. Since

$$G_{x_0} = x_1^2 - x_2^2, \quad G_{x_1} = 2x_0 x_1 - 9x_1^2, \quad G_{x_2} = -2x_0 x_2,$$

if $\mathcal{D}$ is the curve of equation $G = 0$, the points P, R, S are simple points of $\mathcal{D}$. Since $\mathcal{C} = r + \mathcal{D}$, it follows that P, R, S are double points of $\mathcal{C}$ (cf. Exercise 53). Moreover, by computing the partial derivatives of G it is immediate to get the equations of the tangent lines r_P, r_R, r_S to $\mathcal{D}$ at P, R, S. These lines are given by $r_P = \{x_0 = 0\}$, $r_R = \{3x_0 + 11x_1 + 4x_2 = 0\}$, $r_S = \{3x_0 + 11x_1 - 4x_2 = 0\}$. Then the principal tangents to $\mathcal{Q}$ at P (at R, S, respectively) are given by r, r_P (by r, r_R, and by r, r_S, respectively) (cf. Exercise 53).

Exercise 72. Consider the curve $\mathcal{C}$ of $\mathbb{C}^2$ of equation

$$f(x, y) = x^2(y - 5)^2 + y(bx + a^2y) = 0,$$

where the parameters a, b vary in $\mathbb{C}$.

(a) Determine the number of asymptotes of $\mathcal{C}$.
(b) Say whether the point $(0, 0) \in \mathbb{C}^2$ is singular for $\mathcal{C}$, compute its multiplicity and say whether it is an ordinary point.
(c) Say whether there exist values of the parameters $a, b \in \mathbb{C}$ such that the curve $\mathcal{C}$ passes through the point $(-1, 1)$ and it is tangent at that point to the line $3x + 5y - 2 = 0$.

Solution. (a) The projective closure $\overline{\mathcal{C}}$ of $\mathcal{C}$ in $\mathbb{P}^2(\mathbb{C})$ has equation

$$F(x_0, x_1, x_2) = x_1^2(x_2 - 5x_0)^2 + x_0x_2(bx_1 + a^2x_2) = 0.$$

Hence the points at infinity of $\mathcal{C}$, i.e. the intersection points of $\overline{\mathcal{C}}$ with the line $x_0 = 0$, are $P = [0, 1, 0]$ and $Q = [0, 0, 1]$.

In order to determine the principal tangents to $\overline{\mathcal{C}}$ at P, we use the affine coordinates $u = \frac{x_0}{x_1}, v = \frac{x_2}{x_1}$ defined on the affine chart U_1. With this choice of coordinates, the point P corresponds to the point $(0, 0)$ and the affine part $\overline{\mathcal{C}} \cap U_1$ has equation $f_1(u, v) = (v - 5u)^2 + uv(b + a^2v) = 0$. Since f_1 does not contain monomials of degree 1, P is singular. Moreover, the homogeneous part of f_1 of degree 2 is given by $v^2 + uv(b - 10) + 25u^2$: it is never identically zero, and it is the square of a polynomial of degree 1 if and only if $b = 0$ or $b = 20$. It follows that P is a non-ordinary double point if $b = 0$ or $b = 20$, and is an ordinary double point otherwise. Moreover, since the monomial u does not divide the quadratic homogeneous part of f_1, the line $x_0 = 0$ is never one of the principal tangents to $\overline{\mathcal{C}}$ at P. Therefore, there are two asymptotes of $\mathcal{C}$ whose projective closures contain P for $b \neq 0$ and $b \neq 20$, and only one otherwise.

In order to study the point $Q \in \mathcal{C}$, we now choose the affine coordinates $s = \frac{x_0}{x_2}, t = \frac{x_1}{x_2}$ defined on U_2. With this choice of coordinates, Q corresponds to the point $(0, 0)$ and the affine part $\overline{\mathcal{C}} \cap U_2$ has equation $f_2(s, t) = t^2(1 - 5s)^2 + s(bt + a^2) = 0$.

If $a \neq 0$, then the point Q is smooth, and the (principal) tangent to $\overline{\mathcal{C}}$ at Q has equation $x_0 = 0$. In this case $\mathcal{C}$ has no asymptotes whose projective closure passes through Q.

If instead $a = 0$, then the quadratic homogeneous part of f_2 is given by $t(t + bs)$: it is always non-zero, and it is the square of a polynomial of degree 1 if and only if $b = 0$. It follows that P is a non-ordinary double point if $b = 0$, and it is an ordinary double point otherwise. In addition, since the monomial s does not divide the quadratic homogeneous part of f_2, the line $x_0 = 0$ is never one of the principal tangents to $\overline{C}$ at P. Therefore, there are two asymptotes of C whose projective closures contain Q for $b \neq 0$, and one otherwise.

In summary, the number of asymptotes of C is:

- 4 if $a = 0$ and $b \notin \{0, 20\}$;
- 3 if $a = 0$ and $b = 20$;
- 2 if $a = b = 0$, or $a \neq 0$ and $b \notin \{0, 20\}$;
- 1 if $a \neq 0$, $b \in \{0, 20\}$.

(b) Since the equation of C contains no term of degree 1, $(0, 0)$ is singular. Moreover, the quadratic homogeneous part of the equation of C is given by $25x^2 + bxy + a^2y^2$, that is always non-zero, and is equal to the square of a polynomial of degree 1 if and only if $b = \pm 10a$. Hence $(0, 0)$ is a double point, and it is ordinary if and only if $b \neq \pm 10a$.

(c) Imposing $f(-1, 1) = 0$ one obtains $b = a^2 + 16$. One has also $f_x(-1, 1) = b - 32$, $f_y(-1, 1) = 2a^2 - b - 8$. So, the line of equation $3x + 5y - 2 = 0$ is tangent to C at $(-1, 1)$ if and only if $b = a^2 + 16$ and

$$0 = \det \begin{pmatrix} b - 32 & 2a^2 - b - 8 \\ 3 & 5 \end{pmatrix} = 8b - 6a^2 - 136.$$

Since the system $b - a^2 - 16 = 8b - 6a^2 - 136 = 0$ has solutions $a = \pm 2, b = 20$, it follows that the line $3x + 5y - 2 = 0$ is tangent to C at $(-1, 1)$ if and only if $a = \pm 2$ and $b = 20$.

Exercise 73. Consider the curve C of $\mathbb{C}^2$ of equation

$$f(x, y) = x^3(x^2 + a) + y(x^3 - x^2y + by + c) = 0,$$

where the parameters a, b, c vary in $\mathbb{C}$.

(a) Compute the multiplicity of the point $O = (0, 0)$ and the principal tangents to C at O. When O is a simple point, determine the multiplicity of intersection at O between C and the tangent to C at O.
(b) Determine the improper points and the asymptotes of C.
(c) Find the values of a, b, c for which the point $Q = (0, 2)$ is singular for C; for these values, compute the multiplicity of Q for C, and say whether Q is an ordinary point.

Solution. (a) By reordering the monomials of f, one gets $f(x, y) = cy + by^2 + ax^3 + yx^2(x - y) + x^5$. Hence O is simple if and only if $c \neq 0$. If this is the case, then the tangent r to C at O has equation $y = 0$, and therefore the map $\gamma \colon \mathbb{C} \to \mathbb{C}^2$ given by $\gamma(t) = (t, 0)$ is a parametrization of r. Since $f(\gamma(0)) = O$, then $I(C, r, O)$ is equal to the multiplicity of 0 as a root of $f(\gamma(t))$. Since $f(\gamma(t)) = t^3(t^2 + a)$, one has $I(C, r, O) = 3$ if $a \neq 0$, and $I(C, r, O) = 5$ otherwise.

If instead $c = 0$ and $b \neq 0$, the point O is a non-ordinary double point, with $y = 0$ as the only principal tangent. If $c = b = 0$ and $a \neq 0$, then O is a non-ordinary triple point, with the line of equation $x = 0$ as the only principal tangent. Finally, if $b = c = a = 0$, then O is a non-ordinary quadruple point, and the principal tangents to C at O have equations $x = 0$ (of multiplicity 2), $y = 0$ and $x = y$ (each of multiplicity 1).

(b) The projective closure of C has equation

$$F(x_0, x_1, x_2) = x_1^3(x_1^2 + ax_0^2) + x_0 x_2(x_1^3 - x_1^2 x_2 + bx_2 x_0^2 + cx_0^3) = 0.$$

Solving the system $F(x_0, x_1, x_2) = x_0 = 0$ one sees that $P = [0, 0, 1]$ is the only point at infinity of C. Now let $u = \frac{x_0}{x_2}$, $v = \frac{x_1}{x_2}$ be affine coordinates on U_2. With this choice of coordinates P coincides with $(0, 0)$ and the affine part $\overline{C} \cap U_2$ of the projective closure of C has equation

$$g(u, v) = v^3(v^2 + au^2) + u(v^3 - v^2 + bu^2 + cu^3) = 0.$$

The homogeneous component of g of smallest degree is $u(bu^2 - v^2)$, so that if α is a square root of b, then the principal tangents to $\overline{C}$ at P intersect U_2 in the affine lines of equations $u = 0$, $v = \alpha u$, $v = -\alpha u$ (in particular, if $b = 0$ then P is not ordinary, because the tangents $v = \alpha u$, $v = -\alpha u$ coincide). Therefore, the principal tangents to $\overline{C}$ at P have equations $x_0 = 0$, $x_1 = \alpha x_0$, $x_1 = -\alpha x_0$. It follows that, if $b = 0$, then the curve C has only one asymptote, of equation $x = 0$, otherwise C has two asymptotes, of equations $x = \alpha$ and $x = -\alpha$.

(c) In order for Q to be singular for C, one must have $f(Q) = f_x(Q) = f_y(Q) = 0$. Since $f(0, 2) = 4b + 2c$, $f_x(0, 2) = 0$ and $f_y(0, 2) = 4b + c$, this happens if and only if $b = c = 0$. Therefore let us assume $b = c = 0$ and study the nature of the point Q. To this aim, observe that the translation $\tau \colon \mathbb{C}^2 \to \mathbb{C}^2$ defined by $\tau(x, y) = (x, y - 2)$ satisfies $\tau(Q) = (0, 0)$. Moreover

$$f(\tau^{-1}(x, y)) = x^2(x(x^2 + a) + (y + 2)(x - y - 2)),$$

so that the homogeneous component of $f \circ \tau^{-1}$ of smallest degree is $-4x^2$, and the origin is a non-ordinary double point of $\tau(C)$. It follows that Q is a non-ordinary double point of C.

Exercise 74. Consider the affine curve $C_{a,b}$ of equation

$$f(x, y) = x^3 - 2ay^2 + bxy^2 = 0$$

where the parameters a, b vary in $\mathbb{C}$.

(a) Find the values of a, b for which the line at infinity is tangent to the projective closure $\overline{C_{a,b}}$ of $C_{a,b}$ and, for each of those values, say whether $\overline{C_{a,b}}$ is singular at the point of tangency.
(b) Find $a, b \in \mathbb{C}$ such that $C_{a,b}$ passes through the point $(1, 2)$ with tangent line $x - y + 1 = 0$.

Solution. (a) The equation of $\overline{C_{a,b}}$ is

$$F(x_0, x_1, x_2) = x_1^3 - 2ax_0 x_2^2 + bx_1 x_2^2 = 0.$$

The line at infinity is tangent to $\overline{C_{a,b}}$ at a point P if and only if $x_0(P) = F(P) = F_{x_1}(P) = F_{x_2}(P) = 0$. Therefore, as $F_{x_1} = 3x_1^2 + bx_2^2$, $F_{x_2} = -4ax_0 x_2 + 2bx_1 x_2$, the line at infinity is tangent to $\overline{C_{a,b}}$ if and only if the system

$$\begin{cases} x_0 = 0 \\ x_1^3 - 2ax_0 x_2^2 + bx_1 x_2^2 = 0 \\ 3x_1^2 + bx_2^2 = 0 \\ -4ax_0 x_2 + 2bx_1 x_2 = 0 \end{cases}$$

has a non-trivial solution. It is easy to see that this fact occurs if and only if $b = 0$. In addition, if $b = 0$ the only non-trivial homogeneous solution is $[0, 0, 1]$. Finally, since $F_{x_0}(0, 0, 1) = -2a, \overline{C_{a,b}}$ is singular at the point of tangency if and only if $a = 0$.

(b) The curve $C_{a,b}$ passes through $(1, 2)$ with tangent $x - y + 1 = 0$ if and only if $f(1, 2) = 0$ and the vector $(f_x(1, 2), f_y(1, 2))$ is proportional to $(1, -1)$. The first condition is equivalent to $1 - 8a + 4b = 0$, while the second one is equivalent to $f_x(1, 2) = -f_y(1, 2)$, i.e. to $3 - 8a + 8b = 0$. The unique solution of the system consisting of these two equations is $a = -\frac{1}{8}, b = -\frac{1}{2}$, so $C_{a,b}$ passes through $(1, 2)$ with tangent $x - y + 1 = 0$ if and only if $a = -\frac{1}{8}, b = -\frac{1}{2}$.

Exercise 75. Let $k \geq 1$ be an integer and consider the curve $\mathcal{C}$ of $\mathbb{C}^2$ of equation $f(x, y) = x^k y^2 - x^5 + a = 0$, where the parameter a varies in $\mathbb{C}$.

(a) For every k and a, determine the singular points of $\mathcal{C}$ and compute their multiplicities. In addition, say whether these singularities are ordinary.
(b) For every k and a, determine the improper points and the asymptotes of $\mathcal{C}$.
(c) If $k = 3$, say which improper points of $\mathcal{C}$ are inflection points.

Solution. (a) The singular points of $\mathcal{C}$ are given by the solutions of the system

$$\begin{cases} f(x, y) = x^k y^2 - x^5 + a = 0 \\ f_x(x, y) = kx^{k-1} y^2 - 5x^4 = 0 \\ f_y(x, y) = 2x^k y = 0 \end{cases}.$$

The third equation implies that necessarily either $x = 0$ or $y = 0$. Though, if $y = 0$ the second equation yields that $x = 0$, so every singular point must fulfil the condition $x = 0$. Then from the first equation we deduce that no singular point exists if $a \neq 0$.

Therefore we may suppose $a = 0$.

If $k > 1$, all points of the form $(0, y_0)$, $y_0 \in \mathbb{C}$, are singular, while if $k = 1$ the only singular point of $\mathcal{C}$ is $O = (0, 0)$.

At first let us investigate the nature of the point O.

If $k < 3$, the homogeneous component of f of smallest degree is $x^k y^2$, and hence $m_O(\mathcal{C}) = 2 + k$. The point O is non-ordinary, because the principal tangent of equation $y = 0$ has multiplicity 2.

If $k = 3$, we have $f(x, y) = x^3(y^2 - x^2)$, $m_O(\mathcal{C}) = 5$ and O is not ordinary, because the principal tangent of equation $x = 0$ has multiplicity 3.

If instead $k > 3$, the homogeneous component of f of smallest degree is $-x^5$, so O is clearly a non-ordinary singular point of multiplicity 5.

In summary, for any value of k the point O is not ordinary of multiplicity $\min\{5, k + 2\}$.

As already remarked, if $k = 1$ no additional singular point exists. So assume $k > 1$, and let us investigate the nature of the point $(0, y_0)$, with $y_0 \neq 0$. The translation $\tau \colon \mathbb{C}^2 \to \mathbb{C}^2$ defined by $\tau(x, y) = (x, y + y_0)$ fulfils the conditions $\tau(0, 0) = (0, y_0)$ and

$$f(\tau(x, y)) = x^k(y + y_0)^2 - x^5 = y_0^2 x^k - x^5 + 2y_0 x^k y + x^k y^2.$$

It follows that, if $k < 5$, the homogeneous component of smallest degree of $f \circ \tau$ is $y_0^2 x^k$, so the point $(0, y_0)$ has multiplicity k for $\mathcal{C}$ and it is non-ordinary (because its only principal tangent is the line $x = 0$). If $k > 5$, the homogeneous component of smallest degree of $f \circ \tau$ is $-x^5$, so $(0, y_0)$ is a non-ordinary singular point for $\mathcal{C}$ of multiplicity 5. Finally, if $k = 5$ two cases can occur: if $y_0 \neq \pm 1$, the homogeneous component of smallest degree of $f \circ \tau$ is $(y_0^2 - 1)x^5$, so $(0, y_0)$ is a non-ordinary singular point for $\mathcal{C}$ of multiplicity 5, while if $y_0 = \pm 1$ we have $f(\tau(x, y)) = 2y_0 x^5 y + x^5 y^2$, and hence $(0, \pm 1)$ is a non-ordinary singular point for $\mathcal{C}$ of multiplicity 6.

In summary, if $k > 1$ each point of the form $(0, y_0)$ is a non-ordinary singular point, and its multiplicity is $\min\{5, k\}$, except when $k = 5$ and $y_0 = \pm 1$, because in this case the multiplicity is 6.

(b) Let us study separately the cases $k < 3$, $k = 3$, $k > 3$.

If $k < 3$, the projective closure of $\overline{\mathcal{C}}$ of $\mathcal{C}$ has equation

$$F(x_0, x_1, x_2) = x_0^{3-k} x_1^k x_2^2 - x_1^5 + a x_0^5 = 0.$$

It readily follows that the only improper point of $\mathcal{C}$ is $P = [0, 0, 1]$. Using on U_2 the affine coordinates $u = \frac{x_0}{x_2}$, $v = \frac{x_1}{x_2}$, we see that the affine part $\overline{\mathcal{C}} \cap U_2$ of $\overline{\mathcal{C}}$ has equation $u^{3-k} v^k - v^5 + a u^5 = 0$. Moreover, P has affine coordinates $(0, 0)$. Since the principal tangents to $\overline{\mathcal{C}} \cap U_2$ at $(0, 0)$ are evidently the lines of equations $u = 0$ and $v = 0$, it follows that the principal tangents at P to $\overline{\mathcal{C}}$ have equations $x_0 = 0$ and $x_1 = 0$. The first of these lines does not yield any asymptote, so the only asymptote for $\mathcal{C}$ is the line $x = 0$.

If $k > 3$, the projective closure $\overline{C}$ of C has equation

$$F(x_0, x_1, x_2) = x_1^k x_2^2 - x_0^{k-3} x_1^5 + a x_0^{k+2} = 0.$$

So the improper points of C are $P = [0, 0, 1]$ and $Q = [0, 1, 0]$. Using on U_2 the affine coordinates u, v defined in the previous paragraph in such a way that $P = (0, 0)$, the affine part $\overline{C} \cap U_2$ of $\overline{C}$ has equation $v^k - u^{k-3} v^5 + a u^{k+2} = 0$, whose homogenous component of smallest degree is v^k. Therefore the line $v = 0$ is the only principal tangent to $\overline{C} \cap U_2$ at $(0, 0)$, and consequently the only asymptote of C whose projective closure passes through P is the line $x = 0$.

In order to study the asymptotes passing through Q, let us use on U_1 the affine coordinates $s = \frac{x_0}{x_1}$, $t = \frac{x_2}{x_1}$. Thus Q is identified with the origin of $\mathbb{C}^2$, and the affine part $\overline{C} \cap U_1$ of $\overline{C}$ has equation $g(s, t) = t^2 - s^{k-3} + a s^{k+2} = 0$. In the following analysis it may be useful to distinguish different cases according to the value of k.

If $k > 5$, the homogenous component of smallest degree of $g(s, t)$ is t^2; so the line $t = 0$ is the only principal tangent to $\overline{C} \cap U_1$ at $(0, 0)$, which easily implies that the only asymptote of C whose projective closure passes through Q is the line $y = 0$.

If $k = 5$, the homogenous component of smallest degree of $g(s, t)$ is $t^2 - s^2$, hence the principal tangents to $\overline{C} \cap U_1$ at $(0, 0)$ are the lines $t - s = 0$ and $t + s = 0$; therefore there are two asymptotes of C whose projective closures pass through Q, that is the lines $y - 1 = 0$ and $y + 1 = 0$.

If $k = 4$, the homogenous component of smallest degree of $g(s, t)$ is s; then the line $s = 0$ is the only principal tangent to $\overline{C} \cap U_1$ at $(0, 0)$ and so in this case there exists no asymptote of C whose projective closure passes through Q.

Consider now the case $k = 3$. The equation of the projective closure of C is then $F(x_0, x_1, x_2) = x_1^3 x_2^2 - x_1^5 + a x_0^5 = 0$. The system $F(x_0, x_1, x_2) = x_0 = 0$ is evidently equivalent to the system $x_1^3 (x_1^2 - x_2^2) = x_0 = 0$, so that the improper points of C are $P = [0, 0, 1], M = [0, -1, 1], N = [0, 1, 1]$. Using on U_2 the affine coordinates u, v introduced above, the affine part $\overline{C} \cap U_2$ of $\overline{C}$ has equation $f(u, v) = v^3 - v^5 + a u^5 = 0$. Moreover, P, M, N are identified with the points $(0, 0), (0, -1), (0, 1)$ of $\mathbb{C}^2$, respectively. It follows immediately that the only principal tangent to $\overline{C}$ at P has equation $x_1 = 0$, so the only asymptote of C whose projective closure passes through P has equation $x = 0$.

In order to determine the principal tangents to $\overline{C}$ at M and N, we compute the gradient of F. As $F_{x_0} = 5 a x_0^4, F_{x_1} = 3 x_1^2 x_2^2 - 5 x_1^4, F_{x_2} = 2 x_1^3 x_2$, then $\nabla F(0, -1, 1) = (0, -2, -2), \nabla F(0, 1, 1) = (0, -2, 2)$, so M, N are non-singular for $\overline{C}$, and the tangents to $\overline{C}$ at M and at N have equations $x_1 + x_2 = 0$ and $x_1 - x_2 = 0$, respectively. It follows that, if $k = 3$, the asymptotes of C are the lines $x = y, x = -y, x = 0$.

(c) Using the notation introduced in the final part of the solution of (b), first of all we observe that P, being singular for $\overline{C}$, cannot be an inflection point. We check that, instead, M and N are both inflection points of $\overline{C}$. As seen above, the tangent τ_M to the affine part $\overline{C} \cap U_2$ at $(0, -1)$ has equation $v = -1$, and so it can be parametrized by means of the map $\gamma \colon \mathbb{C} \to \mathbb{C}^2$ defined by $\gamma(r) = (r, -1)$. Since $\gamma(0) = (0, -1)$, then the multiplicity of intersection $I(\overline{C}, \tau_M, M)$ is equal to the the multiplicity of

0 as a root of the polynomial $f(\gamma(r))$. As $f(\gamma(r)) = ar^5$, this multiplicity is greater than or equal to 5, so it is greater than 2. Moreover, as seen in the solution of (b), M is non-singular for $\overline{C}$, so M is an inflection point of it.

In the same way we get that also N is an inflection point of $\overline{C}$: the tangent τ_N to $\overline{C} \cap U_2$ at $(0, 1)$ is parametrized by $\eta(r) = (r, 1)$, and $f(\eta(r)) = ar^5$, so $I(\overline{C}, \overline{\tau_N}, N) \geq 5$; this implies the thesis, since N is non-singular for $\overline{C}$.

Exercise 76. Find a cubic $\mathcal{D}$ of $\mathbb{R}^2$ fulfilling the following conditions:

  (i) $(0, 1)$ is an ordinary double point with principal tangents of equations $y = 2x + 1$ and $y = -2x + 1$;
 (ii) the only improper points of $\mathcal{D}$ are $[0, 1, 0]$ and $[0, 0, 1]$;
(iii) the line $y = 5$ is an asymptote for $\mathcal{D}$.

In addition, say whether $\mathcal{D}$ has other asymptotes.

Solution. Consider the translation $\tau \colon (X, Y) \to (x, y) = (X, Y + 1)$ such that $\tau(0, 0) = (0, 1)$. Using the coordinates X, Y we must impose that the principal tangents have equations $Y = \pm 2X$. Therefore, the cubic must have an equation of the form

$$Y^2 - 4X^2 + aX^3 + bX^2Y + cXY^2 + dY^3 = 0,$$

i.e., coming back to the coordinates x, y, an equation of the form

$$(y - 1)^2 - 4x^2 + ax^3 + bx^2(y - 1) + cx(y - 1)^2 + d(y - 1)^3 = 0.$$

By imposing that $[0, 1, 0]$ and $[0, 0, 1]$ be improper points, we get the conditions $a = 0$ and $d = 0$. So the equation of the projective closure $\overline{\mathcal{D}}$ of $\mathcal{D}$ is

$$F(x_0, x_1, x_2) = x_0(x_2 - x_0)^2 - 4x_0x_1^2 + bx_1^2(x_2 - x_0) + cx_1(x_2 - x_0)^2 = 0.$$

The projective closure of the line $y = 5$ has equation $x_2 - 5x_0 = 0$ and its improper point is $[0, 1, 0]$. Since $\nabla F(0, 1, 0) = (-4 - b, 0, b)$, the point $[0, 1, 0]$ is smooth for every b; if we impose that the line $x_2 - 5x_0 = 0$ be tangent at $[0, 1, 0]$, we find the additional condition $b = 1$. Finally observe that necessarily $c = 0$, because otherwise $\mathcal{D}$ would have additional improper points distinct from $[0, 1, 0]$ and $[0, 0, 1]$. Thus we can conclude that the equation of $\mathcal{D}$ must be

$$(y - 1)^2 - 4x^2 + x^2(y - 1) = 0,$$

which fulfils all the requested conditions.

Finally we easily check that $\mathcal{D}$ has no other asymptotes, because the other improper point $[0, 0, 1]$ is smooth for $\overline{\mathcal{D}}$ with tangent $x_0 = 0$.

Exercise 77. Find a cubic $\mathcal{D}$ of $\mathbb{P}^2(\mathbb{C})$ fulfilling the following conditions:

  (i) $[1, 0, 0]$ is an ordinary double point with principal tangents of equations $x_1 = 0$ and $x_2 = 0$;

(ii) $[0, 1, 1]$ is an inflection point with tangent $x_0 = 0$;

(iii) $\mathcal{D}$ passes through the point $[1, 4, 2]$.

In addition, say whether the cubic $\mathcal{D}$ is reducible.

Solution. In order that condition (i) be satisfied, in the affine chart U_0 the equation of $\mathcal{D} \cap U_0$ must be of the form

$$xy + ax^3 + bx^2y + cxy^2 + dy^3 = 0;$$

so $\mathcal{D}$ has equation

$$x_0x_1x_2 + ax_1^3 + bx_1^2x_2 + cx_1x_2^2 + dx_2^3 = 0,$$

with $a, b, c, d \in \mathbb{C}$ not simultaneously zero.

We want now to impose condition (ii), so we observe that in the chart U_2, with respect to the affine coordinates $u = \frac{x_0}{x_2}$, $v = \frac{x_1}{x_2}$, the affine part $\mathcal{D} \cap U_2$ has equation

$$f(u, v) = uv + av^3 + bv^2 + cv + d = 0.$$

Moreover, the affine part of the line $r = \{x_0 = 0\}$ has equation $u = 0$, while the point $[0, 1, 1]$ has coordinates $(u_0, v_0) = (0, 1)$. First of all we note that $f_u(0, 1) = 1$, so the point $[0, 1, 1]$, provided it belongs to $\mathcal{D}$, is non-singular and, in this case, condition (ii) is satisfied if and only if $I(\mathcal{D}, r, [0, 1, 1]) \geq 3$. Since the function $\gamma \colon \mathbb{C} \to \mathbb{C}^2$, $\gamma(t) = (0, t + 1)$ defines a parametrization of $r \cap U_2$ with $\gamma(0) = (0, 1)$, the multiplicity of intersection $I(\mathcal{D}, r, [0, 1, 1])$ coincides with the multiplicity of 0 as a root of the polynomial $f(\gamma(t)) = a(t + 1)^3 + b(t + 1)^2 + c(t + 1) + d$. Therefore, $I(\mathcal{D}, r, [0, 1, 1]) \geq 3$ if and only if

$$a + b + c + d = 3a + 2b + c = 3a + b = 0.$$

Finally $\mathcal{D}$ passes through $[1, 4, 2]$ if and only if $1 + 8a + 4b + 2c + d = 0$. Since the only solution of the system

$$\begin{cases} a + b + c + d = 0 \\ 3a + 2b + c = 0 \\ 3a + b = 0 \\ 1 + 8a + 4b + 2c + d = 0 \end{cases}$$

is $a = -1, b = 3, c = -3, d = 1$, we obtain that the only cubic $\mathcal{D}$ satisfying the requested conditions has equation

$$x_0x_1x_2 - x_1^3 + 3x_1^2x_2 - 3x_1x_2^2 + x_2^3 = 0.$$

As the point $[0, 1, 1] \in \mathcal{D}$ is an inflection point with tangent $x_0 = 0$, by Exercise 62 the curve $\mathcal{D}$ is reducible if and only if it contains the line $x_0 = 0$. By looking at the equation of $\mathcal{D}$ we immediately see that this fact does not happen, and consequently $\mathcal{D}$ is irreducible.

Exercise 78. (a) Determine the equation of a cubic $\mathcal{C}$ of $\mathbb{P}^2(\mathbb{C})$ passing through $[1, 6, 2]$, having at $Q = [1, 0, 0]$ an inflection point with tangent $x_1 + x_2 = 0$ and having at $P = [0, 1, 0]$ a cusp with principal tangent $x_0 = 0$.
(b) Say whether $\mathcal{C}$ has any additional singular points and inflection points.
(c) Say whether $\mathcal{C}$ is irreducible.
(d) Determine all the lines H of $\mathbb{P}^2(\mathbb{C})$ such that the affine curve $\mathcal{C} \cap (\mathbb{P}^2(\mathbb{C}) \setminus H)$ has no asymptotes.

Solution. (a) Let $F(x_0, x_1, x_2) = 0$ be an equation of a cubic $\mathcal{C}$ with the requested properties. Set the affine coordinates $u = \frac{x_0}{x_1}$, $v = \frac{x_2}{x_1}$ on U_1, and let $f_1(u, v) = F(u, 1, v)$ be the equation of the affine part $\mathcal{C} \cap U_1$ of $\mathcal{C}$. Since P must be a cusp of $\mathcal{C}$ with principal tangent $x_0 = 0$, the origin $(0, 0) \in \mathbb{C}^2$ must be a cusp of $\mathcal{C} \cap U_1$ with principal tangent $u = 0$, so that, up to non-zero scalars, $f_1(u, v) = u^2 + au^3 + bu^2v + cuv^2 + dv^3$ for some $a, b, c, d \in \mathbb{C}$ not all equal to 0. Hence $F(x_0, x_1, x_2) = x_0^2 x_1 + a x_0^3 + b x_0^2 x_2 + c x_0 x_2^2 + d x_2^3$.

So, with respect to the affine coordinates $s = \frac{x_1}{x_0}$, $t = \frac{x_2}{x_0}$ on U_0, the equation of the affine part $\mathcal{C} \cap U_0$ of $\mathcal{C}$ is $f_0(s, t) = F(1, s, t) = s + a + bt + ct^2 + dt^3 = 0$. In order for $\mathcal{C}$ to have an inflection point with tangent $x_1 + x_2 = 0$ at Q, the curve defined by f_0 must have an inflection point with tangent $s + t = 0$ at $(0, 0)$. A parametrization $\gamma \colon \mathbb{C} \to \mathbb{C}^2$ of this line is defined by $\gamma(r) = (r, -r)$, and for this parametrization one has $\gamma(0) = (0, 0)$. Therefore, 0 must be a root of multiplicity at least 3 of the polynomial $f_0(\gamma(r))$. Since $f_0(\gamma(r)) = a + (1 - b)r + cr^2 - dr^3$, one must have $a = c = 0$, $b = 1$. Therefore $f_0(s, t) = s + t + dt^3$. Note that Q is automatically non-singular for $\mathcal{C}$, and so it is an inflection point. In addition, the condition $[1, 6, 2] \in \mathcal{C}$ implies that $f_0(6, 2) = 0$, so $8 + 8d = 0$ and $d = -1$, and finally $F(x_0, x_1, x_2) = x_0^2 x_1 + x_0^2 x_2 - x_2^3$.

(b) It is immediate to check that the system

$$\begin{cases} F_{x_0} = 2x_0 x_1 + 2x_0 x_2 = 0 \\ F_{x_1} = x_0^2 = 0 \\ F_{x_2} = x_0^2 - 3x_2^2 = 0 \end{cases}$$

has $(0, 1, 0)$ as the only non-trivial solution up to multiplication by non-zero scalars, and so P is the only singular point of $\mathcal{C}$. Moreover, the determinant of the Hessian matrix of F is given by

$$H_F(X) = \det \begin{pmatrix} 2x_1 + 2x_2 & 2x_0 & 2x_0 \\ 2x_0 & 0 & 0 \\ 2x_0 & 0 & -6x_2 \end{pmatrix} = 24x_0^2 x_2.$$

It is immediate to check that the only non-zero homogeneous triples that satisfy the system $F(x_0, x_1, x_2) = H_F(x_0, x_1, x_2) = 0$ are given by $[0, 1, 0]$, $[1, 0, 0]$, i.e., by the coordinates of P and Q. Since P is singular, it follows that Q is the only inflection point of C.

(c) Since C has a unique singular point and a unique inflection point, we deduce that C is irreducible from Exercise 62 (d).

(d) Note that the affine curve $C \cap (\mathbb{P}^2(\mathbb{C}) \backslash H)$ has no asymptotes if and only if for every $T \in C \cap H$ there exists a unique principal tangent to C at T, and this tangent is equal to H. So let H be a line of $\mathbb{P}^2(\mathbb{C})$. Since C is irreducible and has degree 3, by Bézout's Theorem the intersection $C \cap H$ consists of one, two, or three points. If $C \cap H$ contains at least two points, the multiplicity of intersection of C and H is equal to 1 at at least one of these points, that we denote by S. It follows that H is not tangent to C at S, so that S is non-singular and the (principal) tangent to C at S defines an asymptote of $C \cap (\mathbb{P}^2(\mathbb{C}) \backslash H)$. Therefore, if $C \cap (\mathbb{P}^2(\mathbb{C}) \backslash H)$ has no asymptotes, there must be a point $R \in \mathbb{P}^2(\mathbb{C})$ such that $C \cap H = \{R\}$ and $I(C, H, R) = 3$. It follows that either R is singular for C, or H is an inflection tangent to C at R. In the former case, by what we have seen in (b) one must have $R = P$, and H must be the only principal tangent to C at P, hence $H = \{x_0 = 0\}$. In addition, it is immediate to check that actually $C \cap \{x_0 = 0\} = P$, so that H coincides with the only principal tangent to C at the only point of $C \cap H$. In the latter case, by what we have seen in (b) one must have $R = Q$ and $H = \{x_0 + x_1 = 0\}$. Bézout's Theorem guarantees also in this case that $C \cap H = \{Q\}$. In addition, the (principal) tangent to C at Q is precisely H by construction. Summing up, the affine curve $C \cap (\mathbb{P}^2(\mathbb{C}) \backslash H)$ has no asymptotes if either $H = \{x_0 = 0\}$ or $H = \{x_0 + x_1 = 0\}$.

Exercise 79. Let P, Q be distinct points of $\mathbb{P}^2(\mathbb{C})$ and let r be the line joining them. Determine the integers k for which there exists a quartic C of $\mathbb{P}^2(\mathbb{C})$ that satisfies the following conditions:

(i) P is an inflection point of C with the line r as tangent;
(ii) Q is an ordinary double point of C;
(iii) C has k irreducible components.

Solution. If C is a quartic that satisfies properties (i) and (ii), then $I(C, r, P) \geq 3$ and $I(C, r, Q) \geq 2$. By Bézout's Theorem, the line r must be an irreducible component of the quartic; more precisely, it must be a component of multiplicity 1, since P would be singular otherwise. In particular, C must be reducible i.e. $k \geq 2$.

On the other hand we see that for each of the values $k = 2, 3, 4$ we can find a quartic that satisfies properties (i) and (ii) and that has k irreducible components.

If $k = 2$, it is enough to take a quartic whose irreducible components are the line r and a non-singular (and therefore irreducible by Exercise 55) cubic not passing through P and passing through Q with a tangent different from r. In order to show that such a cubic exists, we choose a homogeneous coordinate system x_0, x_1, x_2 such that $Q = [1, 0, 0]$ and $P = [0, 1, 0]$, so that r has equation $x_2 = 0$. The cubic defined in this coordinate system by the equation $x_0 x_2^2 - x_1^3 - x_1 x_0^2 = 0$ has the requested properties.

If $k = 3$, it is enough to take a quartic whose irreducible components are r, another line $s \neq r$ passing through Q and an irreducible conic passing neither through P nor through Q.

If $k = 4$, it is enough to take a quartic whose irreducible components are r, another line $s \neq r$ passing through Q and two additional distinct lines passing neither through P nor through Q.

Exercise 80. Prove that, if C is an affine curve of $\mathbb{C}^2$ of degree n, then C has at most n distinct asymptotes.

Solution. Let $\overline{C}$ be the projective closure of C in $\mathbb{P}^2(\mathbb{C})$, and let $r \subseteq \mathbb{P}^2(\mathbb{C})$ be the line of equation $x_0 = 0$. Since r is not a component of $\overline{C}$ (cf. Sect. 1.7.4), by Bézout's Theorem r and $\overline{C}$ intersect at finitely many points $P_1, \ldots, P_k$. Let a_i be the number of asymptotes of C whose projective closure passes through P_i. Of course, the integer $a = \sum_{i=1}^{k} a_i$ is the total number of asymptotes of C. In addition, since every asymptote passing through P_i is a principal tangent to $\overline{C}$ at P_i, and the number of principal tangents to a curve at any of its points is bounded by the multiplicity of the point, for every i one has $a_i \leq m_{P_i}(\overline{C})$.

Furthermore, from Bézout's Theorem and from the fact that one has by definition $I(\overline{C}, r, P_i) \geq m_{P_i}(\overline{C})$ for every $i = 1, \ldots, k$, we deduce that

$$ n = \sum_{i=1}^{k} I(\overline{C}, r, P_i) \geq \sum_{i=1}^{k} m_{P_i}(\overline{C}). $$

Hence

$$ a = \sum_{i=1}^{k} a_i \leq \sum_{i=1}^{k} m_{P_i}(\overline{C}) \leq n, $$

and therefore the claim holds.

Exercise 81. Let C be a non-singular projective curve of $\mathbb{P}^2(\mathbb{C})$ of degree $n > 1$ and, for every point $P \in C$, denote by τ_P the tangent to C at P. Given $Q \in \mathbb{P}^2(\mathbb{C})$, prove that the set $C_Q = \{P \in C \mid Q \in \tau_P\}$ is non-empty and contains at most $n(n-1)$ points.

Solution. Let $F = 0$ be an equation of C, and let $Q = [q_0, q_1, q_2]$. The set C_Q coincides with the set of points whose homogeneous coordinates $[x_0, x_1, x_2]$ satisfy the system

$$ \begin{cases} F_{x_0}(x_0, x_1, x_2)q_0 + F_{x_1}(x_0, x_1, x_2)q_1 + F_{x_2}(x_0, x_1, x_2)q_2 = 0 \\ F(x_0, x_1, x_2) = 0 \end{cases}. $$

At first let us show that the first equation of this system is not trivial. Otherwise, the polynomial $q_0 F_{x_0} + q_1 F_{x_1} + q_2 F_{x_2}$ would be identically zero, and hence $F_{x_0}, F_{x_1}, F_{x_2}$ would be linearly dependent polynomials. By Bézout's Theorem, since each F_{x_i} is either homogeneous of degree $n - 1$ or zero, the set of solutions of the system

$F_{x_0} = F_{x_1} = 0$ is not empty as $n > 1$. On the other hand, because of the hypothesis we are assuming by contradiction, the points corresponding to the solutions of these latter two equations would annihilate F_{x_2} too, so they would be singular points of C. A contradiction, because C is non-singular by hypothesis.

The previous considerations imply that the set C_Q coincides with the set of points of intersection between C and a curve of degree $n - 1$. Then, by Bézout's Theorem, C_Q is non-empty. Moreover, the curve C, which is non-singular, is irreducible by Exercise 55. Using again Bézout's Theorem, it follows that the cardinality of C_Q is at most $n(n-1)$.

Note. A particular case of the general situation considered in this exercise is described in Exercise 67.

Arguing in a similar but slightly subtler way, one can prove the more general result that, given an irreducible plane curve C of $\mathbb{P}^2(\mathbb{C})$ of degree $n > 1$ and a point $Q \notin \mathrm{Sing}(C)$, the set of lines passing through Q and tangent to C at some point is non-empty and has cardinality $\leq n(n-1)$.

Exercise 82. Assume that C is an irreducible quartic of $\mathbb{P}^2(\mathbb{C})$ having 3 cusps. Show that the three principal tangents to C at the cusps belong to a pencil.

Solution. Denote by A, B, C the three cusps of the quartic C and by τ_A, τ_B, τ_C the principal tangents at these points. Observe that $B \notin \tau_A$ because otherwise, by Bézout's Theorem, the line τ_A would intersect C in at least 5 points (counted with multiplicity) and consequently it would be a component of C, a contradiction. By the same reason $C \notin \tau_A$, $A \notin \tau_B$ and $C \notin \tau_B$. Therefore the lines τ_A and τ_B meet at a point D which is distinct from A, B, C. On the other hand, the points A, B, C cannot lie on a line r, because otherwise r would be a component of C, contradicting the hypothesis. So A, B, C, D are in general position.

Then we can choose a system of homogeneous coordinates where $A = [1, 0, 0]$, $B = [0, 1, 0]$, $C = [0, 0, 1]$, $D = [1, 1, 1]$. In these coordinates τ_A has equation $x_1 - x_2 = 0$ and τ_B has equation $x_0 - x_2 = 0$.

A quartic having a cusp in A with principal tangent $x_1 - x_2 = 0$ has an equation of the form

$$x_0^2(x_1 - x_2)^2 + x_0(ax_1^3 + bx_1^2x_2 + cx_1x_2^2 + dx_2^3) +$$
$$+ ex_1^4 + fx_1^3x_2 + gx_1^2x_2^2 + hx_1x_2^3 + kx_2^4 = 0.$$

By imposing that C have a cusp in B with principal tangent $x_0 - x_2 = 0$, we obtain the relations
$$a = 0, \quad e = 0, \quad f = 0, \quad g = 1, \quad b = -2.$$

Finally, by imposing that C have a cusp in C, we get

$$d = 0, \quad k = 0, \quad h = 0, \quad c = \pm 2.$$

Therefore C has an equation of the form

$$x_0^2 x_1^2 + x_0^2 x_2^2 + x_1^2 x_2^2 - 2x_0^2 x_1 x_2 - 2x_0 x_1^2 x_2 + cx_0 x_1 x_2^2 = 0.$$

If $c = 2$, the equation of C would be $(x_0 x_1 - x_0 x_2 - x_1 x_2)^2 = 0$, so C would be reducible, in contradiction with the hypothesis. Hence $c = -2$ and then the principal tangent τ_C at C to C has equation $x_0 - x_1 = 0$; as a consequence the point D lies on the third cuspidal tangent τ_C too.

Exercise 83. Let P be a non-singular point of a curve C of $\mathbb{P}^2(\mathbb{K})$ of equation $F(x_0, x_1, x_2) = 0$ and assume that the tangent τ_P to the curve at P is not contained in C. Setting $m = I(C, \tau_P, P)$, prove that:

(a) The Hessian polynomial H_F of F is non-zero.
(b) $I(H(C), \tau_P, P) = m - 2$.
(c) If $m = 3$ (i.e. if P is an ordinary inflection point of C), then $I(H(C), C, P) = 1$.

Solution. (a), (b) Recall at first (cf. Sect. 1.9.4) that one can check whether H_F is the zero polynomial or not working in any system of homogeneous coordinates of $\mathbb{P}^2(\mathbb{K})$; in addition, provided that $H_F \neq 0$ and hence that the Hessian curve is defined, $I(H(C), \tau_P, P)$ can be computed in any system of coordinates.

So choose a system of homogeneous coordinates where $P = [1, 0, 0]$ and τ_P has equation $x_2 = 0$; for the sake of simplicity, still denote by $F(x_0, x_1, x_2) = 0$ an equation of C in this system of coordinates.

The affine part $C \cap U_0$ of the curve in the chart U_0 has equation $f(x, y) = 0$, where $f(x, y) = F(1, x, y)$ is the dehomogenized polynomial of F with respect to x_0, and the tangent to $C \cap U_0$ at the origin has equation $y = 0$. It follows that

$$f(x, y) = x^m \varphi(x) + y \psi(x, y)$$

with $\varphi \in \mathbb{K}[x]$, $\psi \in \mathbb{K}[x, y]$, $\varphi(0) \neq 0$ and $\psi(0, 0) \neq 0$ (recall that $m \geq 2$ by the definition of tangent line). Then we have

$$\begin{aligned}
f_x &= x^{m-1} h(x) + y\psi_x(x, y) \\
f_y &= \psi(x, y) + y\psi_y(x, y) \\
f_{xx} &= x^{m-2} k(x) + y\psi_{xx}(x, y) \\
f_{xy} &= \psi_x(x, y) + y\psi_{xy}(x, y) \\
f_{yy} &= 2\psi_y(x, y) + y\psi_{yy}(x, y)
\end{aligned} \qquad (3.2)$$

with $h(x) = m\varphi(x) + x\varphi_x(x)$ and $k(x) = (m - 1)h(x) + xh_x(x)$. In particular $h(0) \neq 0$ and $k(0) \neq 0$.

Let us now compute the dehomogenized polynomial $H_F(X)$ with respect to x_0. If we denote by d the degree of F, necessarily we have $d \geq 2$ because by hypothesis C does not contain the tangent line τ_P. Observe that by Euler's identity we have

$$
\begin{aligned}
(d-1)F_{x_0} &= x_0 F_{x_0 x_0} + x_1 F_{x_0 x_1} + x_2 F_{x_0 x_2} \\
(d-1)F_{x_1} &= x_0 F_{x_0 x_1} + x_1 F_{x_1 x_1} + x_2 F_{x_1 x_2} \\
(d-1)F_{x_2} &= x_0 F_{x_0 x_2} + x_1 F_{x_1 x_2} + x_2 F_{x_2 x_2} \\
dF &= x_0 F_{x_0} + x_1 F_{x_1} + x_2 F_{x_2}
\end{aligned}
$$

which implies that

$$
\begin{aligned}
x_0 F_{x_0 x_0} &= (d-1)F_{x_0} - x_1 F_{x_0 x_1} - x_2 F_{x_0 x_2} \\
x_0 F_{x_0 x_1} &= (d-1)F_{x_1} - x_1 F_{x_1 x_1} - x_2 F_{x_1 x_2} \\
x_0 F_{x_0 x_2} &= (d-1)F_{x_2} - x_1 F_{x_1 x_2} - x_2 F_{x_2 x_2} \\
x_0 F_{x_0} &= dF - x_1 F_{x_1} - x_2 F_{x_2}.
\end{aligned}
\tag{3.3}
$$

By using the relations (3.3) and the properties of the determinant, we obtain

$$
\begin{aligned}
x_0 H_F(X) &= \det \begin{pmatrix} x_0 F_{x_0 x_0} & F_{x_0 x_1} & F_{x_0 x_2} \\ x_0 F_{x_0 x_1} & F_{x_1 x_1} & F_{x_1 x_2} \\ x_0 F_{x_0 x_2} & F_{x_1 x_2} & F_{x_2 x_2} \end{pmatrix} = \\
&= \det \begin{pmatrix} (d-1)F_{x_0} - x_1 F_{x_0 x_1} - x_2 F_{x_0 x_2} & F_{x_0 x_1} & F_{x_0 x_2} \\ (d-1)F_{x_1} - x_1 F_{x_1 x_1} - x_2 F_{x_1 x_2} & F_{x_1 x_1} & F_{x_1 x_2} \\ (d-1)F_{x_2} - x_1 F_{x_1 x_2} - x_2 F_{x_2 x_2} & F_{x_1 x_2} & F_{x_2 x_2} \end{pmatrix} = \\
&= \det \begin{pmatrix} (d-1)F_{x_0} & F_{x_0 x_1} & F_{x_0 x_2} \\ (d-1)F_{x_1} & F_{x_1 x_1} & F_{x_1 x_2} \\ (d-1)F_{x_2} & F_{x_1 x_2} & F_{x_2 x_2} \end{pmatrix} = \\
&= (d-1)\det \begin{pmatrix} F_{x_0} & F_{x_0 x_1} & F_{x_0 x_2} \\ F_{x_1} & F_{x_1 x_1} & F_{x_1 x_2} \\ F_{x_2} & F_{x_1 x_2} & F_{x_2 x_2} \end{pmatrix}.
\end{aligned}
$$

By using once more the relations (3.3) and the properties of the determinant we obtain

$$
\begin{aligned}
x_0^2 H_F(X) &= (d-1)\det \begin{pmatrix} x_0 F_{x_0} & x_0 F_{x_0 x_1} & x_0 F_{x_0 x_2} \\ F_{x_1} & F_{x_1 x_1} & F_{x_1 x_2} \\ F_{x_2} & F_{x_1 x_2} & F_{x_2 x_2} \end{pmatrix} = \\
&= (d-1)\det \begin{pmatrix} dF & (d-1)F_{x_1} & (d-1)F_{x_2} \\ F_{x_1} & F_{x_1 x_1} & F_{x_1 x_2} \\ F_{x_2} & F_{x_1 x_2} & F_{x_2 x_2} \end{pmatrix}.
\end{aligned}
$$

Thus, if we dehomogenize with respect to x_0, we get

$$H_F(1, x, y) = (d - 1) \det \begin{pmatrix} df & (d-1)f_x & (d-1)f_y \\ f_x & f_{xx} & f_{xy} \\ f_y & f_{xy} & f_{yy} \end{pmatrix}. \qquad (3.4)$$

By replacing the relations (3.2) in (3.4) we obtain

$$H_F(1, x, y) =$$
$$= (d-1) \det \begin{pmatrix} d(x^m \varphi + y\psi) & (d-1)(x^{m-1}h + y\psi_x) & (d-1)(\psi + y\psi_y) \\ x^{m-1}h + y\psi_x & x^{m-2}k + y\psi_{xx} & \psi_x + y\psi_{xy} \\ \psi + y\psi_y & \psi_x + y\psi_{xy} & 2\psi_y + y\psi_{yy} \end{pmatrix}.$$

In order to compute the multiplicity of intersection at $P = (0, 0)$ with the line of equation $y = 0$, it is sufficient to compute $H_F(1, x, 0)$. Then we observe that

$$H_F(1, x, 0) = (d-1) \det \begin{pmatrix} d\, x^m \varphi & (d-1)x^{m-1}h & (d-1)\psi(x, 0) \\ x^{m-1}h & x^{m-2}k & \psi_x(x, 0) \\ \psi(x, 0) & \psi_x(x, 0) & 2\psi_y(x, 0) \end{pmatrix} =$$
$$= x^{m-2} g(x)$$

with $g(0) = -(d-1)k(0)(\psi(0, 0))^2$. As $k(0) \neq 0$ and $\psi(0, 0) \neq 0$, then $g(0) \neq 0$. This implies that H_F is non-zero (so that the Hessian curve $H(\mathcal{C})$ of $\mathcal{C}$ is defined), and also that $I(H(\mathcal{C}), \tau_P, P) = m - 2$.

(c) If $m = 3$, then $I(H(\mathcal{C}), \tau_P, P) = 1$ by part (b). It follows that P is a smooth point of $H(\mathcal{C})$ and that the line τ_P is not tangent to $H(\mathcal{C})$ at P. Therefore, by Exercise 109, the curves $\mathcal{C}$ and $H(\mathcal{C})$ are not tangent at P, and hence $I(H(\mathcal{C}), \mathcal{C}, P) = 1$.

 Exercise 84. Let $\mathcal{C}$ be a reduced curve of $\mathbb{P}^2(\mathbb{C})$ of equation $F(x_0, x_1, x_2) = 0$. Show that:

(a) If $\mathcal{C}$ is not a union of lines, than the Hessian polynomial $H_F(X)$ is non-zero.
(b) If $\mathcal{C}$ is irreducible and has infinitely many inflection points, then $\mathcal{C}$ is a line.
(c) If $H_F(X) \neq 0$, then the only common irreducible components of $\mathcal{C}$ and of the Hessian $H(\mathcal{C})$ are the lines contained in $\mathcal{C}$.
(d) If $\mathcal{C}$ is a union of lines all passing through one point, then $H_F(X) = 0$.

Solution. (a) Since $\mathcal{C}$ is reduced, $\mathrm{Sing}(\mathcal{C})$ is a finite set (cf. Exercise 56). Moreover, by assumption there exists an irreducible component $\mathcal{C}_1$ of $\mathcal{C}$ which is not a line. So the claim can be proven by choosing a point $P \in \mathcal{C}_1$ which is non-singular for $\mathcal{C}$ and by using Exercise 83 (a).

(b) Assume by contradiction that $\mathcal{C}$ is not a line and denote by $d \geq 2$ its degree. Let $P \in \mathcal{C}$ be a non-singular point, let τ_P be the tangent to $\mathcal{C}$ at P and let $m = I(\mathcal{C}, \tau_P, P)$. Since $\mathcal{C}$ is irreducible, τ_P cannot be a component of the curve and so $m < \infty$. By what we proved in Exercise 83, the polynomial $H_F(X)$ is non-zero and the Hessian curve $H(\mathcal{C})$ is such that $I(H(\mathcal{C}), \tau_P, P) = m - 2$. Since $\mathcal{C}$ has infinitely many inflection points by assumption, one has $H_F(Q) = 0$ for infinitely many points $Q \in C$. Hence, by Bézout's Theorem, C is an irreducible component of $H(C)$ and therefore $I(C, \tau_P, P) \leq I(H(C), \tau_P, P)$. So one obtains $m \leq m - 2$, a contradiction.

(c) Any line contained in C is also contained in $H(C)$, since all its points that are non-singular for C (there are infinitely many of these, because C is reduced) are inflection points. Conversely, if C_1 is a common irreducible component of C and $H(C)$, then any point of C_1 that is smooth for C is an inflection point (both for C_1 and for C). Since there exist infinitely many points with this property, C_1 is a line by (b).

(d) If C is a union of lines all passing through one point, then up to a change of coordinates we may assume that they all pass through the point $[1, 0, 0]$ and so they have an equation of the form $ax_1 + bx_2 = 0$. Then C is defined by a homogeneous polynomial F not depending on x_0; therefore the Hessian matrix of F has a zero column and, as a consequence, its determinant is zero.

Exercise 85. Let C be the cubic of $\mathbb{P}^2(\mathbb{K})$ defined by the equation:

$$F(x_0, x_1, x_2) = x_0 x_2^2 - x_1^3 - ax_1 x_0^2 - bx_0^3 = 0,$$

with $a, b \in \mathbb{K}$. Show that:

(a) The point $P = [0, 0, 1]$ is an inflection point of C.
(b) C is irreducible.
(c) C is non-singular if and only if $g(x) = x^3 + ax + b$ has no multiple roots or, equivalently, iff $4a^3 + 27b^2 \neq 0$.
(d) If $\mathbb{K} = \mathbb{C}$ and C is non-singular, then there are precisely 4 lines passing through P and tangent to C.

Solution. (a) One has:

$$\nabla F(x_0, x_1, x_2) = (x_2^2 - 2ax_0 x_1 - 3bx_0^2, -3x_1^2 - ax_0^2, 2x_0 x_2).$$

So it is immediate to check that P is a smooth point of C and that $T_P(C)$ is the line $x_0 = 0$. Since $F(0, x_1, x_2) = -x_1^3$, the line $x_0 = 0$ intersects C at P with multiplicity 3 and so P is an inflection point.

(b) Note first of all that it suffices to consider the case $\mathbb{K} = \mathbb{C}$, since if $\mathbb{K} = \mathbb{R}$ and the complexified curve $C_{\mathbb{C}}$ of C is irreducible, then C is obviously irreducible too.

So assume $\mathbb{K} = \mathbb{C}$. Since P is an inflection point and the inflection tangent is the line $x_0 = 0$ which is not contained in C, the curve is irreducible by Exercise 62.

(c) Since P is the only intersection point of C with the line at infinity $x_0 = 0$, it is enough to examine the affine part $C \cap U_0$ of C. In the affine coordinates $x = \frac{x_1}{x_0}, y = \frac{x_2}{x_0}$, the curve $C \cap U_0$ is defined by the equation $f(x, y) = y^2 - g(x) = 0$, where $g(x) = x^3 + ax + b$. The singular points of C_0 are the solutions of the system

$$2y = 0, \quad g'(x) = 0, \quad y^2 - g(x) = 0,$$

i.e., they are the points $(\alpha, 0)$ with $g(\alpha) = g'(\alpha) = 0$. So $\mathcal{C}$ is singular if and only if $g(x)$ has a multiple root α. This happens if and only if the resultant $\text{Ris}(g, g')$ of the polynomials $g(x)$ and $g'(x) = 3x^2 + a$ vanishes (cf. Sect. 1.9.2). By definition $\text{Ris}(g, g')$ is the determinant of the Sylvester matrix:

$$S(g, g') = \begin{pmatrix} b & a & 0 & 1 & 0 \\ 0 & b & a & 0 & 1 \\ a & 0 & 3 & 0 & 0 \\ 0 & a & 0 & 3 & 0 \\ 0 & 0 & a & 0 & 3 \end{pmatrix},$$

and so it is equal to $4a^3 + 27b^2$. Alternatively, one can observe that the pair of equations $g(x) = x^3 + ax + b = 0$, $g'(x) = 3x^2 + a = 0$ is equivalent to the pair of equations $2ax + 3b = 0$, $3x^2 + a = 0$; then it is easy to check that the latter two equations have a common solution if and only if $4a^3 + 27b^2 = 0$.

(d) We have already checked that the line at infinity $x_0 = 0$ is tangent to $\mathcal{C}$ at P. The affine part of the tangent line to $\mathcal{C}$ at a proper point R with affine coordinates (α, β) has equation $-g'(\alpha)(x - \alpha) + 2\beta(y - \beta) = 0$, so that the point P belongs to $T_R(\mathcal{C})$ if and only if $\beta = 0$ (and therefore $g(\alpha) = 0$).

So the proper points R such that $P \in T_R(\mathcal{C})$ are in one-to-one correspondence with the roots of $g(x)$ and the tangent lines to $\mathcal{C}$ at these points are distinct. Since $\mathcal{C}$ is non-singular by assumption, the roots of g are distinct as explained in the solution of (c). Since g has degree 3 and $\mathbb{K} = \mathbb{C}$, it follows from this argument that there are in all 4 lines that are tangent to $\mathcal{C}$ and pass through P.

Exercise 86. (*Weierstrass equation of a plane cubic*) Let $\mathcal{C}$ be a non-singular irreducible cubic of $\mathbb{P}^2(\mathbb{K})$. Show that:

(a) $\mathcal{C}$ has at least one inflection point.
(b) If $P \in \mathcal{C}$ is an inflection point, then there exists a system of homogeneous coordinates x_0, x_1, x_2 of $\mathbb{P}^2(\mathbb{K})$ such that P has coordinates $[0, 0, 1]$ and $\mathcal{C}$ is defined by the equation

$$x_0 x_2^2 - x_1^3 - ax_1 x_0^2 - bx_0^3 = 0,$$

where $a, b \in \mathbb{K}$ and $4a^3 + 27b^2 \neq 0$.
(c) If $\mathbb{K} = \mathbb{C}$, there exists a system of homogeneous coordinates x_0, x_1, x_2 of $\mathbb{P}^2(\mathbb{C})$ such that P has coordinates $[0, 0, 1]$ and $\mathcal{C}$ is defined by the equation

$$x_0 x_2^2 - x_1(x_1 - x_0)(x_1 - \lambda x_0) = 0,$$

with $\lambda \in \mathbb{C} \setminus \{0, 1\}$.

Solution. (a) First of all note that by Exercise 64 the support of $\mathcal{C}$ contains infinitely many points also in the case $\mathbb{K} = \mathbb{R}$.

Let $F(x_0, x_1, x_2) = 0$ be an equation of $\mathcal{C}$ and let $H_F(X)$ be the Hessian polynomial of F. If $H_F = 0$, then all the points of $\mathcal{C}$, being smooth by assumption, are inflection points (actually, using Exercises 61 and 84, one can show that this case does not occur). If H_F is non-zero, then it has degree 3; so, by Bézout's Theorem if $\mathbb{K} = \mathbb{C}$, or by Exercise 64 if $\mathbb{K} = \mathbb{R}$, there is at least one point $P \in C$ such that $H_F(P) = 0$. Since C is non-singular, P is an inflection point.

(b) Choose homogeneous coordinates w_0, w_1, w_2 such that $P = [0, 0, 1]$ and the line $T_P(C)$ has equation $w_0 = 0$. Since P is an inflection point, the curve C is defined by an equation of the form:

$$w_0 w_2^2 + 2 w_2 w_0 A(w_0, w_1) + B(w_0, w_1) = 0,$$

where A and B are homogeneous (or zero) polynomials of degrees 1 and 3, respectively. In the homogeneous coordinate system $y_0 = w_0, y_1 = w_1, y_2 = w_2 + A(w_0, w_1)$ the equation of C takes the form $y_0 y_2^2 + C(y_0, y_1) = 0$, where C is homogeneous of degree 3. More precisely, since C is irreducible and therefore does not have the line $y_0 = 0$ as a component, one has $C(y_0, y_1) = -c y_1^3 + y_0 D(y_0, y_1)$, with $c \in \mathbb{K}^*$ and D either zero or a homogeneous polynomial of degree 2. Changing the coordinate system to $z_0 = y_0, z_1 = \frac{y_1}{c}, z_2 = \frac{y_2}{c^2}$, the curve C is defined by an equation of the form $z_0 z_2^2 - z_1^3 - \alpha z_1^2 z_0 - \beta z_1 z_0^2 - \gamma z_0^3 = 0$, with $\alpha, \beta, \gamma \in \mathbb{K}$. Finally, the change of coordinates $x_0 = z_0, x_1 = z_1 + \frac{\alpha}{3} z_0, x_2 = z_2$ transforms the equation of C into $x_0 x_2^2 - x_1^3 - a x_1 x_0^2 - b x_0^3 = 0$, with $a, b \in \mathbb{K}$. Since in all the coordinate systems that we have used the point P has coordinates $[0, 0, 1]$, we have obtained an equation of C of the requested form. In addition one has $4a^3 + 27b^2 \neq 0$ by Exercise 85.

(c) Consider the homogeneous coordinates y_0, y_1, y_2 and the polynomial $C(y_0, y_1)$ introduced in the solution of (b). Let $c(y) = C(1, y)$, let $\alpha_1, \alpha_2, \alpha_3 \in \mathbb{C}$ be the roots of $c(y)$ and, for $i = 1, 2, 3$, let $Q_i = [1, \alpha_i, 0]$. Note that the Q_i are distinct, because if α_i were a multiple root of c then the point Q_i would be singular. There exists a change of homogeneous coordinates $z_0 = y_0, z_1 = \mu y_0 + \nu y_1, z_2 = y_2$, with $\mu, \nu \in \mathbb{C}, \nu \neq 0$, such that in the coordinates z_0, z_1, z_2 one has $Q_1 = [1, 0, 0], Q_2 = [1, 1, 0]$ and $Q_3 = [1, \lambda, 0]$, for a suitable $\lambda \neq 0, 1$. In this coordinate system C is given by an equation of the form $z_0 z_2^2 - G(z_0, z_1) = 0$. Since $C \cap \{z_2 = 0\} = \{Q_1, Q_2, Q_3\}$, there exists $\tau \in \mathbb{C}^*$ such that $G(z_0, z_1) = \tau z_1(z_1 - z_0)(z_1 - \lambda z_0)$. Let $\delta \in \mathbb{C}$ be such that $\delta^2 = \tau$; in the coordinates $x_0 = z_0, x_1 = z_1, x_2 = \frac{z_2}{\delta}$ the curve C is defined by the equation $x_0 x_2^2 - x_1(x_1 - x_0)(x_1 - \lambda x_0) = 0$

Exercise 87. Let $\mathcal{D} \subset \mathbb{P}^2(\mathbb{R})$ be an irreducible cubic. Show that $\mathcal{D}$ is non-singular if and only if $\mathcal{D}_\mathbb{C} \subset \mathbb{P}^2(\mathbb{C})$ is non-singular.

Solution (1). Note that the singular points of $\mathcal{D}$ are precisely the real singular points of $\mathcal{D}_\mathbb{C}$. So, if $\mathcal{D}_\mathbb{C}$ is non-singular, then also $\mathcal{D}$ is non-singular.

Conversely, let us show that, if $\mathcal{D}$ is non-singular, then also $\mathcal{D}_\mathbb{C}$ is non-singular. So assume by contradiction that $\mathcal{D}$ is smooth and that there exists a singular point $P \in \mathcal{D}_\mathbb{C}$. The point $Q = \sigma(P)$ is distinct from P and it is a singular point of $\mathcal{D}_\mathbb{C}$, too. The line $r = L(P, Q)$ intersects $\mathcal{D}_\mathbb{C}$ with multiplicity ≥ 2 both at P and at Q and therefore is a component of $\mathcal{D}_\mathbb{C}$ by Bézout's Theorem. Since $r = \sigma(r)$, the line r is real (cf. Exercise 59) and so it is a component of $\mathcal{D}$, against the assumption.

Solution (2). As in Solution (1) we observe that, if $\mathcal{D}_{\mathbb{C}}$ is non-singular, then also $\mathcal{D}$ is non-singular.

Assume now that the curve $\mathcal{D}$ is smooth. By Exercise 86 there exists a homogeneous coordinate system x_0, x_1, x_2 of $\mathbb{P}^2(\mathbb{R})$ in which $\mathcal{D}$ is defined by an equation of the form $F(x_0, x_1, x_2) = x_0 x_2^2 - x_1^3 - a x_1 x_0^2 - b x_0^3 = 0$ with $a, b \in \mathbb{R}$ such that $4a^3 + 27b^2 \neq 0$. Then the curve $\mathcal{D}_{\mathbb{C}}$, being defined by the same equation $F(x_0, x_1, x_2) = 0$ in the corresponding homogeneous coordinate system of $\mathbb{P}^2(\mathbb{C})$, is non-singular by Exercise 85.

Note. The assumption that $\mathcal{D}$ is irreducible cannot be removed. Indeed, if $\mathcal{D} = r + \mathcal{Q}$, where r is a line and $\mathcal{Q}$ is a non-degenerate conic such that $\mathcal{Q} \cap r = \emptyset$, then the curve $\mathcal{D}$ is non-singular but the complexified curve $\mathcal{D}_{\mathbb{C}}$ is singular at the points of $r_{\mathbb{C}} \cap \mathcal{Q}_{\mathbb{C}}$ (cf. Exercise 54).

Exercise 88. (*Nodal complex cubics*) If C is an irreducible projective cubic of $\mathbb{P}^2(\mathbb{C})$ with one node, show that:

(a) C is projectively equivalent to the cubic of equation

$$x_0 x_1 x_2 + x_0^3 + x_1^3 = 0.$$

(b) C has three inflection points, which are collinear.

Solution. (a) Observe at first that two curves are projectively equivalent if and only if there exist systems of projective coordinates where they are described by the same equation. Denote by P the node of C and by r, s the principal tangents to C at P. Since the curve of equation $x_0 x_1 x_2 + x_0^3 + x_1^3 = 0$ has a node at $[0, 0, 1]$ with principal tangents of equations $x_0 = 0$ and $x_1 = 0$, on $\mathbb{P}^2(\mathbb{C})$ we can choose coordinates y_0, y_1, y_2 such that $P = [0, 0, 1]$, $r = \{y_0 = 0\}$, $s = \{y_1 = 0\}$ (we can readily check that such a system of coordinates does exist). In the affine coordinates $u = \frac{y_0}{y_2}$, $v = \frac{y_1}{y_2}$ defined on the affine chart U_2 the equation of $C \cap U_2$ is of the form $uv + au^3 + bu^2v + cuv^2 + dv^3 = 0$ for some $a, b, c, d \in \mathbb{C}$ not simultaneously zero. So C has equation $y_0 y_1 (y_2 + by_0 + cy_1) + ay_0^3 + dy_1^3 = 0$. In addition, observe that, in order that C be irreducible, necessarily $a \neq 0, d \neq 0$.

Now it is not difficult to construct a new system of coordinates where C is described by the required equation. Namely, assume that α, δ are cube roots of a, d, respectively. As $a \neq 0, d \neq 0$, clearly $\alpha \neq 0, \delta \neq 0$. The matrix

$$\begin{pmatrix} \alpha & 0 & \frac{b}{\alpha\delta} \\ 0 & \delta & \frac{c}{\alpha\delta} \\ 0 & 0 & \frac{1}{\alpha\delta} \end{pmatrix}$$

is invertible, so there exists a well-defined system of homogeneous coordinates z_0, z_1, z_2 on $\mathbb{P}^2(\mathbb{C})$ such that $z_0 = \alpha y_0$, $z_1 = \delta y_1$, $z_2 = \frac{y_2 + by_0 + cy_1}{\alpha\delta}$. As

$$z_0 z_1 z_2 + z_0^3 + z_1^3 = \alpha \delta y_0 y_1 \frac{y_2 + b y_0 + c y_1}{\alpha \delta} + \alpha^3 y_0^3 + \delta^3 y_1^3 =$$
$$= y_0 y_1 (y_2 + b y_0 + c y_1) + a y_0^3 + d y_1^3,$$

in the system of coordinates z_0, z_1, z_2 the curve $\mathcal{C}$ is defined by the equation given in the statement, so the thesis follows.

(b) By using that projectivities preserve the multiplicity of intersection between a line and a curve, it is immediate to check that, if $\mathcal{C}$ is a projective curve, $P \in \mathcal{C}$ and $f : \mathbb{P}^2(\mathbb{C}) \to \mathbb{P}^2(\mathbb{C})$ is a projectivity, then P is an inflection point of $\mathcal{C}$ if and only if $f(P)$ is an inflection point of $f(\mathcal{C})$. Therefore, in order to prove statement (b) we may assume that $\mathcal{C}$ is defined by the equation $G(x_0, x_1, x_2) = x_0 x_1 x_2 + x_0^3 + x_1^3 = 0$. The equation of the Hessian curve of $\mathcal{C}$ is

$$H_G(x_0, x_1, x_2) = \det \begin{pmatrix} 6x_0 & x_2 & x_1 \\ x_2 & 6x_1 & x_0 \\ x_1 & x_0 & 0 \end{pmatrix} = 2x_0 x_1 x_2 - 6(x_0^3 + x_1^3) = 0.$$

It can be immediately checked that the system $G(x_0, x_1, x_2) = H_G(x_0, x_1, x_2) = 0$ is equivalent to $x_0 x_1 x_2 = x_0^3 + x_1^3 = 0$, having as solutions the points $[0, 0, 1]$, $[1, -1, 0]$, $[1, \omega, 0]$, $[1, \omega^2, 0]$, where $\omega \neq -1$ is a cube root of -1 in $\mathbb{C}$. By construction, $[0, 0, 1]$ is singular for $\mathcal{C}$, while it is not difficult to check that $[1, -1, 0]$, $[1, \omega, 0]$, $[1, \omega^2, 0]$ are non-singular for $\mathcal{C}$, so they are inflection points of $\mathcal{C}$. These inflection points lie on the line $x_2 = 0$, so they are collinear.

Note. Point (a) of the exercise ensures that two irreducible cubics of $\mathbb{P}^2(\mathbb{C})$, each with one node, are projectively equivalent. So, in particular, every complex irreducible cubic with one node is projectively equivalent to the cubic of equation $x_0 x_2^2 - x_1^3 - x_0 x_1^2 = 0$.

An explicit case of statement (b) is considered in Exercise 69.

Note. (*Nodal real cubics*) Denote by $\mathcal{D}$ an irreducible cubic of $\mathbb{P}^2(\mathbb{R})$ having one node P. Also the complexification $\mathcal{C} = \mathcal{D}_{\mathbb{C}}$ has a node at P and, having an odd degree, it is irreducible by Exercise 61. Therefore, $\mathcal{C}$ has exactly three inflection points by Exercise 88. Since $\mathcal{C}$ is the complexification of a real curve, the set of its inflection points is invariant under conjugation. This observation implies that at least one of the inflection points of $\mathcal{C}$ is a real point Q, and consequently it is an inflection point of $\mathcal{D}$. Then, arguing as in Exercise 86, it is not difficult to show there exists a system of projective coordinates x_0, x_1, x_2 of $\mathbb{P}^2(\mathbb{R})$ such that $P = [1, 0, 0]$, $Q = [0, 0, 1]$ and $\mathcal{D}$ is defined by one of the following equations:

(1) $x_0 x_2^2 - x_1^3 + x_0 x_1^2 = 0$;
(2) $x_0 x_2^2 - x_1^3 - x_0 x_1^2 = 0$.

Note that the curves of $\mathbb{P}^2(\mathbb{R})$ defined by the Eqs. (1) and (2) are not projectively isomorphic, because the point P is the only singular point of both curves, but the tangent cones at P to the two curves are not projectively isomorphic. If we compute the Hessian explicitly, we can check that in case (1) the curve $\mathcal{D}$ has three inflection points, while in case (2) only one inflection point exists.

Exercise 89. (*Cuspidal complex cubics*) If C is an irreducible cubic of $\mathbb{P}^2(\mathbb{C})$ with a cusp, show that:

(a) C has exactly one inflection point.
(b) C is projectively equivalent to the cubic of equation $x_0 x_1^2 = x_2^3$.

Solution. (a) Denote by P the cusp of C and by r the principal tangent at that point. Let Q be another point of the cubic distinct from P; necessarily Q is non-singular, because otherwise the line $L(P, Q)$ would meet the cubic in at least 4 points (counted with multiplicity) and hence it would be an irreducible component of C, which instead is irreducible by hypothesis. Denote by s the tangent to C at Q. If s should pass through P, then it would meet the cubic in at least 4 points (counted with multiplicity), again a contradiction. In particular $r \neq s$, so the point $R = r \cap s$ is well defined. As the points P, Q, R are not collinear, we can choose a projective frame in $\mathbb{P}^2(\mathbb{C})$ having them as fundamental points, so that $P = [1, 0, 0]$, $Q = [0, 1, 0]$, $R = [0, 0, 1]$.

In order that $[1, 0, 0]$ be a cusp with tangent $L(P, R) = \{x_1 = 0\}$, the equation of C must be of the form

$$x_0 x_1^2 + a x_1^3 + b x_1^2 x_2 + c x_1 x_2^2 + d x_2^3 = 0,$$

with $a, b, c, d \in \mathbb{C}$ not simultaneously zero. The curve defined by the latter equation passes through Q if and only if $a = 0$ and it has $x_0 = 0$ as tangent at Q iff $b = 0$. Therefore, the equation of C is of the form

$$F(x_0, x_1, x_2) = x_0 x_1^2 + c x_1 x_2^2 + d x_2^3 = 0$$

with $d \neq 0$ because otherwise C would be reducible. Since we aim at determining the inflection points of C, observe that

$$H_F(x_0, x_1, x_2) = \det \begin{pmatrix} 0 & 2x_1 & 0 \\ 2x_1 & 2x_0 & 2cx_2 \\ 0 & 2cx_2 & 2cx_1 + 6dx_2 \end{pmatrix} = -8x_1^2(3dx_2 + cx_1).$$

Note that the line $x_1 = 0$ intersects the curve C only at the singular point P, coherently with the fact that P is singular and $x_1 = 0$ is the (only) principal tangent to C. As $d \neq 0$, the line $3dx_2 + cx_1 = 0$ is not a principal tangent and hence it intersects C at P with multiplicity 2. It follows that it intersects C with multiplicity 1 at a point $T \neq P$, which necessarily is non-singular for C. (In fact, an easy computation shows that $T = [-2c^3, 27d^2, -9cd]$ and $F_{x_0}(T) \neq 0$). Therefore T is the only inflection point of C.

(b) Observe at first that two curves are projectively equivalent if and only if there exist systems of projective coordinates where they are described by the same equation. Observe also that the curve of equation $x_0 x_1^2 = x_2^3$ has a cusp at $[1, 0, 0]$ with principal tangent of equation $x_1 = 0$, one inflection point at $[0, 1, 0]$ with tangent of equation $x_0 = 0$ and it passes through $[1, 1, 1]$.

Denote by P, T the cusp and the inflection point of C, respectively; moreover denote by r the principal tangent to C at P and by s the tangent to C at the inflection point T. If $R = r \cap s$, as remarked in the solution of (a) the points P, T, R are in general position. In addition, as C is irreducible, the support of C is not contained in $L(P, T) \cup L(P, R) \cup L(T, R)$, so we can choose a point $S \in C$ in such a way that P, T, R, S form a projective frame of $\mathbb{P}^2(\mathbb{C})$. Let us fix on $\mathbb{P}^2(\mathbb{C})$ the homogeneous coordinates y_0, y_1, y_2 induced by the frame $\{P, T, R, S\}$.

As shown above, the equation of C is of the form

$$F(y_0, y_1, y_2) = y_0 y_1^2 + c y_1 y_2^2 + d y_2^3 = 0$$

with $d \neq 0$, and $H_F(y_0, y_1, y_2) = -8y_1^2(3dy_2 + cy_1)$. By imposing that C pass through S and that T be an inflection point of C, we obtain the conditions $F(1, 1, 1) = H_F(0, 1, 0) = 0$, which imply that $c = 0$ and $d = -1$. Hence $F(y_0, y_1, y_2) = y_0 y_1^2 - y_2^3$, and the thesis is proved.

Note. In particular the exercise proves that two irreducible cubics of $\mathbb{P}^2(\mathbb{C})$, each having one cusp, are projectively equivalent.

Explicit cases of the property just proved are described in Exercise 78 and in Exercise 105.

Note. (*Cuspidal real cubics*) Let $\mathcal{D}$ be an irreducible cubic of $\mathbb{P}^2(\mathbb{R})$ having a cusp P. The complexified curve $C = \mathcal{D}_{\mathbb{C}}$ also has a cusp at P and, having an odd degree, is irreducible by Exercise 61. So C has precisely one inflection point T by Exercise 89. Since C is the complexified curve of a real curve, the set of inflection points of C is invariant under conjugation. It follows that T is a real point and therefore is an inflection point of $\mathcal{D}$.

Since the support of $\mathcal{D}$ contains infinitely many points by Exercise 64, one can show as in the solution of (b) the existence of a projective coordinate system x_0, x_1, x_2 of $\mathbb{P}^2(\mathbb{R})$ such that $P = [1, 0, 0]$, $T = [0, 1, 0]$, $[1, 1, 1] \in \mathcal{D}$ and $\mathcal{D}$ is defined by the equation $x_0 x_1^2 - x_2^3 = 0$. So the irreducible real cubics with a cusp are all projectively equivalent.

Finally note that the change of coordinates $y_0 = x_0$, $y_1 = x_2$, $y_2 = x_1$ transforms $\mathcal{D}$ into the cubic $y_0 y_2^2 - y_1^3 = 0$, which is in Weierstrass form.

Exercise 90. Let C be a smooth irreducible cubic of $\mathbb{P}^2(\mathbb{K})$ and let $P_1, P_2 \in C$ be two distinct inflection points. Show that:

(a) The line $L(P_1, P_2)$ intersects C at a third point P_3, different from P_1 and P_2, which is also an inflection point of C.

(b) There exists a projectivity g of $\mathbb{P}^2(\mathbb{K})$ such that $g(C) = C$ and $g(P_1) = P_2$.

Solution. (a) By Exercise 86 there exists a system of homogenous coordinates x_0, x_1, x_2 on $\mathbb{P}^2(\mathbb{K})$ such that $P_1 = [0, 0, 1]$ and C has equation

$$x_0 x_2^2 - x_1^3 - a x_1 x_0^2 - b x_0^3 = 0,$$

where $a, b \in \mathbb{K}$ and $4a^3 + 27b^2 \neq 0$. Since P_1 is the only point of C on the line $x_0 = 0$, the point P_2 has coordinates $[1, \alpha, \beta]$, with $\alpha, \beta \in \mathbb{K}$. Since $\deg C = 3$ and P_2 is an inflection point, the line $T_{P_2}(C)$ intersects C only at P_2, and therefore it does not pass through P_1. A calculation analogous to those carried out in the solution of Exercise 85 (d) shows that this fact is equivalent to the condition $\beta \neq 0$.

Now let $f : \mathbb{P}^2(\mathbb{K}) \to \mathbb{P}^2(\mathbb{K})$ be the projectivity defined by $[x_0, x_1, x_2] \mapsto [x_0, x_1, -x_2]$. It is easy to check that $f(P_1) = P_1$ and $f(C) = C$. The point $f(P_2)$ has homogeneous coordinates $[1, \alpha, -\beta]$ and so it is distinct from P_2 and collinear with P_1 and P_2, and therefore it is the third point P_3 at which $L(P_1, P_2)$ intersects C. Moreover $P_3 = f(P_2)$ is an inflection point, since it is the image of an inflection point under a projectivity that preserves the curve C.

(b) Let now y_0, y_1, y_2 be a system of homogeneous coordinates such that $P_3 = [0, 0, 1]$ and C has equation $y_0 y_2^2 - y_1^3 - a' y_1 y_0^2 - b' y_0^3 = 0$, with $a', b' \in \mathbb{K}$ (cf. Exercise 86). By what we have seen in the solution of (a), the coordinates of P_1 and P_2 in this system are $[1, \alpha', \beta']$ and $[1, \alpha', -\beta']$, respectively, for some $\alpha' \in \mathbb{K}$, $\beta' \in \mathbb{K}^*$. It is now immediate to check that the projectivity $g : \mathbb{P}^2(\mathbb{K}) \to \mathbb{P}^2(\mathbb{K})$ defined by $[y_0, y_1, y_2] \mapsto [y_0, y_1, -y_2]$ has the requested properties.

Exercise 91. (*Salmon's Theorem*) Let C be a smooth cubic of $\mathbb{P}^2(\mathbb{C})$ and let $P \in C$ be an inflection point. Show that:

(a) There are precisely 4 lines r_1, r_2, r_3, r_4 tangent to C and passing through P.

(b) Let $k_P = \beta(r_1, r_2, r_3, r_4)$ and let $j(C, P) = \dfrac{(k_P^2 - k_P + 1)^3}{k_P^2(k_P - 1)^2}$ (note that $j(C, P)$ is independent of the ordering of the lines r_1, r_2, r_3, r_4 by Exercise 21); if $P' \in C$ is another inflection point, then $j(C, P) = j(C, P')$.

Solution. (a) By Exercise 86 there exists a system of homogeneous coordinates x_0, x_1, x_2 such that $P = [0, 0, 1]$ and the equation of C is in Weierstrass form. So the claim follows from Exercise 85.

(b) If $P' \in C$ is an inflection point, then by Exercise 90 there exists a projectivity g of $\mathbb{P}^2(\mathbb{C})$ such that $g(C) = C$ and $g(P) = g(P')$. If r_1', r_2', r_3', r_4' are the lines tangent to C and passing through P', then one has $g(\{r_1, r_2, r_3, r_4\}) = \{r_1', r_2', r_3', r_4'\}$. So by Exercise 21 one has $j(C, P) = j(C, P')$.

Note. (*The j-invariant of a smooth cubic*) Given a smooth cubic C of $\mathbb{P}^2(\mathbb{C})$, we have seen (cf. Exercise 86) that C has at least an inflection point P, and so one can define $j(C, P)$ as in Exercise 91. In addition, again by Exercise 91, $j(C, P)$ does not depend on the choice of the inflection point and therefore it is uniquely determined by C. For this reason it is usual to write simply $j(C)$. The complex number $j(C)$ is called *j-invariant*, or *modulus*, of the cubic; indeed in Exercise 92 it is shown that $j(C)$ determines C up to projective equivalence.

Exercise 92. Given a smooth cubic C of $\mathbb{P}^2(\mathbb{C})$, let $j(C)$ be the invariant defined in Exercise 91 and in the Note following it. Prove the following statements:

(a) If C_λ is the cubic of equation:

$$x_0 x_2^2 - x_1(x_1 - x_0)(x_1 - \lambda x_0) = 0,$$

with $\lambda \in \mathbb{C} \setminus \{0, 1\}$, then $j(C_\lambda) = \dfrac{(\lambda^2 - \lambda + 1)^3}{\lambda^2 (\lambda - 1)^2}$.

(b) For every $\alpha \in \mathbb{C}$ there exists a smooth cubic C of $\mathbb{P}^2(\mathbb{C})$ such that $j(C) = \alpha$.

(c) Two smooth cubics C and C' of $\mathbb{P}^2(\mathbb{C})$ are projectively equivalent if and only if $j(C) = j(C')$.

Solution. (a) Arguing precisely as in Exercise 85 one checks that C_λ is smooth, that $P = [0, 0, 1] \in C_\lambda$ is an inflection point and that the lines tangent to C_λ and passing through P are the line $r_1 = \{x_0 = 0\}$, which is the tangent at P, and the lines $r_2 = \{x_1 = 0\}$, $r_3 = \{x_1 - x_0 = 0\}$ and $r_4 = \{x_1 - \lambda x_0 = 0\}$. So one has $\beta(r_2, r_1, r_3, r_4) = \lambda$, and therefore $j(C_\lambda) = \dfrac{(\lambda^2 - \lambda + 1)^3}{\lambda^2 (\lambda - 1)^2}$.

(b) Fix $\alpha \in \mathbb{C}$; consider the polynomial $p(t) = (t^2 - t + 1)^3 - \alpha t^2 (t - 1)^2$. Since $p(t)$ has a positive degree, there exists $\lambda \in \mathbb{C}$ such that $p(\lambda) = 0$. Since $p(0) = p(1) = 1$, one has $\lambda \neq 0, 1$ and so $\dfrac{(\lambda^2 - \lambda + 1)^3}{\lambda^2 (\lambda - 1)^2} = \alpha$. By (a) the cubic C_λ of equation $x_0 x_2^2 - x_1(x_1 - x_0)(x_1 - \lambda x_0) = 0$ has modulus equal to α.

(c) Assume that C and C' are projectively equivalent and that g is a projectivity of $\mathbb{P}^2(\mathbb{C})$ such that $g(C) = C'$. Let $P \in C$ be an inflection point and let r_1, r_2, r_3, r_4 be the lines passing through P and tangent to C; then $P' = g(P)$ is an inflection point of C' and the lines $r_1' = g(r_1), \ldots, r_4' = g(r_4)$ are the lines passing through P' and tangent to C'. Since g induces a projective isomorphism between the pencil of lines with centre P and the pencil of lines with centre P', one has $\beta(r_1, r_2, r_3, r_4) = \beta(r_1', r_2', r_3', r_4')$, and therefore, a fortiori, $j(C) = j(C')$. So we have shown that $j(C)$ is invariant under projective isomorphisms.

Conversely, let C and C' be two smooth cubics such that $j(C) = j(C') = \alpha$. By Exercise 86, there exist $\lambda, \lambda' \in \mathbb{C} \setminus \{0, 1\}$ such that C is projectively equivalent to the cubic C_λ defined by the equation $x_0 x_2^2 - x_1(x_1 - x_0)(x_1 - \lambda x_0) = 0$ and C' is projectively equivalent to the cubic $C_{\lambda'}$ defined by the equation $x_0 x_2^2 - x_1(x_1 - x_0)(x_1 - \lambda' x_0) = 0$. Since projective transformations preserve j, by part (a) we have

$$\frac{(\lambda'^2 - \lambda' + 1)^3}{\lambda'^2 (\lambda' - 1)^2} = j(C') = j(C) = \frac{(\lambda^2 - \lambda + 1)^3}{\lambda^2 (\lambda - 1)^2}. \tag{3.5}$$

Since projective equivalence is an equivalence relation, in order to conclude it is enough to show that C_λ and $C_{\lambda'}$ are projectively equivalent. Since the relation (3.5) holds, by Exercise 21 there exists a projectivity f of the line $\{x_2 = 0\}$ such that:

$$f\{[1, 0, 0], [0, 1, 0], [1, 1, 0], [1, \lambda, 0]\} = \{[1, 0, 0], [0, 1, 0], [1, 1, 0], [1, \lambda', 0]\}.$$

In addition, since the permutations $(12)(34)$, $(13)(24)$ and $(14)(32)$ preserve the cross-ratio of a quadruple P_1, P_2, P_3, P_4 of distinct points of a projective line (cf. Sect. 1.5.2), we may assume that $[0, 1, 0]$ is fixed by f and so that f is of the form $[x_0, x_1, 0] \mapsto [x_0, ax_0 + bx_1, 0]$, where $a, b \in \mathbb{C}$ and $b \neq 0$. Then, arguing as in the solution of Exercise 86 (c), one sees that the projectivity g of $\mathbb{P}^2(\mathbb{C})$ defined by $[x_0, x_1, x_2] \mapsto [x_0, ax_0 + bx_1, x_2]$ transforms C_λ into a cubic $\mathcal{D}$ of equation $x_0 x_2^2 - \tau x_1 (x_1 - x_0)(x_1 - \lambda' x_0) = 0$, where $\tau \in \mathbb{C}^*$. Let $\delta \in \mathbb{C}$ be such that $\tau = \delta^2$; the projectivity h defined by $[x_0, x_1, x_2] \mapsto \left[x_0, x_1, \frac{x_2}{\delta} \right]$ transforms $\mathcal{D}$ into $C_{\lambda'}$.

Exercise 93. Prove that every smooth cubic C of $\mathbb{P}^2(\mathbb{C})$ has exactly 9 inflection points.

Solution (1). As C has degree 3 and it is irreducible by Exercise 55, for each inflection point P of C we have $I(C, T_P(C), P) = 3$, i.e. all inflection points of C are ordinary. By Exercise 83 the Hessian curve $H(C)$ is well defined and the multiplicity of intersection of C with $H(C)$ is equal to 1 at each inflection point of C. Since $H(C)$ is a cubic, Bézout's Theorem implies that C and $H(C)$ meet exactly at 9 point, so C has exactly 9 inflection points.

Solution (2). By Exercise 85 there exist homogeneous coordinates x_0, x_1, x_2 of $\mathbb{P}^2(\mathbb{C})$ such that C is defined by the equation $F(x_0, x_1, x_2) = x_0 x_2^2 - x_1^3 - ax_1 x_0^2 - bx_0^3 = 0$, with $a, b \in \mathbb{C}$ and $4a^3 + 27b^2 \neq 0$. As we have shown in Exercise 85 and its solution, the point $P = [0, 0, 1]$ is an inflection point of C and C intersects the line $x_2 = 0$ in three distinct points R_1, R_2, R_3 such that P belongs to the tangent lines to C at R_1, R_2, R_3. As C is smooth, and hence irreducible, then $I(C, T_{R_i}(C), R_i) = 2$ for $i = 1, 2, 3$. In other words, the points R_1, R_2, R_3 are not inflection points.

As P is the only improper point of C and it is an inflection point, it suffices to look for inflection points only in the affine part $C \cap U_0$ of C. In the affine coordinates $x = \frac{x_1}{x_0}, y = \frac{x_2}{x_0}$, the curve $C \cap U_0$ is given by the equation $f(x, y) = y^2 - g(x) = 0$, where $g(x) = x^3 + ax + b$. The Hessian matrix of F is:

$$\text{Hess}_F(X) = \begin{pmatrix} -2ax_1 - 6bx_0 & -2ax_0 & 2x_2 \\ -2ax_0 & -6x_1 & 0 \\ 2x_2 & 0 & 2x_0 \end{pmatrix}.$$

If we compute the determinant of $\text{Hess}_F(X)$ and we dehomogenize it with respect to x_0, we obtain that the affine part $H(C) \cap U_0$ of the Hessian curve $H(C)$ is given by the equation $h(x, y) = 3xy^2 + 3ax^2 + 9bx - a^2 = 0$. So the inflection points of $C \cap U_0$ are the solutions of the system of equations

$$y^2 - g(x) = 0, \quad h(x, y) = 0.$$

By eliminating y^2 in the second equation, we get the following system, which is equivalent to the previous one:

$$y^2 - g(x) = 0, \quad 3x^4 + 6ax^2 + 12bx - a^2 = 0.$$

We are now going to show that the polynomial $p(x) = 3x^4 + 6ax^2 + 12bx - a^2 = 0$ has distinct roots. We have $p'(x) = 12x^3 + 12ax + 12b = 12g(x)$, so p and p' have a common root if and only if p and g have a common root. Assume by contradiction that there exists $\alpha \in \mathbb{C}$ such that $p(\alpha) = g(\alpha) = 0$: then the point $R = [1, \alpha, 0]$ belongs to the intersection of C with the line $x_2 = 0$ and it is an inflection point, in contradiction with the observations made at the beginning. Therefore $p(x)$ has 4 distinct roots $\alpha_1, \alpha_2, \alpha_3, \alpha_4$, and none of them is a root of $g(x)$; in correspondence with each root α_i we have two inflection points $[1, \alpha_i, \beta_i]$ and $[1, \alpha_i, -\beta_i]$ of C, where $\beta_i \neq 0$ satisfies the relation $\beta_i^2 = g(\alpha_i)$.

In summary, the affine curve $C \cap U_0$ has 8 inflection points and so C has 9 inflection points.

Note. (*Configuration of the inflection points of a smooth complex cubic*) If C is a smooth cubic of $\mathbb{P}^2(\mathbb{C})$ and $P_1, \ldots, P_9$ are the inflection points of C, by Exercise 90 for each $1 \leq i < j \leq 9$ the line $L(P_i, P_j)$ contains a third inflection point. As a consequence of this fact, it is not difficult to check that the set of lines joining two inflection points of C has exactly 12 elements and the incidence relations among these 12 lines and the points $P_1, \ldots, P_9$ are the same that hold in the affine plane $(\mathbb{Z}/3\mathbb{Z})^2$. An explicit example of this correspondence is described in the Note following Exercise 102.

Exercise 94. (*Inflection points of a smooth real cubic*) Prove that every smooth and irreducible cubic C of $\mathbb{P}^2(\mathbb{R})$ has exactly three inflection points.

Solution. By Exercise 86 the curve C has at least one inflection point P and there exists a system of homogeneous coordinates of $\mathbb{P}^2(\mathbb{R})$ such that $P = [0, 0, 1]$ and the affine curve $C \cap U_0$ is given by an equation of the form $f(x, y) = y^2 - g(x) = 0$, where g is a monic real polynomial of degree 3 with distinct roots.

The polynomial g has at least one real root, because its degree is odd; so, up to performing a change of affine coordinates of the form $(x, y) \mapsto (x + c, y)$, we may assume that $g(0) = 0$ and $g(x) > 0$ for $x > 0$. In other words, we may assume that $g(x) = x(x^2 + mx + q)$ with $m \geq 0$ and $q > 0$. Arguing exactly as in the solution of Exercise 93, one shows that the inflection points of $C \cap U_0$ are the solutions of the system:

$$y^2 - g(x) = 0, \quad p(x) = 3x^4 + 4mx^3 + 6qx^2 - q^2 = 0. \qquad (3.6)$$

Observe also that $p'(x) = 12g(x)$. As $p(0) = -q^2 < 0$ and $p'(x) = 12g(x) > 0$ for every $x > 0$, there exists a unique $\alpha_1 \in (0, +\infty)$ such that $p(\alpha_1) = 0$. By construction $g(\alpha_1) > 0$, so there exists $\beta_1 \in \mathbb{R}$ such that $\beta_1^2 = g(\alpha_1)$ and the points $P_1 = (\alpha_1, \beta_1)$ and $P_2 = (\alpha_1, -\beta_1)$ are inflection points of $C \cap U_0$.

If 0 is the only root of $g(x)$, then $g(x) < 0$ for $x \in (-\infty, 0)$; in this case, since α_1 is the only positive root of $p(x)$, the system (3.6) has no solutions different from P_1 and P_2. If instead g has roots $\lambda_1 < \lambda_2 < 0$ and if there exists a solution $Q = (\alpha_2, \beta_2)$ of (3.6), then $p(\alpha_2) = 0$ and $g(\alpha_2) = \beta_2^2 > 0$ (as explained in the solution of Exercise 93, the polynomials p and g have no common roots). It follows that $\lambda_1 < \alpha_2 < \lambda_2$ and that α_2 is the only root of p in the interval (λ_1, λ_2), because $p(x)$ is increasing in this interval. Therefore in this case, besides P, P_1 and P_2, there are exactly two additional inflection points corresponding to the points $Q_1 = (\alpha_2, \beta_2)$ and $Q_2 = (\alpha_2, -\beta_2)$. In summary, we have shown that C can have either three or five inflection points.

Suppose that C has five inflection points. Then, by Exercise 90, the line $L(P_1, Q_2)$ intersects C in a third inflection point R and it can be readily checked that R does not coincide with any of the points P, P_1, P_2, Q_1 and Q_2. Thus we get to a contradiction, and therefore C has exactly three inflection points.

Exercise 95. Let $d \geq 3$; consider two distinct points $P_1, P_2 \in \mathbb{P}^2(\mathbb{K})$ and, for $i = 1, 2$, a projective line l_i passing through P_i. Denote by Λ_d the space of projective curves of degree d, and let

$$\mathcal{F}_1 = \{C \in \Lambda_d \mid l_1 \text{ is tangent to } C \text{ at } P_1\},$$
$$\mathcal{F}_2 = \{C \in \Lambda_d \mid l_i \text{ is tangent to } C \text{ at } P_i, \ i = 1, 2\}.$$

Show that $\mathcal{F}_i$ is a linear system, and compute its dimension.

Solution. On $\mathbb{P}^2(\mathbb{K})$ consider homogeneous coordinates such that $P_1 = [1, 0, 0]$, $l_1 = \{x_2 = 0\}$, and let

$$F = \sum_{i+j+k=d} a_{i,j,k} x_0^i x_1^j x_2^k = 0$$

be the equation of the generic projective curve of degree d. As remarked in Sect. 1.9.6, the curve defined by F is tangent at $[y_0, y_1, y_2]$ to the line of equation $ax_0 + bx_1 + cx_2 = 0$ if and only if $\nabla F(y_0, y_1, y_2)$ is a multiple of (a, b, c). It follows that, in our situation, F represents a curve in $\mathcal{F}_1$ iff $F_{x_0}(1, 0, 0) = F_{x_1}(1, 0, 0) = 0$. Observe now that $\dfrac{\partial x_0^i x_1^j x_2^k}{\partial x_0}(1, 0, 0) = d$ if $i = d, j = k = 0$, while it is zero otherwise. So $F_{x_0}(1, 0, 0) = 0$ iff $a_{d,0,0} = 0$. Similarly, $\dfrac{\partial x_0^i x_1^j x_2^k}{\partial x_1}(1, 0, 0) = 1$ if $i = d - 1, j = 1$, $k = 0$, and it is zero otherwise, so that $F_{x_1}(1, 0, 0) = 0$ iff $a_{d-1,1,0} = 0$. Since the coefficients $a_{i,j,k}$ give a system of homogeneous coordinates for Λ_d, the previous considerations imply that $\mathcal{F}_1$ is a subspace of Λ_d of codimension 2. As $\dim \Lambda_d = \dfrac{d(d+3)}{2}$, we obtain that $\dim \mathcal{F}_1 = \dfrac{(d+4)(d-1)}{2}$.

Suppose now that $P_1 \notin l_2, P_2 \notin l_1$. In this case, l_1 and l_2 meet at a point P_3 which forms with P_1, P_2 a triple of points in general position. Then we can choose coordinates such that $P_1 = [1, 0, 0], P_2 = [0, 0, 1], P_3 = [0, 1, 0]$, and hence $l_1 = \{x_2 = 0\}, l_2 = \{x_0 = 0\}$. So F defines a curve in $\mathcal{F}_2$ iff

$$F_{x_0}(1, 0, 0) = F_{x_1}(1, 0, 0) = 0, \quad F_{x_1}(0, 0, 1) = F_{x_2}(0, 0, 1) = 0.$$

Arguing as above, one checks that these conditions are equivalent to $a_{d,0,0} = a_{d-1,1,0} = a_{0,1,d-1} = a_{0,0,d} = 0$. Therefore $\mathcal{F}_2$ si defined by 4 independent linear conditions, and hence it has codimension 4 in Λ_d.

Suppose now $P_2 \in l_1$, $P_1 \notin l_2$. In this case, we can choose coordinates such that $P_1 = [1, 0, 0]$, $P_2 = [0, 1, 0]$, $l_1 = \{x_2 = 0\}$, $l_2 = \{x_0 = 0\}$, and $F \in \mathcal{F}_2$ iff $F_{x_0}(1, 0, 0) = F_{x_1}(1, 0, 0) = 0$, i.e. $a_{d,0,0} = a_{d-1,1,0} = 0$, and $F_{x_1}(0, 1, 0) = F_{x_2}(0, 1, 0) = 0$, i.e. $a_{0,d,0} = a_{0,d-1,1} = 0$. Also in this case $\mathcal{F}_2$ is defined by 4 independent linear conditions, so it has codimension 4 in Λ_d. Of course, the same result holds also when $P_1 \in l_2$, $P_2 \notin l_1$.

Finally, suppose $l_1 = l_2 = L(P_1, P_2)$, and choose coordinates such that $P_1 = [1, 0, 0]$, $P_2 = [0, 1, 0]$. Arguing as above, we obtain that $F \in \mathcal{F}_2$ iff $a_{d,0,0} = a_{d-1,1,0} = a_{1,d-1,0} = a_{0,d,0} = 0$. As $d \geq 3$ we have $d - 1 \neq 1$, so that also in this case $\mathcal{F}_2$ is defined by 4 independent linear conditions.

Therefore, in any case $\dim \mathcal{F}_2 = \dim \Lambda_d - 4 = \dfrac{d^2 + 3d - 8}{2}$.

Note. We briefly sketch two alternative methods for computing the dimension of $\mathcal{F}_1$, that can also be used to compute $\dim \mathcal{F}_2$.

Using the same notations and coordinates as in the above solution, if $f(u, v) = F(1, u, v)$ is the equation of the affine part $\mathcal{C} \cap U_0$ of the generic curve $\mathcal{C} \in \Lambda_d$, then $f(u, v) = a_{d,0,0} + a_{d-1,1,0}u + a_{d-1,0,1}v + g(u, v)$, where $g(u, v)$ is a sum of monomials all having degree greater than or equal to 2. Since the affine part $l_1 \cap U_0$ of l_1 has equation $v = 0$ and P_1 has affine coordinates $(0, 0)$, it follows that $\mathcal{C}$ passes through P_1 with tangent l_1 if and only if $a_{d,0,0} = a_{d-1,1,0} = 0$. This allows us to conclude that $\mathcal{F}_1$ has codimension 2 in Λ_d.

Alternatively, one can proceed as follows. As noted above, F represents a curve of $\mathcal{F}_1$ if and only if $F_{x_0}(1, 0, 0) = F_{x_1}(1, 0, 0) = 0$. Since the conditions that we have just described are linear, it is clear that $\mathcal{F}_1$ is a linear system of codimension at most 2. In order to show that codim $\mathcal{F}_1 = 2$, it suffices to produce a curve $\mathcal{C}_1 \in \Lambda_d$ not passing through $P_1 = [1, 0, 0]$ and a curve $\mathcal{C}_2 \in \Lambda_d$ passing through P_1 having at P_1 only one tangent different from $l_1 = \{x_2 = 0\}$. If this is the case, then, denoting by $\mathcal{F}_3$ the linear system consisting of the curves passing through P_1, one has $\mathcal{F}_1 \subsetneq \mathcal{F}_3 \subsetneq \Lambda_d$, so that $\dim \mathcal{F}_1 < \dim \mathcal{F}_3 < \dim \Lambda_d$ and $\dim \Lambda_2 - \dim \mathcal{F}_1 \geq 2$, as requested. Therefore, the proof can be concluded by setting, for instance, $\mathcal{C}_i = [F_i]$ with $F_1(x_0, x_1, x_2) = x_0^d$ and $F_2(x_0, x_1, x_2) = x_0^{d-1}x_1$.

Note in addition that the considerations made to compute the dimension of $\mathcal{F}_1$ apply also when $d = 2$; therefore the conics tangent to l_1 at P_1 form a linear system of dimension 3. On the other hand, when $d = 2$ the dimension of the linear system $\mathcal{F}_2$ depends on the position of the points P_1, P_2 with respect to the lines l_1, l_2; a case is studied in Exercise 125.

Exercise 96. Let $h, k \geq 1$ and let $\mathcal{W}$ be a linear system of curves of degree h in $\mathbb{P}^2(\mathbb{K})$. Let $\mathcal{C}$ be a projective curve of degree k, and set

$$\mathcal{H} = \{C + C' \mid C' \in \mathcal{W}\}.$$

Show that $\mathcal{H}$ is a linear system of curves of degree $k + h$, and that dim $\mathcal{H} = $ dim $\mathcal{W}$.

Solution. Let $V = \mathbb{K}[x_0, x_1, x_2]_h$ (respectively $V' = \mathbb{K}[x_0, x_1, x_2]_{h+k}$) be the vector space consisting of the zero polynomial and of the homogeneous polynomials of degree h (respectively $h + k$), and let $F(x_0, x_1, x_2)$ be a homogeneous polynomial of degree k that represents C. It is immediate to check that the function

$$\varphi \colon V \to V', \quad \varphi(G) = FG$$

is linear and injective. Moreover, if $\mathcal{W} = \mathbb{P}(S)$, clearly one has $\mathcal{H} = \mathbb{P}(\varphi(S))$, and the claim can be immediately deduced from this fact.

Exercise 97. Let Λ_3 be the set of all cubics of $\mathbb{P}^2(\mathbb{C})$, and let $\mathcal{S} \subseteq \Lambda_3$ be the set of the cubics C that satisfy the following conditions:

(i) $[1, 0, 0]$ is either a non-ordinary double point with principal tangent $x_2 = 0$ or a triple point for C;
(ii) C passes through the points $[0, 0, 1]$ and $[1, 1, -1]$.

Verify that $\mathcal{S}$ is a linear system and write down one of its projective frames.

Solution. Consider the curve $C \in \mathcal{S}$ of equation $F(x_0, x_1, x_2) = 0$. Set affine coordinates $u = \frac{x_1}{x_0}, v = \frac{x_2}{x_0}$ on U_0, and let $f(u, v) = F(1, u, v)$. In order that C satisfy (i), there must be $a, b, c, d, e \in \mathbb{C}$ not all zero and such that $f(u, v) = av^2 + bu^3 + cu^2v + duv^2 + ev^3$, and so

$$F(x_0, x_1, x_2) = ax_0x_2^2 + bx_1^3 + cx_1^2x_2 + dx_1x_2^2 + ex_2^3.$$

The fact that C passes through $[0, 0, 1]$ and through $[1, 1, -1]$ is equivalent to the conditions $e = 0$ and $a + b - c + d - e = 0$, hence the generic cubic of $\mathcal{S}$ has equation

$$ax_0x_2^2 + bx_1^3 + (a + b + d)x_1^2x_2 + dx_1x_2^2 = 0, \quad [a, b, d] \in \mathbb{P}^2(\mathbb{C}).$$

It follows that $\mathcal{S}$ is a linear system of dimension 2, and a projective frame of $\mathcal{S}$ is given by the curves determined by the parametres $[1, 0, 0], [0, 1, 0], [0, 0, 1], [1, 1, 1]$, that is by the curves of equation

$$x_0x_2^2 + x_1^2x_2 = 0, \quad x_1^3 + x_1^2x_2 = 0,$$
$$x_1^2x_2 + x_1x_2^2 = 0, \quad x_0x_2^2 + x_1^3 + 3x_1^2x_2 + x_1x_2^2 = 0.$$

Exercise 98. Given two distinct points P and Q of $\mathbb{P}^2(\mathbb{K})$, consider the set Ω of cubics of $\mathbb{P}^2(\mathbb{K})$ which are singular at Q and tangent at P to the line $L(P, Q)$.

(a) Prove that Ω is a linear system and compute its dimension.
(b) If R and S are points such that P, Q, R, S are in general position, show that

$$\Lambda = \{C \in \Omega \mid S \in C, \ C \text{ is tangent to the line } L(R, S) \text{ at } R\}$$

is a linear system, compute its dimension and determine the number of irreducible components of every cubic of Λ.

Solution. (a) Observe that all cubics of Ω are reducible. Namely, if $r = L(P, Q)$, for every $C \in \Omega$ we have $I(C, r, P) \geq 2$ and $I(C, r, Q) \geq 2$, so by Bézout's Theorem r is a component of C. Therefore Ω consists exactly of the cubics of the form $r + \mathcal{D}$ as $\mathcal{D}$ varies in the set of conics passing through Q (which must be singular); since such conics form a linear system of dimension 4, it is easy to check (cf. Exercise 96) that Ω is a linear system of dimension 4 too.

(b) Assume that $C = r + \mathcal{D}$ is a cubic of Ω. As P, Q, R, S are in general position, neither S nor R lies on r; therefore $C \in \Lambda$ if and only if the conic $\mathcal{D}$, which passes through Q, passes through S too and is tangent at R to the line $L(R, S)$. Then $\mathcal{D}$ intersects $L(R, S)$ at least in 3 points (counted with multiplicity), so $L(R, S)$ is a component of $\mathcal{D}$.

If t is any line passing through Q, the cubic $r + L(R, S) + t$ satisfies the required conditions; then Λ coincides with the set of cubics of the form $r + L(R, S) + t$ where t varies in the pencil of lines with centre Q. Therefore (cf. Exercise 96) Λ is a linear system of dimension 1.

Every cubic of Λ has 3 distinct irreducible components, except the cubic $2r + L(R, S)$ for which the line t of the pencil coincides with r.

Exercise 99. Consider in $\mathbb{P}^2(\mathbb{C})$ the curve C of equation

$$F(x_0, x_1, x_2) = 2x_1^4 - x_0^2 x_1^2 + 2x_0^3 x_1 + 3x_0^3 x_2 = 0$$

and the curve $\mathcal{D}$ of equation

$$G(x_0, x_1, x_2) = ax_1^2 x_2^2 + 2bx_0^2 x_1 x_2 + 4x_0^3 x_1 + cx_0^3 x_2 = 0,$$

as a, b vary in $\mathbb{C}$ and c varies in $\mathbb{C}^*$.

(a) Find the values of the parameters a, b, c for which the pencil generated by C and $\mathcal{D}$ contains a curve $\mathcal{G}$ singular at the point $P = [1, 0, 0]$. For those values, determine the multiplicity and the principal tangents to $\mathcal{G}$ at P.
(b) Find the values of the parameters a, b, c for which the pencil generated by C and $\mathcal{D}$ contains a curve $\mathcal{G}$ such that its affine part $\mathcal{G} \cap U_0$ has the line $x - y = 0$ as an asymptote and the corresponding point at infinity is non-singular for $\mathcal{G}$.

Solution. (a) The affine part $C \cap U_0$ of C is defined by the equation $f(x, y) = F(1, x, y) = 2x^4 - x^2 + 2x + 3y = 0$ and the affine part $\mathcal{D} \cap U_0$ of $\mathcal{D}$ is defined by the equation $g(x, y) = G(1, x, y) = ax^2 y^2 + 2bxy + 4x + cy = 0$. The affine part of the generic curve of the pencil generated by C and D has equation $\alpha f + \beta g = 0$ when $[\alpha, \beta]$ varies in $\mathbb{P}^1(\mathbb{C})$. The homogeneous linear term of $\alpha f + \beta g$ is equal to $x(2\alpha + 4\beta) + y(3\alpha + c\beta)$; so the pencil contains a curve which is singular at $P = (0, 0)$ if and only if the system $2\alpha + 4\beta = 3\alpha + c\beta = 0$ has non-trivial solutions, that is iff $c = 6$. In addition, in this case the curve G of the pencil which is singular at P has equation $2f - g = 0$. Since for $c = 6$ the first non-zero homogeneous component of $2f - g$ is $2x(-x - by)$, then the origin is a double point for G, with principal tangents of equations $x = 0$ and $x + by = 0$. In particular, P is ordinary for G if and only if $b \neq 0$.

(b) The generic curve $G_{\alpha,\beta}$ of the pencil generated by C and D has equation $H^{\alpha,\beta} = \alpha F + \beta G = 0$, $[\alpha, \beta] \in \mathbb{P}^1(\mathbb{C})$.

Assume that $x - y = 0$ is an asymptote of $G_{\alpha,\beta} \cap U_0$, and assume that the point at infinity of that line is non-singular for $G_{\alpha,\beta}$. The projective closure of $x - y = 0$ has equation $x_1 - x_2 = 0$, and intersects the improper line at $[0, 1, 1]$. A necessary and sufficient condition in order that $[0, 1, 1]$ be a non-singular point of $G_{\alpha,\beta}$ and the (principal) tangent to $G_{\alpha,\beta}$ at $[0, 1, 1]$ have equation $x_1 - x_2 = 0$ is that $H^{\alpha,\beta}_{x_0}(0, 1, 1) = 0, H^{\alpha,\beta}_{x_1}(0, 1, 1) = -H^{\alpha,\beta}_{x_2}(0, 1, 1) \neq 0$. An easy computation shows that the first equation is always satisfied, while the second condition is equivalent to $8\alpha + 2\beta = -2a\beta \neq 0$. It follows that the pencil generated by C and D contains a curve G satisfying the required conditions if and only if $a \neq 0$. Moreover, if $a \neq 0$ then $G = G_{-a,2}$, so that G has equation $H^{-a,2} = -aF + 2G = 0$.

Exercise 100. Consider the cubic D_1 of $\mathbb{P}^2(\mathbb{C})$ of equation

$$F(x_0, x_1, x_2) = x_1 x_2^2 - 2x_0 x_1 x_2 - 6x_0^2 x_2 + x_0 x_2^2 + 8x_0^3 = 0.$$

(a) Check that D_1 has exactly two singular points A and B.
(b) Say whether D_1 is irreducible.
(c) Find a cubic D_2 in $\mathbb{P}^2(\mathbb{C})$ such that all the cubics of the pencil generated by D_1 and D_2 have infinitely many inflection points, have A and B as singular points and pass through $P = [-1, 2, 4]$.

Solution. (a) If we compute the gradient of the polynomial F

$$\nabla F = (-2x_1 x_2 - 12x_0 x_2 + x_2^2 + 24x_0^2, x_2(x_2 - 2x_0), 2x_1 x_2 - 2x_0 x_1 - 6x_0^2 + 2x_0 x_2),$$

we see that the only singularities of D_1 are the points $A = [0, 1, 0]$ and $B = [1, 1, 2]$.

(b) Any cubic having two singular points is surely reducible by Bézout's Theorem, because the line joining the two singularities intersects the cubic in at least 4 points (counted with multiplicity). In this case the line $r = L(A, B)$ has equation $x_2 - 2x_0 = 0$ and in fact the polynomial F factors as $F = (x_2 - 2x_0)(x_1 x_2 + x_0 x_2 - 4x_0^2)$. Therefore D_1 is the union of the line r and the conic C of equation $x_1 x_2 + x_0 x_2 - 4x_0^2 = 0$, which is non-singular and hence irreducible.

(c) As already observed in the solution of (b), every cubic which is singular both at A and at B has the line r as a component. On the other hand, if we choose as D_2

the sum of r and a conic passing through A, B and P and distinct from C, all cubics of the pencil generated by $\mathcal{D}_1$ and $\mathcal{D}_2$ are singular at A and at B. For instance we can take $\mathcal{D}_2 = r + L(A, P) + L(B, P)$, so that every cubic of the pencil generated by $\mathcal{D}_1$ and $\mathcal{D}_2$ is the sum of a line r and a conic of the pencil $\mathcal{F}$ generated by the conic C and by the reducible conic $L(A, P) + L(B, P)$.

Moreover, given a cubic $\mathcal{D}$ of this pencil, all points of $r \backslash \{A, B\}$ are inflection points of $\mathcal{D}$ unless r is a double component of the cubic. Namely, if the latter situation occurs, all points of r are singular and so they are not inflection points; however in this case $\mathcal{D}$ consists, in addition to the double line r, of the tangent to C at P, so that all points of this line (except the intersection point with r) are inflection points of the cubic.

Exercise 101. Consider the points $A = [1, 0, -1]$, $B = [0, 1, 0]$, $C = [1, 0, 0]$ and $D = [0, 1, 1]$ of $\mathbb{P}^2(\mathbb{C})$. Consider also the lines r of equation $x_0 + x_2 = 0$, s of equation $x_1 = 0$ and t of equation $2x_1 - 2x_2 - x_0 = 0$. Denote by $\mathcal{W}$ the set of projective quartics $\mathcal{Q}$ of $\mathbb{P}^2(\mathbb{C})$ that satisfy the following conditions:

(i) A is a singular point;
(ii) $I(\mathcal{Q}, r, B) \geq 3$;
(iii) C is a singular point and $I(\mathcal{Q}, s, C) \geq 3$;
(iv) the line t is tangent to $\mathcal{Q}$ at D.

(a) Say whether $\mathcal{W}$ is a linear system of curves and, if so, compute its dimension.
(b) Say whether there exists in $\mathcal{W}$ a quartic having B as a triple point.
(c) Write the equation of a quartic $\mathcal{Q} \in \mathcal{W}$ such that the affine line $t \cap U_0$ is not an asymptote of the affine part $\mathcal{Q} \cap U_0$ of the quartic $\mathcal{Q}$.

Solution. (a) Observe that the points A and B lie on the line r. If the quartic $\mathcal{Q}$ belongs to $\mathcal{W}$, then $I(\mathcal{Q}, r, A) \geq 2$ and $I(\mathcal{Q}, r, B) \geq 3$; so, by Bézout's Theorem, r must be an irreducible component of $\mathcal{Q}$. Moreover, the points A and C lie on the line s; since $I(\mathcal{Q}, s, A) \geq 2$ and $I(\mathcal{Q}, s, C) \geq 3$, in the same way we deduce that s must be an irreducible component of $\mathcal{Q}$ too. Therefore the quartic $\mathcal{Q}$ is necessarily given by $r + s + C$ with C a curve of degree 2. A necessary and sufficient condition in order that this quartic satisfy all the conditions defining the set $\mathcal{W}$ is that the conic C passes through the point C (so that C is singular) and is tangent at D to the line t. So the set $\mathcal{W}$ consists of the quartics $r + s + C$, when C varies in the linear system of conics passing through C and tangent at D to the line t. As these conics form a linear system of dimension 2, then also $\mathcal{W}$ is a linear system of dimension 2 (cf. Exercise 96).

(b) Assume by contradiction that there exists a quartic $\mathcal{Q} = r + s + C \in \mathcal{W}$ having B as a triple point. Then B has to be a double point for the conic C, i.e. C must have the lines $L(B, C)$ and $L(B, D)$ as components. Since $L(B, D)$ is distinct from the line t, the quartic is not tangent to t at D. This contradicts the hypothesis that $\mathcal{Q} \in \mathcal{W}$.

(c) Every quartic of $\mathcal{W}$ is tangent to t at D. Therefore, the affine line $t \cap U_0$ is not an asymptote if and only if the point D is singular and t is not a principal tangent to the curve at that point. As $L(C, D) \neq t$, all quartics $\mathcal{Q}$ of the form $r + s + L(C, D) + l$, where l is a line passing through D and distinct from t, have the required property. If for instance we choose $l = \{x_0 = 0\}$, then we have $\mathcal{Q} = [x_0 x_1 (x_0 + x_2)(x_1 - x_2)]$.

Exercise 102. (*Hesse pencil of cubics*) Consider the pencil $\mathcal{F} = \{C_{\lambda,\mu} \mid [\lambda, \mu] \in \mathbb{P}^1(\mathbb{C})\}$, where $C_{\lambda,\mu}$ is the cubic of $\mathbb{P}^2(\mathbb{C})$ of equation

$$F_{\lambda,\mu}(x_0, x_1, x_2) = \lambda(x_0^3 + x_1^3 + x_2^3) + \mu x_0 x_1 x_2 = 0.$$

(a) Find the base points of $\mathcal{F}$ and show that all of them are non-singular for every $C_{\lambda,\mu}$.
(b) Show that for every $[\lambda, \mu] \in \mathbb{P}^1(\mathbb{C})$ the Hessian curve $H(C_{\lambda,\mu})$ of $C_{\lambda,\mu}$ is defined and belongs to the pencil $\mathcal{F}$, so that there is a map $H: \mathcal{F} \to \mathcal{F}$ which associates $H(C_{\lambda,\mu})$ to $C_{\lambda,\mu}$.
(c) Prove that $H: \mathcal{F} \to \mathcal{F}$ has four fixed points, and that it is not a projectivity.
(d) Find the inflection points of $C_{\lambda,\mu}$ when $C_{\lambda,\mu}$ is not a fixed point of H.

Solution. (a) Let $\omega = \dfrac{1 + i\sqrt{3}}{2}$ be a cube root of -1. If we consider the system formed by the equations of $C_{1,0}$ and of $C_{0,1}$, it is immediate to check that the base points of $\mathcal{F}$ are

$$\begin{aligned}
P_1 &= [0, 1, -1], \; P_2 = [0, 1, \omega], \; P_3 = [0, 1, \overline{\omega}], \\
P_4 &= [1, 0, -1], \; P_5 = [1, 0, \omega], \; P_6 = [1, 0, \overline{\omega}], \\
P_7 &= [1, -1, 0], \; P_8 = [1, \omega, 0], \; P_9 = [1, \overline{\omega}, 0].
\end{aligned}$$

In addition, we have

$$\nabla F_{\lambda,\mu}(x_0, x_1, x_2) = 3\lambda(x_0^2, x_1^2, x_2^2) + \mu(x_1 x_2, x_0 x_2, x_0 x_1),$$

so that $C_{1,0}$ is clearly non-singular (in particular no P_i is singular for $C_{1,0}$). So take $\mu \neq 0$. For $\{i, j, k\} = \{1, 2, 3\}$, if $y_i = 0$, $y_j \neq 0$, $y_k \neq 0$ it turns out that $\dfrac{\partial F_{\lambda,\mu}}{\partial x_i}(y_0, y_1, y_2) = \mu y_j y_k \neq 0$, which immediately implies that P_i is non-singular for every cubic of $\mathcal{F}$.

(b) The Hessian curve of $C_{\lambda,\mu}$ has equation

$$H_{F_{\lambda,\mu}} = \det \begin{pmatrix} 6\lambda x_0 & \mu x_2 & \mu x_1 \\ \mu x_2 & 6\lambda x_1 & \mu x_0 \\ \mu x_1 & \mu x_0 & 6\lambda x_2 \end{pmatrix} = -6\lambda\mu^2(x_0^3 + x_1^3 + x_2^3) + (216\lambda^3 + 2\mu^3)x_0 x_1 x_2 = 0.$$

As $(\lambda, \mu) \neq (0, 0)$ clearly implies $(-6\lambda\mu^2, 216\lambda^3 + 2\mu^3) \neq (0, 0)$, we have $H_{F_{\lambda,\mu}} \neq 0$ and $H(C_{\lambda,\mu}) \in \mathcal{F}$ for every $C_{\lambda,\mu} \in \mathcal{F}$.

(c) As seen in the previous part, $H(C_{\lambda,\mu}) = C_{\lambda,\mu}$ if and only if $[\lambda, \mu] = [-6\lambda\mu^2, 216\lambda^3 + 2\mu^3]$, i.e. iff $\det \begin{pmatrix} -6\lambda\mu^2 & 216\lambda^3 + 2\mu^3 \\ \lambda & \mu \end{pmatrix} = 0$. This equation is equivalent to $\lambda(27\lambda^3 + \mu^3) = 0$, whose non-trivial solutions are $[0, 1], [1, -3], [1, 3\omega]$, $[1, 3\overline{\omega}]$. It follows that the fixed points of H are the cubics $C_{0,1}, C_{1,-3}, C_{1,3\omega}, C_{1,3\overline{\omega}}$. If H were a projectivity of the 1-dimensional projective space $\mathcal{F}$, then it should coincide with the identity map, since it has at least 3 fixed points. This contradicts the fact that H has exactly 4 fixed points, so H is not a projectivity.

(d) By the computations seen in (b), the intersection points between $\mathcal{C}_{\lambda,\mu}$ and $H(\mathcal{C}_{\lambda,\mu})$ are given by the solutions of the system

$$\begin{cases} \lambda(x_0^3 + x_1^3 + x_2^3) + \mu x_0 x_1 x_2 = 0 \\ -6\lambda\mu^2(x_0^3 + x_1^3 + x_2^3) + (216\lambda^3 + 2\mu^3)x_0 x_1 x_2 = 0 \end{cases}.$$

As explained in (c), if $\mathcal{C}_{[\lambda,\mu]}$ is not a fixed point of H then

$$\det\begin{pmatrix} -6\lambda\mu^2 & 216\lambda^3 + 2\mu^3 \\ \lambda & \mu \end{pmatrix} \neq 0,$$

so that the previous system is equivalent to $x_0^3 + x_1^3 + x_2^3 = x_0 x_1 x_2 = 0$. The solutions of this system exactly correspond to the base points $P_1, \ldots, P_9$ of $\mathcal{F}$, which, as seen in (a), are all non-singular for $\mathcal{C}_{\lambda,\mu}$. It follows that, if $H(\mathcal{C}_{\lambda,\mu}) \neq \mathcal{C}_{\lambda,\mu}$, then the inflection points of $\mathcal{C}_{\lambda,\mu}$ are exactly the base points of $\mathcal{F}$.

Note. If $\mathcal{C}_{\lambda_0,\mu_0}$ is a fixed point of H, then all non-singular points of $\mathcal{C}_{\lambda_0,\mu_0}$ are inflection points. Since every cubic of $\mathcal{F}$ is reduced by part (a) of the Exercise, it follows from Exercise 84 that $\mathcal{C}_{\lambda_0,\mu_0}$ is the union of three lines. In fact, if $[\lambda_0, \mu_0] = [0, 1]$ then the cubic $\mathcal{C}_{\lambda_0,\mu_0}$ decomposes as the union of the lines of equation $x_0 = 0$ (containing P_1, P_2, P_3), $x_1 = 0$ (containing P_4, P_5, P_6), $x_2 = 0$ (containing P_7, P_8, P_9).

If $[\lambda_0, \mu_0] = [1, -3]$, then $\mathcal{C}_{\lambda_0,\mu_0}$ has equation $x_0^3 + x_1^3 + x_2^3 - 3x_0 x_1 x_2 = 0$, and decomposes as the union of the lines of equation $x_0 + x_1 + x_2 = 0$ (containing P_1, P_4, P_7), $\overline{\omega}x_0 + \omega x_1 - x_2 = 0$ (containing P_2, P_6, P_8), and $\omega x_0 + \overline{\omega}x_1 - x_2 = 0$ (containing P_3, P_5, P_9).

If $[\lambda_0, \mu_0] = [1, 3\omega]$, then $\mathcal{C}_{\lambda_0,\mu_0}$ has equation $x_0^3 + x_1^3 + x_2^3 + 3\omega x_0 x_1 x_2 = 0$, and decomposes as the union of the lines of equation $\omega x_0 - x_1 - x_2 = 0$ (containing P_1, P_5, P_8), $x_0 - \omega x_1 + x_2 = 0$ (containing P_2, P_4, P_9), and $\overline{\omega}x_0 + \overline{\omega}x_1 - x_2 = 0$ (containing P_3, P_6, P_7).

Finally, if $[\lambda_0, \mu_0] = [1, 3\overline{\omega}]$, then $\mathcal{C}_{\lambda_0,\mu_0}$ has equation $x_0^3 + x_1^3 + x_2^3 + 3\overline{\omega}x_0 x_1 x_2 = 0$, and decomposes as the union of the lines of equation $\overline{\omega}x_0 - x_1 - x_2 = 0$ (containing P_1, P_6, P_9), $x_0 - \overline{\omega}x_1 + x_2 = 0$ (containing P_3, P_4, P_8), and $\omega x_0 + \omega x_1 - x_2 = 0$ (containing P_2, P_5, P_7).

Now define a bijection Φ between the set $\{P_1, \ldots, P_9\}$ and $(\mathbb{Z}/3\mathbb{Z})^2$:

$$\begin{aligned} P_1 &\mapsto (0,0), & P_2 &\mapsto (1,0), & P_3 &\mapsto (2,0), \\ P_4 &\mapsto (0,1), & P_5 &\mapsto (2,1), & P_6 &\mapsto (1,1), \\ P_7 &\mapsto (0,2), & P_8 &\mapsto (1,2), & P_9 &\mapsto (2,2). \end{aligned}$$

It is immediate to check, using the computations that we have just made, that three points P_i, P_j and P_k are collinear in $\mathbb{P}^2(\mathbb{C})$ if and only if $\Phi(P_i)$, $\Phi(P_j)$ and $\Phi(P_k)$ are collinear in $(\mathbb{Z}/3\mathbb{Z})^2$, and that the four reducible cubics of the pencil correspond to the partition of the set of lines of $(\mathbb{Z}/3\mathbb{Z})^2$ in four subsets of parallel lines.

Exercise 103. Let $\mathcal{D}$ and $\mathcal{G}$ be the cubics of $\mathbb{P}^2(\mathbb{C})$ of equations $x_0(x_1^2 - x_2^2) = 0$ and $x_1^2(x_0 + x_2) = 0$, respectively. For $[\lambda, \mu] \in \mathbb{P}^1(\mathbb{C})$, consider the cubics $\mathcal{C}_{\lambda, \mu}$ of the pencil generated by $\mathcal{D}$ and $\mathcal{G}$.

(a) Determine the base points of the pencil.
(b) Show that there exists a unique point P which is singular for all the cubics of the pencil and, as $[\lambda, \mu]$ varies in $\mathbb{P}^1(\mathbb{C})$, compute the multiplicity of P and the principal tangents to $\mathcal{C}_{\lambda, \mu}$ at P.
(c) Determine all the reducible cubics of the pencil.

Solution. (a) Both generators of the pencil are reducible: $\mathcal{D}$ is the union of three distinct lines, while $\mathcal{G}$ has two distinct irreducible components, one of which is double. The base points of the pencil are the points of $\mathcal{D} \cap \mathcal{G}$; hence, considering all the possible intersections of an irreducible component of $\mathcal{D}$ with an irreducible component of $\mathcal{G}$, we find that the pencil has 5 base points, namely the points $A = [1, -1, -1]$, $B = [1, 1, -1]$, $C = [0, 1, 0]$ (that lie on the line of equation $x_0 + x_2 = 0$) and the points $P = [1, 0, 0]$, $Q = [0, 0, 1]$ (that lie on the double line $x_1 = 0$ contained in $\mathcal{G}$).

(b) The points which are singular for all the cubics of the pencil are precisely the points which are singular for both generators; looking at the five base points of the pencil, it is easy to see that only $P = [1, 0, 0]$ is singular both for $\mathcal{D}$ and for $\mathcal{G}$.

The generic cubic $\mathcal{C}_{\lambda, \mu}$ of the pencil has equation $\lambda x_0(x_1^2 - x_2^2) + \mu x_1^2(x_0 + x_2) = 0$. In order to determine the multiplicity of P and the principal tangents to $\mathcal{C}_{\lambda, \mu}$ at P, we work in the affine chart U_0 with the affine coordinates $x = \frac{x_1}{x_0}, y = \frac{x_2}{x_0}$. With respect to these coordinates the affine curve $\mathcal{C}_{\lambda, \mu} \cap U_0$ has equation

$$\lambda(x^2 - y^2) + \mu x^2(1 + y) = (\lambda + \mu)x^2 - \lambda y^2 + \mu x^2 y = 0$$

and P has coordinates $(0, 0)$. Since for $[\lambda, \mu]$ varying in $\mathbb{P}^1(\mathbb{C})$ it is not possible that λ and $\lambda + \mu$ vanish simultaneously, the homogeneous part $(\lambda + \mu)x^2 - \lambda y^2$ is never identically zero and so P is always a double point. In particular:

(b1) If $\lambda = 0$, P is a non-ordinary double point with principal tangent $x_1 = 0$ (in that case we find again the cubic $\mathcal{C}_{0,1} = \mathcal{G}$ which, as we already knew, has P as a non-ordinary double point since the double line $x_1^2 = 0$ is a component of $\mathcal{G}$).
(b2) If $\lambda + \mu = 0$, P is a non-ordinary double point with principal tangent $x_2 = 0$ (in that case we obtain the cubic $\mathcal{C}_{1,-1}$ of equation $x_2(x_0x_2 + x_1^2) = 0$ whose components $x_0x_2 + x_1^2 = 0$ and $x_2 = 0$ are mutually tangent at P).
(b3) If $\lambda \neq 0$ and $\lambda + \mu \neq 0$, then P is ordinary double; denoting by α a square root of $\frac{\lambda}{\lambda + \mu}$, the distinct principal tangents to the cubic at P are the lines of equations $x_1 - \alpha x_2 = 0$ and $x_1 + \alpha x_2 = 0$.

(c) By what we have seen in Exercise 62, if $C_{\lambda,\mu}$ is reducible then it contains a line passing through P which of course will be a principal tangent at P.

As we have seen in part (b), P turns out to be non-ordinary only in the cases $\lambda = 0$ or $\lambda + \mu = 0$, namely in the cases corresponding to the reducible cubics $\mathcal{G}$ and $C_{1,-1}$.

If P is ordinary double for a reducible cubic $C_{\lambda,\mu}$, then by the considerations of part (b3) the line through P contained in $C_{\lambda,\mu}$ must be either the line $x_1 - \alpha x_2 = 0$ or the line $x_1 + \alpha x_2 = 0$. On the other hand, since only even powers of x_1 appear in the equation of $C_{\lambda,\mu}$, we deduce immediately that $C_{\lambda,\mu}$ contains both these lines: indeed, if substituting $x_1 = \alpha x_2$ in the equation of $C_{\lambda,\mu}$ we obtain the zero polynomial, then the same happens also substituting $x_1 = -\alpha x_2$, and conversely.

Since $\alpha \neq 0$, these two lines pass neither through $C = [0, 1, 0]$ nor through $Q = [0, 0, 1]$. Hence the line $x_0 = 0$ must be the third irreducible component of the cubic, i.e. $C_{\lambda,\mu}$ must have equation $x_0(x_1^2 - \alpha^2 x_2^2) = 0$. Imposing that this cubic pass through the points $A = [1, -1, -1]$ and $B = [1, 1, -1]$, we get $\alpha^2 = 1$, i.e. $\frac{\lambda}{\lambda + \mu} = 1$ and so $\mu = 0$. Therefore, besides the two reducible cubics found above, the only other reducible cubic of the pencil is the cubic $C_{1,0} = \mathcal{D}$.

It is also possible to conclude the proof of (c) in a slightly different way. After observing that, whenever $C_{\lambda,\mu}$ is reducible and has P as an ordinary double point, the degenerate conic of equation $x_1^2 = \alpha^2 x_2^2$ must be contained in $C_{\lambda,\mu}$, by substituting $x_1^2 = \frac{\lambda}{\lambda + \mu} x_2^2$ in the equation of $C_{\lambda,\mu}$ one obtains the equation $\frac{\lambda\mu}{\lambda + \mu} x_2^3 = 0$. In order that this equation be identically zero, one must have $\lambda = 0$ (but in this case P is not an ordinary double point of $C_{\lambda,\mu}$) or $\mu = 0$, case that corresponds to the reducible cubic $C_{1,0} = \mathcal{D}$.

Exercise 104. Consider the cubics $\mathcal{D}_1$ and $\mathcal{D}_2$ of $\mathbb{P}^2(\mathbb{C})$ of equations

$$F_1(x_0, x_1, x_2) = x_2(x_1^2 - 2x_0x_1 + x_2^2) = 0,$$
$$F_2(x_0, x_1, x_2) = (x_2 - x_1)(x_2 + x_1)(x_1 - x_0) = 0,$$

respectively. Determine the base points and the reducible cubics of the pencil $\mathcal{F}$ generated by $\mathcal{D}_1$ and $\mathcal{D}_2$.

Solution. The cubic $\mathcal{D}_1$ has as irreducible components the line $r_1 = \{x_2 = 0\}$ and an irreducible conic; denote by r_2, r_3, r_4 the lines of equations $x_2 - x_1 = 0, x_2 + x_1 = 0$ and $x_1 - x_0 = 0$, respectively, which are the irreducible components of $\mathcal{D}_2$.

It can be easily computed that the base points of the pencil, i.e. the intersection points between $\mathcal{D}_1$ and $\mathcal{D}_2$, are the points $A = [1, 1, -1]$, $B = [1, 1, 0]$, $C = [1, 1, 1]$ and $Q = [1, 0, 0]$. Note that the points A, B, C are collinear and lie on the line r_4; moreover, Q is a singular point both for $\mathcal{D}_1$ and for $\mathcal{D}_2$, so it is singular for all cubics of the pencil.

Since the base locus of $\mathcal{F}$ is a finite set, two distinct cubics of $\mathcal{F}$ have no common components. Assume now that $C_{\lambda,\mu}$ is reducible. As Q is singular for $C_{\lambda,\mu}$ (because it is singular for every cubic of the pencil), there exists a line r contained in $C_{\lambda,\mu}$ and passing through Q (cf. Exercise 62). Since $Q \notin r_4$, necessarily $r \neq r_4$.

If $r = r_1$, then the previous observation implies that $C_{\lambda,\mu} = \mathcal{D}_1$.

If either $r = r_2$ or $r = r_3$, then $C_{\lambda,\mu} = \mathcal{D}_2$ for the same reason.

If $r \neq r_i$ for $i = 1, 2, 3$, then $C_{\lambda,\mu} = r + \mathcal{G}$ where $\mathcal{G}$ is a conic that necessarily passes through the points $A \in r_4, B \in r_4, C \in r_4$. Hence $\mathcal{G}$ must be reducible and must contain the line r_4. Again from the previous observation it follows that $C_{\lambda,\mu} = \mathcal{D}_2$.

Therefore we obtain that the only reducible cubics of the pencil are the generators $\mathcal{D}_1$ and $\mathcal{D}_2$.

Exercise 105. Denote by C the curve of $\mathbb{P}^2(\mathbb{C})$ of equation

$$F(x_0, x_1, x_2) = x_1^3 - 2x_0x_2^2 + x_1x_2^2 = 0.$$

(a) Find the singular points of C, their multiplicities and their principal tangents. In addition, for each singular point of C, compute the multiplicity of intersection between the curve and each principal tangent at that point.
(b) Find the inflection points of C.
(c) Find a projective cubic $\mathcal{D}$ of $\mathbb{P}^2(\mathbb{C})$ such that the pencil $\mathcal{F}$ generated by C and $\mathcal{D}$ satisfies the following properties:

  (i) all cubics of $\mathcal{F}$ have at least a non-ordinary double point or a triple point;
  (ii) the points $[0, 0, 1]$ and $[1, 1, 1]$ belong to the base locus of the pencil $\mathcal{F}$.

Solution. (a) As
$$\nabla F = (-2x_2^2, 3x_1^2 + x_2^2, 2x_2(x_1 - 2x_0)),$$

the only singular point of C is $P = [1, 0, 0]$. If we set on U_0 the affine coordinates $x = \frac{x_1}{x_0}, y = \frac{x_2}{x_0}$, the point P has coordinates $(0, 0)$, while the affine part $C \cap U_0$ of C has equation $x^3 - 2y^2 + xy^2 = 0$. So P is a cusp with principal tangent r of equation $x_2 = 0$. Moreover $I(C, r, P) = 3$.

  (b) As
$$H_F(X) = \det \begin{pmatrix} 0 & 0 & -4x_2 \\ 0 & 6x_1 & 2x_2 \\ -4x_2 & 2x_2 & 2x_1 - 4x_0 \end{pmatrix} = -96x_1x_2^2,$$

the only points of intersection between C and its Hessian are P and $Q = [0, 0, 1]$. Since P is singular and Q is not, the only inflection point is Q.

  (c) If we choose as $\mathcal{D}$ a cubic having P as a non-ordinary double point with the same principal tangent as C, i.e. the line $r = \{x_2 = 0\}$, then all cubics of the pencil generated by C and $\mathcal{D}$ fulfil property (i). If in addition $\mathcal{D}$ passes through the points $[0, 0, 1]$ and $[1, 1, 1]$, these points turn out to be base points of the pencil generated by C and $\mathcal{D}$. A cubic $\mathcal{D}$ with these properties is, for instance, the one defined by the equation $x_2^2(x_0 - x_1) = 0$.

Note. In the solution of the previous exercise we proved, in particular, that C has exactly one inflection point. This fact can also be deduced without performing any computation. Namely, C has an ordinary cusp as its unique singularity, so it is irreducible by Exercise 62. Then C has exactly one inflection point by Exercise 89.

Exercise 106. Assume that C_1 and C_2 are curves of $\mathbb{P}^2(\mathbb{K})$ of degree n that intersect exactly at N distinct points $P_1, \ldots, P_N$. Let $\mathcal{D}$ be an irreducible curve of degree $d < n$ passing through the points $P_1, \ldots, P_{nd}$. Prove that there exists a curve $\mathcal{G}$ of degree $n - d$ passing through the points $P_{nd+1}, \ldots, P_N$.

Solution. Observe that $P_j \notin \mathcal{D}$ for every j such that $nd < j \leq N$: otherwise, $\mathcal{D}$ would have more than nd points in common both with C_1 and with C_2 and consequently it would be an irreducible component of both curves, while by hypothesis C_1 and C_2 meet at finitely many points.

Denote by Q a point of $\mathcal{D}$ such that $Q \neq P_j$ for each $j = 1, \ldots, nd$. In the pencil generated by C_1 and C_2 there is a curve $\mathcal{W}$ passing through Q; in particular $\mathcal{W}$ has degree n and passes through $P_1, \ldots, P_N, Q$. Since $\mathcal{W}$ shares with $\mathcal{D}$ at least $nd + 1$ distinct points and $\mathcal{D}$ is irreducible, then by Bézout's Theorem $\mathcal{D}$ is an irreducible component of $\mathcal{W}$, i.e. $\mathcal{W} = \mathcal{D} + \mathcal{G}$, where $\mathcal{G}$ is a curve of degree $n - d$. As $\mathcal{W}$ contains the points $P_1, \ldots, P_N$ and $\mathcal{D}$ contains $P_1, \ldots, P_{nd}$ but, as seen above, it does not contain any point among $P_{nd+1}, \ldots, P_N$, these latter points must be contained in the curve $\mathcal{G}$.

Exercise 107. (*Poncelet's Theorem*) Assume that C is an irreducible projective curve of $\mathbb{P}^2(\mathbb{C})$ of degree $n \geq 3$. Let r be a line that intersects C at n distinct points $P_1, \ldots, P_n$ and, for each $i = 1, \ldots, n$, let τ_i be a line tangent to C at P_i. Denote by $Q_1, \ldots, Q_k$ the additional points of intersection between C and the lines $\tau_1, \ldots, \tau_n$. Prove that $Q_1, \ldots, Q_k$ belong to the support of a curve of degree $n - 2$.

Solution. First of all observe that each point P_i is simple for C: otherwise the line r would meet the curve in more than n points (counted with multiplicity) and hence it would be an irreducible component of C, which instead is irreducible by hypothesis. For the same reason r is not tangent to C at any P_i.

Consider the curve $\mathcal{D}$ of degree n defined by $\mathcal{D} = \tau_1 + \cdots + \tau_n$ and take $Q \in r \setminus \{P_1, \ldots, P_n\}$. Assume that $\mathcal{G}$ is a curve of the pencil generated by C and $\mathcal{D}$ passing through Q. Since $\mathcal{G}$ has degree n and has at least $n + 1$ distinct points in common with r, then r is a component of $\mathcal{G}$, that is $\mathcal{G} = r + \mathcal{G}'$.

The curve $\mathcal{G}$, like all the curves of the pencil generated by C and $\mathcal{D}$, is tangent at P_i to the line τ_i; as r has not this property, then $\mathcal{G}'$ has to pass through the points $P_1, \ldots, P_n$. Consequently, as $\mathcal{G}'$ has degree $n - 1$, the line r is an irreducible component of $\mathcal{G}'$ too, i.e. $\mathcal{G} = 2r + \mathcal{G}''$ with $\deg \mathcal{G}'' = n - 2$. Then the points $Q_1, \ldots, Q_k$, which do not belong to r, must belong to the curve $\mathcal{G}''$.

Note. If $n = 3$ the statement of Exercise 107 turns into the following result. If C is an irreducible cubic of $\mathbb{P}^2(\mathbb{C})$, let r be a line meeting C in three distinct points P_1, P_2, P_3 which are not inflection points of C and, for $i = 1, 2, 3$, let τ_i be a line tangent to C at P_i. If Q_i is the additional point of intersection between τ_i and C, then the points Q_1, Q_2, Q_3 are collinear.

Exercise 108. Consider the curves C and D of $\mathbb{P}^2(\mathbb{C})$ defined by

$$F(x_0, x_1, x_2) = x_0^2 - x_1^2 + x_2^2 = 0, \quad G(x_0, x_1, x_2) = x_0 x_1 - x_2^2 + x_1^2 = 0$$

respectively. For every $P \in C \cap D$, compute $I(C, D, P)$.

Solution. Note that $[0, 0, 1] \notin C \cup D$; so, if $[q_0, q_1, q_2] \in C \cap D$, then $(q_0, q_1) \neq (0, 0)$ annihilates the resultant

$$\text{Ris}(F, G, x_2)(x_0, x_1) = \det \begin{pmatrix} x_0^2 - x_1^2 & 0 & 1 & 0 \\ 0 & x_0^2 - x_1^2 & 0 & 1 \\ x_1^2 + x_0 x_1 & 0 & -1 & 0 \\ 0 & x_1^2 + x_0 x_1 & 0 & -1 \end{pmatrix} = x_0^2 (x_0 + x_1)^2.$$

Since $\text{Ris}(F, G, x_2) \neq 0$, the curves C and D have no component in common. If $x_0 = 0$, from $F = G = 0$ it follows that $x_2^2 - x_1^2 = 0$, a condition that identifies the points $Q_1 = [0, 1, 1] \in C \cap D$, $Q_2 = [0, 1, -1] \in C \cap D$. If instead we set $x_0 + x_1 = 0$, then we find the point $Q_3 = [1, -1, 0] \in C \cap D$.

Observe now that $Q_1, Q_2, [0, 0, 1]$ are collinear, while the line $L(Q_3, [0, 0, 1])$ does not contain any point of $C \cap D$ distinct from Q_3. Therefore, the multiplicity of intersection $I(C, D, Q_3)$ is equal to the multiplicity of $[1, -1]$ as a root of $\text{Ris}(F, G, x_2)$, that is 2.

By Bézout's Theorem $\sum_{i=1}^{3} I(C, D, Q_i) = 4$; since of course $I(C, D, Q_i) \geq 1$ for $i = 1, 2$, then we can conclude that $I(C, D, Q_1) = I(C, D, Q_2) = 1$.

Exercise 109. Consider two curves C, D in $\mathbb{P}^2(\mathbb{K})$ without common components and let $P \in C \cap D$. Assume that P is non-singular both for C and for D, and denote by τ_C, τ_D the tangents at P to C, D, respectively. Show that $I(C, D, P) \geq 2$ if and only if $\tau_C = \tau_D$.

Solution. Let $Q \notin C \cup D \cup \tau_C$ be a point such that there exists no point of $C \cap D$ distinct from P and lying on the line joining P and Q. It is immediate to check that there are homogeneous coordinates such that $P = [1, 0, 0]$, $Q = [0, 0, 1] \notin C \cup D$ and $\tau_C = \{x_2 = 0\}$. Let $F(x_0, x_1, x_2) = 0$, $G(x_0, x_1, x_2) = 0$ be the equations of C and D, respectively, and denote by m the degree of F and by d the degree of G. By definition $I(C, D, P)$ is equal to the multiplicity of $[1, 0]$ as a root of the resultant $\text{Ris}(F, G, x_2)$.

As $Q \notin C$, the coefficient of x_2^m in F is not zero, so the degree of F with respect to the variable x_2 is m. In particular, if $f(x_1, x_2) = F(1, x_1, x_2)$ is the dehomogenized polynomial of F with respect to x_0, then $\deg f = \deg F = m$. Similarly, if $g(x_1, x_2) = G(1, x_1, x_2)$, then $\deg g = \deg G = d$. Moreover, since the specialization to $x_0 = 1$ does not lower the degrees of F and G, this specialization commutes with computing the resultant (cf. Sect. 1.9.2). Therefore $\mathrm{Ris}(F, G, x_2)(1, x_1) = \mathrm{Ris}(f, g, x_2)(x_1)$, so that $I(\mathcal{C}, \mathcal{D}, P)$ is equal to the order $\mathrm{ord}(\mathrm{Ris}(f, g, x_2)(x_1))$ (i.e. to the multiplicity of 0 as a root of $\mathrm{Ris}(f, g, x_2)(x_1)$, or in other words to the largest $i \geq 0$ such that x_1^i divides $\mathrm{Ris}(f, g, x_2)(x_1)$).

Now, as $\tau_{\mathcal{C}} = \{x_2 = 0\}$, up to multiplying f by a non-zero constant we have

$$f(x_1, x_2) = x_2 + \varphi_0(x_1) + \varphi_1(x_1)x_2 + \cdots + \varphi_{m-1}(x_1)x_2^{m-1} + ax_2^m,$$

with $\mathrm{ord}\,(\varphi_0(x_1)) \geq 2$, $\mathrm{ord}\,(\varphi_1(x_1)) \geq 1$ and $a \neq 0$. In the same way, if $\tau_{\mathcal{D}} = \{\alpha x_1 + \beta x_2 = 0\}$ then

$$g(x_1, x_2) = \alpha x_1 + \beta x_2 + \psi_0(x_1) + \psi_1(x_1)x_2 + \cdots + \psi_{d-1}(x_1)x_2^{d-1} + bx_2^d,$$

with $\mathrm{ord}\,(\psi_0(x_1)) \geq 2$, $\mathrm{ord}\,(\psi_1(x_1)) \geq 1$ and $b \neq 0$.

The resultant $\mathrm{Ris}(f, g, x_2)(x_1)$ is given by the determinant of the Sylvester matrix

$$S(x_1) = \begin{pmatrix} \varphi_0(x_1) & 1 + \varphi_1(x_1) & \varphi_2(x_1) & \ldots & a & 0 \ldots \\ 0 & \varphi_0(x_1) & 1 + \varphi_1(x_1) & \ldots & \ldots & a \ldots \\ \vdots & \vdots & \vdots & \vdots & \vdots & \vdots \\ \alpha x_1 + \psi_0(x_1) & \beta + \psi_1(x_1) & \psi_2(x_1) & \ldots & b & 0 \ldots \\ 0 & \alpha x_1 + \psi_0(x_1) & \beta + \psi_1(x_1) & \ldots & \ldots & b \ldots \\ \vdots & \vdots & \vdots & \vdots & \vdots & \vdots \end{pmatrix}.$$

For $i, j = 1, \ldots, m + d$, $i < j$, denote by $D_{i,j}$ the determinant of the 2×2 submatrix obtained by taking the first two columns and the ith and jth rows of $S(x_1)$, and denote by $D'_{i,j}$ the determinant of the corresponding cofactor, that is of the $(m + d - 2) \times (m + d - 2)$ submatrix obtained by deleting in $S(x_1)$ the first two columns and the ith and jth rows. By Laplace rule we get

$$\det S(x_1) = \sum_{i<j} (-1)^{i+j+1} D_{i,j}(x_1) D'_{i,j}(x_1).$$

We are going to estimate the order of $\det S(x_1)$ by analysing how the various summands contribute to the previous summation. Since $\varphi_0(x_1), \psi_0(x_1)$ are divisible by x_1^2, it readily follows that if $i \neq 1$ or $j \neq d + 1$ then we have $\mathrm{ord}(D_{ij}(x_1)) \geq 2$. So $\mathrm{ord}(\mathrm{Ris}(f, g, x_2)(x_1)) \geq 2$ if and only if

$$\mathrm{ord}(D_{1,d+1}(x_1)D'_{1,d+1}(x_1)) = \mathrm{ord}(D_{1,d+1}(x_1)) + \mathrm{ord}(D'_{1,d+1}(x_1)) \geq 2.$$

Now, all terms of

$$D_{1,d+1}(x_1) = \varphi_0(x_1)(\beta + \psi_1(x_1)) - (1 + \varphi_1(x_1))(\alpha x_1 + \psi_0(x_1)),$$

apart from αx_1, have order at least two, so that $\mathrm{ord}(D_{1,d+1}(x_1)) \geq 2$ if and only if $\alpha = 0$.

Let us now check that $D'_{1,d+1}(0) \neq 0$. It is immediate to observe that $D'_{1,d+1}(0)$ is the determinant of the Sylvester matrix of the polynomials $\widehat{f}(x_2) = \dfrac{f(0, x_2)}{x_2}$ and $\widehat{g}(x_2) = \dfrac{g(0, x_2)}{x_2}$, which have no common roots: namely, since the line $L(P, Q)$ does not contain any point of $\mathcal{C} \cap \mathcal{D}$ different from P, then 0 is the only common root of $f(0, x_2)$ and $g(0, x_2)$; on the other hand, $\tau_{\mathcal{C}}$ has equation $x_2 = 0$, so that the order of $f(0, x_2)$ is equal to 1 and 0 cannot be a root of $\widehat{f}$. Therefore $\mathrm{ord}(D'_{1,d+1}(x_1)) = 0$, and $I(\mathcal{C}, \mathcal{D}, P) = \mathrm{ord}(\mathrm{Ris}(f, g, x_2)(x_1)) \geq 2$ if and only if $\alpha = 0$, i.e. iff $\tau_{\mathcal{C}} = \tau_{\mathcal{D}}$.

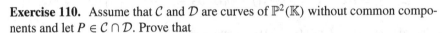

Exercise 110. Assume that $\mathcal{C}$ and $\mathcal{D}$ are curves of $\mathbb{P}^2(\mathbb{K})$ without common components and let $P \in \mathcal{C} \cap \mathcal{D}$. Prove that

$$I(\mathcal{C}, \mathcal{D}, P) \geq m_P(\mathcal{C})m_P(\mathcal{D}).$$

Solution. Let $Q \notin \mathcal{C} \cup \mathcal{D}$ be a point such that there exists no point of $\mathcal{C} \cap \mathcal{D}$ distinct from P and lying on the line joining P and Q. One can readily check that there exist homogeneous coordinates such that $P = [1, 0, 0]$ and $Q = [0, 0, 1]$.

Denote by $F(x_0, x_1, x_2) = 0, G(x_0, x_1, x_2) = 0$ equations of $\mathcal{C}$ and $\mathcal{D}$, respectively, with $\deg F = d$ and $\deg G = d'$, and set $m = m_P(\mathcal{C}), m' = m_P(\mathcal{D})$.

By definition $I(\mathcal{C}, \mathcal{D}, P)$ is equal to the multiplicity of $[1, 0]$ as a root of the resultant $\mathrm{Ris}(F, G, x_2)$. Moreover, arguing as in the solution of Exercise 109 we obtain that, if we set $f(x_1, x_2) = F(1, x_1, x_2)$ and $g(x_1, x_2) = G(1, x_1, x_2)$, the mentioned multiplicity coincides with the order $\mathrm{ord}(\mathrm{Ris}(f, g, x_2)(x_1))$ of $\mathrm{Ris}(f, g, x_2)(x_1)$ (that is with the multiplicity of 0 as a root of $\mathrm{Ris}(f, g, x_2)(x_1)$, or also with the largest $i \geq 0$ such that x_1^i divides $\mathrm{Ris}(f, g, x_2)(x_1)$).

As $Q \notin \mathcal{C} \cup \mathcal{D}$, then

$$f(x_1, x_2) = \varphi_0(x_1) + \varphi_1(x_1)x_2 + \cdots + \varphi_{d-1}(x_1)x_2^{d-1} + ax_2^d,$$
$$g(x_1, x_2) = \psi_0(x_1) + \psi_1(x_1)x_2 + \cdots + \psi_{d'-1}(x_1)x_2^{d'-1} + bx_2^{d'},$$

with $a \neq 0, b \neq 0$. In addition, the fact that $m = m_P(\mathcal{C}), m' = m_P(\mathcal{D})$ implies that φ_i (resp. ψ_i) is divisible by x_1^{m-i} (resp. by $x_1^{m'-i}$) for each $i = 0, \ldots, m$ (resp. for each $i = 0, \ldots, m'$).

The resultant $\mathrm{Ris}(f, g, x_2)(x_1)$ is the determinant of the Sylvester matrix

$$S(x_1) = \begin{pmatrix} \varphi_0(x_1) & \varphi_1(x_1) & \varphi_2(x_1) & \ldots & \ldots & a & 0 & \ldots \\ 0 & \varphi_0(x_1) & \varphi_1(x_1) & \ldots & \ldots & \ldots & a & \ldots \\ \vdots & \vdots & \vdots & \vdots & \vdots & \vdots & \vdots & \vdots \\ \psi_0(x_1) & \psi_1(x_1) & \psi_2(x_1) & \ldots & b & 0 & \ldots & \ldots \\ 0 & \psi_0(x_1) & \psi_1(x_1) & \ldots & \ldots & b & 0 & \ldots \\ \vdots & \vdots & \vdots & \vdots & \vdots & \vdots & \vdots \end{pmatrix}.$$

Consider now the matrix $\widehat{S}(x_1)$ obtained by multiplying by $x_1^{m'-i+1}$ the ith row of $S(x_1)$ for each $i = 1, \ldots, m'$, and by multiplying by x_1^{m-i+1} the $(d+i)$th row of $S(x_1)$ for each $i = 1, \ldots, m$ (since $m \leq d$ and $m' \leq d'$, the operations just described actually make sense). Then we have

$$\mathrm{ord}(\det \widehat{S}(x_1)) = \mathrm{ord}(\det S(x_1)) + \frac{m(m+1) + m'(m'+1)}{2}.$$

Moreover, since $\mathrm{ord}(\varphi_i(x_1)) \geq m - i$ (resp. $\mathrm{ord}(\psi_i(x_1)) \geq m' - i$) for each $i = 0, \ldots, m$ (resp. for each $i = 0, \ldots, m'$), then we obtain that, for each $i = 1, \ldots, m + m'$ the ith column of $\widehat{S}(x_1)$ is divisible by the monomial $x_1^{m+m'-i+1}$, so that

$$\mathrm{ord}(\det \widehat{S}(x_1)) \geq \frac{(m+m')(m+m'+1)}{2}.$$

Therefore $\mathrm{ord}(\det S(x_1)) \geq mm'$, and hence $I(\mathcal{C}, \mathcal{D}, P) \geq mm'$, as required.

Exercise 111. Prove that, if two cubics of $\mathbb{P}^2(\mathbb{K})$ intersect exactly at 9 points, then every cubic passing through 8 of those 9 points passes through the ninth point too.

Solution. Denote by $P_1, \ldots, P_9$ the nine points where the two cubics $\mathcal{C}_1 = [F_1]$ and $\mathcal{C}_2 = [F_2]$ meet.

Some easy consequences immediately follow from the fact that the cubics intersect at finitely many points. First of all, however we choose four out of the nine points, they are not collinear, because otherwise the line joining them would be an irreducible component both of $\mathcal{C}_1$ and of $\mathcal{C}_2$ by Bézout's Theorem, so that the cubics would have infinitely points in common. Similarly, however we choose seven out of the nine points, they cannot lie on a conic.

Assume by contradiction there exists a cubic $\mathcal{D} = [G]$ passing through eight of the nine points, say $P_1, \ldots, P_8$, but not passing through P_9.

The curve $\mathcal{D}$ does not belong to the pencil of cubics generated by $\mathcal{C}_1$ and $\mathcal{C}_2$, which has $\{P_1, \ldots, P_9\}$ as base locus. Hence $\mathcal{C}_1, \mathcal{C}_2$ and $\mathcal{D}$ generate a 2-dimensional linear system $\mathcal{L}$ having $\{P_1, \ldots, P_8\}$ as base locus; as a consequence, however we choose two points in $\mathbb{P}^2(\mathbb{K})$, there exist $\lambda, \mu, \nu \in \mathbb{K}$ not simultaneously zero such that the cubic $[\lambda F_1 + \mu F_2 + \nu G]$ passes through those two points.

We are now going to show that any three points chosen among $P_1, \ldots, P_9$ are not collinear. Namely, assume by contradiction that, for instance, P_1, P_2, P_3 lie on a line r and denote by $\mathcal{Q}$ the only conic passing through $P_4, \ldots, P_8$ (this conic is unique because, as shown above, no four of the five points are collinear – cf. Sect. 1.9.6). Chosen two points $A \in r \backslash \{P_1, P_2, P_3\}$ and $B \notin \mathcal{Q} \cup r$, there exist $\lambda, \mu, \nu \in \mathbb{K}$ not simultaneously zero such that the cubic $\mathcal{W} = [\lambda F_1 + \mu F_2 + \nu G]$ passes through $P_1, \ldots, P_8, A, B$. Since $\mathcal{W}$ intersects the line r at the four points P_1, P_2, P_3, A, by Bézout's Theorem r is an irreducible component of $\mathcal{W}$; hence $\mathcal{W}$ is the sum of r and a conic passing through $P_4, \ldots, P_8$ which, by the uniqueness recalled above, necessarily must be $\mathcal{Q}$. On the other hand, we cannot have $\mathcal{W} = r + \mathcal{Q}$, because $B \in \mathcal{W}$ but B lies neither on r nor on $\mathcal{Q}$.

Similar arguments allow us to prove that any six points chosen among $P_1, \ldots, P_9$ cannot lie on a conic. Namely, assume by contradiction that, for instance, $P_1, \ldots, P_6$ lie on a conic $\mathcal{Q}$ and denote by r the line joining P_7 and P_8. Arguing as above, we arrive at a contradiction by choosing a point $A \in \mathcal{Q} \backslash \{P_1, \ldots, P_6\}$ and a point $B \notin \mathcal{Q} \cup r$, and by considering a cubic $\mathcal{W}$ of the linear system $\mathcal{L}$ passing through $P_1, \ldots, P_8, A, B$.

The previous considerations allow us to get to a contradiction. Namely, consider the line $s = L(P_1, P_2)$ and the only conic $\mathcal{G}$ passing through $P_3, \ldots, P_7$. As shown above, $P_8 \notin s$ and $P_8 \notin \mathcal{G}$. After choosing two distinct points A, B on $s \backslash \{P_1, P_2\}$, let $\mathcal{W}$ be a cubic of the linear system $\mathcal{L}$ passing through A and B. Then necessarily $\mathcal{W} = s + \mathcal{G}$: namely, by Bézout's Theorem s is an irreducible component of $\mathcal{W}$; hence $\mathcal{W}$ is the sum of s and a conic passing through $P_3, \ldots, P_7$ which, by the uniqueness recalled above, necessarily must be $\mathcal{G}$. On the other hand we cannot have $\mathcal{W} = s + \mathcal{G}$, because $P_8 \in \mathcal{W}$ but $P_8 \notin s$ and $P_8 \notin \mathcal{G}$.

Exercise 112. If $n \geq 2$, assume that $\mathcal{I}$ is a hypersurface of degree d of $\mathbb{P}^n(\mathbb{C})$ and let $P \in \mathcal{I}$ be a point. Prove that:

(a) $\mathcal{I}$ is a cone with vertex P if and only if $m_P(\mathcal{I}) = d$.
(b) If $\mathcal{I}$ is a cone, then the set X of vertices of $\mathcal{I}$ is a projective subspace of $\mathbb{P}^n(\mathbb{C})$.
(c) If $\mathcal{I}$ is a cone with vertex P and $Q \in \mathrm{Sing}(\mathcal{I})$, then $L(P, Q) \subseteq \mathrm{Sing}(\mathcal{I})$.

Solution. (a) $\mathcal{I}$ is a cone with vertex P if and only if every line passing through P and not contained in $\mathcal{I}$ intersects $\mathcal{I}$ only at P. Since such a line intersects $\mathcal{I}$ exactly at d points counted with multiplicity (cf. Sect. 1.7.6), $\mathcal{I}$ is a cone with vertex P if and only if for every line passing through P we have $I(\mathcal{I}, r, P) \geq d$, that is iff P is a point of $\mathcal{I}$ of multiplicity d.

(b) If $P_1 \neq P_2$ are vertices of $\mathcal{I}$ and $R \in \mathcal{I}$, then $L(P_1, P_2, R) \subseteq \mathcal{I}$: namely, $P_1 \in X$ implies that $L(P_1, R) \subseteq \mathcal{I}$, and $P_2 \in X$ implies that $L(P_2, S) \subseteq \mathcal{I}$ for every $S \in L(P_1, R)$; so $L(P_1, P_2, R) = L(P_2, L(P_1, R)) \subseteq \mathcal{I}$. It follows that for every $Q \in L(P_1, P_2)$ and for every $R \in \mathcal{I}$ the line $L(Q, R)$ is contained in $\mathcal{I}$, so that every point of $L(P_1, P_2)$ is a vertex of $\mathcal{I}$. For every $P_1, P_2 \in X$, $P_1 \neq P_2$, we so have that $L(P_1, P_2) \subseteq X$, and therefore X is a projective subspace of $\mathbb{P}^n(\mathbb{C})$.

Alternatively, we can argue as follows. By (a), P_1 and P_2 are points of $\mathcal{I}$ of multiplicity d. We can choose homogeneous coordinates $x_0, \ldots, x_n$ on $\mathbb{P}^n(\mathbb{K})$

such that P_1 has coordinates $[1, 0, \ldots, 0]$ and P_2 has coordinates $[0, 1, \ldots 0]$. Let $F(x_0, \ldots, x_n) = 0$ be an equation of $\mathcal{I}$. Since $P_1 \in \mathcal{I}$, then

$$F(x_0, \ldots, x_n) = x_0^{d-1} F_1(x_1, \ldots, x_n) + x_0^{d-2} F_2(x_1, \ldots, x_n) + \cdots + F_d(x_1, \ldots, x_n),$$

where F_i is either zero or homogeneous of degree i for $i = 1, \ldots, d$. From the previous expression it appears that P_1 has multiplicity d for $\mathcal{I}$ if and only if $F_1 = \cdots = F_{d-1} = 0$, that is iff in $F = F_d$ the variable x_0 does not appear. In the same way, since also P_2 is a vertex, the variable x_1 does not appear in F. It is now immediate to check that all points of the line $L(P_1, P_2)$, which in the chosen system of coordinates is described by $x_2 = \cdots = x_n = 0$, are points of multiplicity d for $\mathcal{I}$.

(c) Choose homogeneous coordinates such that $P = [1, 0, \ldots, 0]$, so that, as seen above, $\mathcal{I}$ is defined by a homogeneous polynomial F which does not depend on the variable x_0. In particular $F_{x_0} = 0$, so the point Q of coordinates $[a_0, a_1, \ldots, a_n]$ is singular for $\mathcal{I}$ if and only if $F_{x_i}(a_0, a_1, \ldots, a_n) = 0$ for $i = 1, \ldots, n$. In addition, for $i = 1, \ldots, n$ the polynomial F_{x_i} is either zero or homogeneous and does not depend on the variable x_0, so $F_{x_i}(a_0, a_1, \ldots, a_n) = 0$ iff $F_{x_i}(t, sa_1, \ldots, sa_n) = 0$ for every $[s, t] \in \mathbb{P}^1(\mathbb{C})$. It follows that if Q is singular for $\mathcal{I}$, then all points of $L(P, Q)$ are singular for $\mathcal{I}$.

Note. Statements (b) and (c) of the previous exercise hold also on $\mathbb{R}$, since their proofs do not use properties of the complex numbers.

As regards (a), if $\mathbb{K} = \mathbb{R}$ it is true that, if $P \in \mathcal{I}$ is a point of multiplicity d, then $\mathcal{I}$ is a cone with vertex P, but the opposite implication is false. For instance, the cubic $\mathcal{C}$ of $\mathbb{P}^2(\mathbb{R})$ defined by the equation $x_0(x_0^2 + x_1^2 + x_2^2) = 0$ has as support the line $r = \{x_0 = 0\}$, so it is a cone with vertex Q for every $Q \in r$. Though, it can be immediately checked that all points of r are simple for $\mathcal{C}$.

Note. If $\mathcal{I}$ is a curve of degree d of $\mathbb{P}^2(\mathbb{C})$, then $\mathcal{I}$ is a cone with vertex P if and only if it is the sum of d (non necessarily distinct) lines passing through P. Namely, if $\mathcal{I}$ is a cone with vertex P, from the definition it readily follows that the support of $\mathcal{I}$ is the union of a family of lines passing through P. By applying Bézout's Theorem (or Exercise 57), we easily deduce that every line contained in $\mathcal{I}$ is an irreducible component of the curve, so that $\mathcal{I}$ decomposes as the sum of d lines passing through P (counted with multiplicity).

Therefore statement (a) implies that, if $\mathcal{I}$ is a curve of $\mathbb{P}^2(\mathbb{C})$ of degree d and P is a point of the curve of multiplicity d, then $\mathcal{I}$ decomposes as the sum of d lines passing through P (counted with multiplicity).

Exercise 113. If $n \geq 2$ and $H \subset \mathbb{P}^n(\mathbb{K})$ is a hyperplane, consider a hypersurface $\mathcal{J}$ of H of degree d and a point $P \in \mathbb{P}^n(\mathbb{K}) \setminus H$. Prove that

$$X = \bigcup_{Q \in \mathcal{J}} L(P, Q) \cup \{P\} \subseteq \mathbb{P}^n(\mathbb{K})$$

is the support of a hypersurface $\mathcal{C}_P(\mathcal{J})$ of degree d in $\mathbb{P}^n(\mathbb{K})$, called the "cone over $\mathcal{J}$ with vertex P", which verifies the following properties:

(i) $\mathcal{C}_P(\mathcal{J})$ is either irreducible or reduced if and only if $\mathcal{J}$ is;
(ii) a point $Q \in X \backslash \{P\}$ is singular for $\mathcal{C}_P(\mathcal{J})$ if and only if the point $L(P, Q) \cap \mathcal{J}$ is singular for $\mathcal{J}$.

Solution. Choose homogeneous coordinates $x_0, \ldots, x_n$ on $\mathbb{P}^n(\mathbb{K})$ in such a way that $P = [1, 0, \ldots, 0]$ and H has equation $x_0 = 0$. Note that these coordinates induce homogeneous coordinates $x_1, \ldots, x_n$ on H, and let $G(x_1, \ldots, x_n) = 0$ be an equation of $\mathcal{J}$. Set $F(x_0, x_1, \ldots, x_n) = G(x_1, \ldots, x_n)$, and denote by $\mathcal{C}_P(\mathcal{J})$ the hypersurface of $\mathbb{P}^n(\mathbb{K})$ of equation $F = 0$. It is immediate to check that $\mathcal{C}_P(\mathcal{J})$ has degree d and it is a cone with vertex P.

Let us now check that the support of $\mathcal{C}_P(\mathcal{J})$ (which, thanks to the usual abuse of notation, we will simply denote by $\mathcal{C}_P(\mathcal{J})$) coincides with X. From the definitions it follows that $P \in X \cap \mathcal{C}_P(\mathcal{J})$. If the support of $\mathcal{J}$ is empty (and therefore $\mathbb{K} = \mathbb{R}$), we have $\mathcal{C}_P(\mathcal{J}) = \{P\} = X$. Let now $Q \in \mathcal{J} \neq \emptyset$. Then $Q = [0, b_1, \ldots, b_n]$ for some $[b_1, \ldots, b_n] \in \mathbb{P}^{n-1}(\mathbb{K})$, and in addition $G(b_1, \ldots, b_n) = 0$. As the points of $L(P, Q) \backslash \{P\}$ have coordinates $[t, b_1, \ldots b_n]$, $t \in \mathbb{K}$ and $F(t, b_1, \ldots, b_n) = G(b_1, \ldots, b_n) = 0$ for every $t \in \mathbb{K}$, then $L(P, Q) \subseteq \mathcal{C}_P(\mathcal{J})$. This shows that $X \subseteq \mathcal{C}_P(\mathcal{J})$.

Conversely, let $Q \in \mathcal{C}_P(\mathcal{J}) \backslash \{P\}$ and set $R = L(P, Q) \cap H$. If Q has coordinates $[a_0, \ldots, a_n]$, the point R has coordinates $[0, a_1, \ldots, a_n]$. Since $Q \in \mathcal{C}_P(\mathcal{J})$, then $F(a_0, \ldots, a_n) = 0$, so that $G(a_1, \ldots, a_n) = 0$. Therefore R belongs to the support of $\mathcal{J}$, and $Q \in L(P, R)$ belongs to X. Since Q has been arbitrarily chosen, then $\mathcal{C}_P(\mathcal{J}) \subseteq X$.

The hypersurface $\mathcal{C}_P(\mathcal{J})$ obviously verifies (i): if $G = G_1 \cdot \ldots \cdot G_r$ is the decomposition of G into irreducible factors, if we set $F_k(x_0, x_1, \ldots, x_n) = G_k(x_1, \ldots, x_n)$, $k = 1, \ldots, r$, then $F = F_1 \cdot \ldots \cdot F_r$ is the decomposition of F into irreducible factors.

With regard to (ii), if a point $Q \neq P$ has coordinates $[a_0, a_1, \ldots, a_n]$, the point $R = L(P, Q) \cap \mathcal{J}$ has coordinates $[0, a_1, \ldots, a_n]$. Moreover we have $F_{x_0} = 0$ and $F_{x_i}(a_0, \ldots, a_n) = G_{x_i}(a_1, \ldots, a_n)$ for each $i = 1, \ldots, n$. Therefore Q is singular for $\mathcal{C}_P(\mathcal{J})$ if and only if R is singular for $\mathcal{J}$, as required.

Exercise 114. Consider a cubic surface $\mathcal{S}$ of $\mathbb{P}^3(\mathbb{C})$ such that $\mathrm{Sing}(\mathcal{S})$ is a finite set. Prove that:

(a) $\mathcal{S}$ is irreducible.
(b) If $P \in \mathrm{Sing}(\mathcal{S})$ is a triple point, then $\mathcal{S}$ is a cone with vertex P and P is the only singular point of $\mathcal{S}$.

Solution. (a) Assume by contradiction that $\mathcal{S}$ is reducible. Then $\mathcal{S} = H + \mathcal{Q}$, where H is a plane and $\mathcal{Q}$ is a quadric, possibly reducible or non-reduced. By Exercise 54 all points of the conic $\mathcal{Q} \cap H$ are singular for $\mathcal{S}$, against the assumption that $\mathrm{Sing}(\mathcal{S})$ is a finite set (recall that, if $n \geq 2$, then the support of a hypersurface $\mathbb{P}^n(\mathbb{C})$ contains infinitely many points – cf. Sect. 1.7.2).

(b) By Exercise 112, if $P \in \mathrm{Sing}(S)$ is a triple point, S is a cone of vertex P. In this case, if there existed a point $Q \neq P$ singular for S, by Exercise 112 (c) all points of the line $L(P, Q)$ would be singular for S, contradicting the hypotheses.

Exercise 115. Let S be a cubic surface of $\mathbb{P}^3(\mathbb{C})$ and let r be a line contained in $\mathrm{Sing}(S)$. Prove that, if S is irreducible, then $\mathrm{Sing}(S) = r$.

Solution. Assume by contradiction that there exists a point $Q \in \mathrm{Sing}(S) \backslash r$. Since S is irreducible, the plane $H = L(r, Q)$ is not contained in S (cf. Exercise 57), hence it intersects S in a plane cubic C. Since $r \subseteq C$, by Bézout's Theorem one has $C = r + Q$, with Q a conic of H. In addition, by Exercise 58 (a) one has $r \subseteq \mathrm{Sing}(C)$, so that Exercise 54 implies that $r \subseteq Q$. Then we deduce again from Bézout's Theorem that $Q = r + s$, where s is a line of H. So $C = 2r + s$, and $Q \in s$, because $Q \notin r$; therefore Q is a smooth point of C. On the other hand, again by Exercise 58 (a), Q should be a singular point of C. So we have reached a contradiction, showing that $\mathrm{Sing}(S) = r$.

Exercise 116. (a) Let S be a cubic surface of $\mathbb{P}^3(\mathbb{C})$. Show that, if P is a singular point of S, then there exists a line $r \subset S$ such that $P \in r$.
(b) Let T be a cubic hypersurface of $\mathbb{P}^4(\mathbb{C})$. Prove that for every $P \in T$ there exists a line r such that $P \in r$ and $r \subset T$.

Solution. (a) Let x_0, x_1, x_2, x_3 be a homogeneous coordinate system of $\mathbb{P}^3(\mathbb{C})$ such that P has coordinates $[1, 0, 0, 0]$ and let $y_i = \frac{x_i}{x_0}$, $i = 1, 2, 3$, be the corresponding affine coordinates on U_0. Since S is singular at P by assumption, its affine part $S \cap U_0$ is described by an equation of the form $f(y_1, y_2, y_3) = f_2(y_1, y_2, y_3) + f_3(y_1, y_2, y_3) = 0$, where f_2 is either the zero polynomial or is homogeneous of degree 2 and f_3 is either zero or homogeneous of degree 3 (however note that f_2 and f_3 cannot be both zero). Given $v \in \mathbb{C}^3 \backslash \{0\}$, the affine line r_v passing through P and parallel to v is described in parametric form by $r_v = \{tv \mid t \in \mathbb{C}\}$, and it is immediate to check that r_v is contained in S if and only if $f_2(v) = f_3(v) = 0$. If the polynomials f_2 and f_3 are both not identically zero, then, being homogeneous, they define curves of $\mathbb{P}^2(\mathbb{C})$ of degree 2 and 3, respectively. By Bézout's Theorem, there exists at least one point $R = [v_0]$ such that $f_2(v_0) = f_3(v_0) = 0$. On the other hand, in case f_2 or f_3 is identically zero, the set of points of $\mathbb{P}^2(\mathbb{C})$ at which both f_2 and f_3 vanish is infinite, and therefore it is still possible to choose $v_0 \in \mathbb{C}^3 \backslash \{0\}$ such that $f_2(v_0) = f_3(v_0) = 0$. In any case, the line r_{v_0} is then contained in S and so is its projective closure $r = \overline{r_{v_0}}$.

(b) If P is a non-singular point of T, we denote by H the tangent hyperplane $T_P(T)$. If P is singular, instead, we denote by H any hyperplane containing P. If H is contained in T, then there exist infinitely many lines passing through P and contained in T. So we may assume that H is not contained in T. In this case $S = T \cap H$ is a cubic surface of H and, by Exercise 58, is singular at P. So, by part (a), there exists a line r of H passing through P and contained in $S \subset T$.

Exercise 117. Let $F(x_0, x_1, x_2, x_3) = x_0^3 - x_1 x_2 x_3$, and consider the surface S of $\mathbb{P}^3(\mathbb{R})$ defined by the equation $F(x_0, x_1, x_2, x_3) = 0$.

(a) Determine the singular points of S and describe the tangent cone to S at each of these points.
(b) Determine the lines contained in S.
(c) Say whether S is irreducible.

Solution. (a) Since $\nabla F(x_0, x_1, x_2, x_3) = (3x_0^2, -x_2x_3, -x_1x_3, -x_1x_2)$, the singular points of S are $P_1 = [0, 1, 0, 0]$, $P_2 = [0, 0, 1, 0]$ and $P_3 = [0, 0, 0, 1]$. Let us study the nature of the singular point P_1. An equation of $S \cap U_1$ in the affine coordinates $y_0 = \frac{x_0}{x_1}, y_2 = \frac{x_2}{x_1}, y_3 = \frac{x_3}{x_1}$ is given by $-y_2y_3 + y_0^3 = 0$. So P_1 is a double point and the projective tangent cone to S at P_1 is $x_2x_3 = 0$. Similarly, one checks that P_2 and P_3 are double points whose tangent cones have equation $x_1x_3 = 0$ and $x_1x_2 = 0$, respectively.

(b) The intersection of S with the plane $H_0 = \{x_0 = 0\}$ is the union of the three lines $L(P_1, P_2)$, $L(P_1, P_3)$ and $L(P_2, P_3)$.

Now we check that there are no additional lines contained in S. Assume by contradiction that $r \subseteq S$ is a line not contained in H_0 and let $Q = r \cap H_0$. Up to permuting the coordinates x_1, x_2, x_3, we may assume that Q is on the line $L(P_2, P_3)$. The plane K generated by r and by $L(P_2, P_3)$ intersects S in a reducible cubic C, which is the union of r, $L(P_2, P_3)$ and a line s (possibly not distinct from $L(P_2, P_3)$ and r).

On the other hand, since $L(P_2, P_3) \subseteq K$ and $K \neq H_0$, it is immediate to check that K is defined by the equation $\lambda x_0 + x_1 = 0$, for some $\lambda \in \mathbb{R}$. So, eliminating x_1, one sees that C is defined by the equation $x_0(x_0^2 + \lambda x_2x_3) = 0$ (cf. Sect. 1.7.3 for a more detailed description of the process of elimination of the variable x_1). Therefore C is the union of $L(P_2, P_3)$ and of the conic of equation $x_0^2 + \lambda x_2x_3 = 0$, whose support, by what we have seen above, must contain the line $r \neq L(P_2, P_3)$. Since this conic is irreducible if $\lambda \neq 0$ and is the line $L(P_2, P_3)$ counted twice if $\lambda = 0$, we have a contradiction.

(c) Since S has degree 3, if it were reducible it would contain a plane, and so infinitely many lines, contradicting the fact that S contains precisely three lines.

Exercise 118. For $\lambda \in \mathbb{C}$, consider the surface $S_\lambda \subset \mathbb{P}^3(\mathbb{C})$ of equation

$$F_\lambda(x_0, x_1, x_2, x_3) = x_0^4 + x_1^4 + x_2^4 + x_3^4 - 4\lambda x_0x_1x_2x_3 = 0.$$

(a) Determine the values $\lambda \in \mathbb{C}$ such that S_λ is singular.
(b) For every value of λ as in (a), determine the multiplicity of the singular points of S_λ.
(c) Determine the values $\lambda \in \mathbb{R}$ such that S_λ has at least one real point.

Solution. (a) For $i = 0, \ldots, 3$ denote the coordinate hyperplane of equation $x_i = 0$ by H_i. Note that, in the coordinates induced on H_3 by the standard coordinates of $\mathbb{P}^3(\mathbb{C})$, the curve $S_\lambda \cap H_3$ has equation $G(x_0, x_1, x_2) = x_0^4 + x_1^4 + x_2^4 = 0$ for any $\lambda \in \mathbb{C}$. Since $\nabla G = 4(x_0^3, x_1^3, x_2^3)$, the curve $S_\lambda \cap H_3$ is non-singular. By Exercise 58,

$\mathrm{Sing}(\mathcal{S}_\lambda) \cap H_3 = \emptyset$ for all $\lambda \in \mathbb{C}$. A similar computation shows that $\mathrm{Sing}(\mathcal{S}_\lambda) \cap H_i = \emptyset$ for $i = 0, 1, 2$ and for all $\lambda \in \mathbb{C}$. The gradient of F_λ is

$$\nabla F_\lambda = 4(x_0^3 - \lambda x_1 x_2 x_3, \, x_1^3 - \lambda x_0 x_2 x_3, \, x_2^3 - \lambda x_0 x_1 x_3, \, x_3^3 - \lambda x_0 x_1 x_2).$$

If $\nabla F_\lambda(x_0, x_1, x_2, x_3) = 0$, multiplying the ith component of ∇F_λ by x_i for $i = 0, 1, 2, 3$ we get $x_i^4 = \lambda x_0 x_1 x_2 x_3$. Multiplying the four relations thus obtained one deduces $x_0^4 x_1^4 x_2^4 x_3^4 (1 - \lambda^4) = 0$. Since, as remarked above, regardless of the value of $\lambda \in \mathbb{C}$ no point of $\mathrm{Sing}(\mathcal{S}_\lambda)$ belongs to a coordinate hyperplane, if $\mathcal{S}_\lambda$ is singular then one has $\lambda^4 = 1$, namely $\lambda \in \{1, -1, i, -i\}$. On the other hand, if $\lambda \in \{1, -1, i, -i\}$ then the point $P = [\lambda^3, 1, 1, 1]$ satisfies $\nabla F_\lambda(P) = 0$, and so it is singular for $\mathcal{S}_\lambda$.

(b) Let us consider first of all the case $\lambda = 1$. By what we have seen in (a), no point of the coordinate hyperplane H_0 is singular for $\mathcal{S}_1$, hence it suffices to study the affine part $\mathcal{S}_1 \cap U_0$ of $\mathcal{S}_1$, using the usual affine coordinates $y_i = \frac{x_i}{x_0}$, $i = 1, 2, 3$, on U_0. The singular points of $\mathcal{S}_1$ are given by the solutions of the system of equations

$$y_1 y_2 y_3 = 1, \quad y_1^3 = y_2 y_3, \quad y_2^3 = y_1 y_3, \quad y_3^3 = y_1 y_2. \tag{3.7}$$

Multiplying the second equation of (3.7) by y_1 and substituting $y_1 y_2 y_3 = 1$ one obtains $y_1^4 = 1$. Analogous computations give $y_2^4 = y_3^4 = 1$. It is now easy to check that the singular points of $\mathcal{S}_1$ are the points of affine coordinates $(\alpha, \beta, \alpha^3 \beta^3)$, for $\alpha, \beta \in \{1, -1, i, -i\}$. Hence $\mathrm{Sing}(\mathcal{S}_1)$ consists of 16 points. One has $(F_1)_{x_0 x_0} = 12 x_0^2$ and so $(F_1)_{x_0 x_0}(P) \neq 0$ for all $P \in \mathrm{Sing}(\mathcal{S}_1)$. This proves that all singular points of $\mathcal{S}_1$ have multiplicity 2.

The singularities of the surfaces $\mathcal{S}_\lambda$ for $\lambda = -1, i, -i$ can be analysed in the same way. Alternatively, one can observe that, if $\lambda^4 = 1$, then the projectivity $g_\lambda : \mathbb{P}^3(\mathbb{C}) \to \mathbb{P}^3(\mathbb{C})$ defined by $[x_0, x_1, x_2, x_3] \mapsto [\lambda x_0, x_1, x_2, x_3]$ transforms $\mathcal{S}_\lambda$ into $\mathcal{S}_1$. Therefore, for $\lambda^4 = 1$, $\mathcal{S}_\lambda$ is projectively equivalent to $\mathcal{S}_1$ and $\mathrm{Sing}(\mathcal{S}_\lambda)$ consists of 16 points of multiplicity 2.

(c) Note that the projectivity of $\mathbb{P}^3(\mathbb{C})$ defined by $[x_0, x_1, x_2, x_3] \mapsto [-x_0, x_1, x_2, x_3]$ maps real points to real points and transforms $\mathcal{S}_\lambda$ into $\mathcal{S}_{-\lambda}$ for all $\lambda \in \mathbb{R}$, so that $\mathcal{S}_\lambda$ has real points if and only if $\mathcal{S}_{-\lambda}$ has real points. So it is enough to consider the case $\lambda \geq 0$.

For $\lambda = 1$, it is immediate to check that $\mathcal{S}_1$ contains, for instance, the real points $Q_1 = [1, 1, 1, 1]$, $Q_2 = [1, 1, -1, -1]$.

Let us now study the intersections of $\mathcal{S}_\lambda$ with the line $r = L(Q_1, Q_2)$ as λ varies in $\mathbb{R}$. Since $[0, 0, 1, 1] \notin \mathcal{S}_\lambda$ for every $\lambda \in \mathbb{R}$, it is enough to consider the points of the form $[1, 1, t, t]$, $t \in \mathbb{R}$. One of these points belongs to $\mathcal{S}_\lambda$ if and only if $g(t) = t^4 + 1 - 2\lambda t^2 = 0$. Let now $\lambda > 1$. Then one has $g(0) = 1$, $g(1) = 2(1 - \lambda) < 0$, and therefore there exists a real value $t_0 \in (0, 1)$ such that $g(t_0) = 0$. So $[1, 1, t_0, t_0] \in \mathcal{S}_\lambda$, and $\mathcal{S}_\lambda$ has real points.

To conclude, we show that $\mathcal{S}_\lambda$ has no real points if $0 \leq \lambda < 1$. For every $\lambda \in \mathbb{R}$ the curve $\mathcal{C}_\lambda = \mathcal{S}_\lambda \cap H_0$ is defined by $x_1^4 + x_2^4 + x_3^4 = 0$, and so it does not contain any real point. Therefore it is enough to determine the support of the real affine hypersurface $S_\lambda \cap U_0$. Consider the affine coordinates $y_i = \frac{x_i}{x_0}$, $i = 1, 2, 3$, on U_0 and let $f_\lambda(y_1, y_2, y_3)$ be the polynomial obtained by dehomogenizing F_λ with respect to x_0. Let $\Omega = \{(a, b, c) \in \mathbb{R}^3 \mid (b, c) \neq (0, 0)\}$, fix $(a, b, c) \in \Omega$ and restrict f_λ to the line $r_{a,b,c} = \{(a, tb, tc) \mid t \in \mathbb{R}\}$. One obtains a polynomial $g_\lambda(t) = (b^4 + c^4)t^4 - 4\lambda abct^2 + a^4 + 1$. If $abc \leq 0$, then $g_\lambda(t) \geq a^4 + 1 > 0$ for every $t \in \mathbb{R}$ and so $r_{a,b,c}$ does not contain points of S_λ. Assume now $abc > 0$, set $z = t^2$ and study the function $h(z) = (b^4 + c^4)z^2 - 4\lambda abcz + a^4 + 1$ for $z \geq 0$. The minimum point of h is given by $z_0 = \dfrac{2\lambda abc}{b^4 + c^4}$. One has

$$(b^4 + c^4)h(z_0) = (a^4 + 1)(b^4 + c^4) - 4\lambda^2(abc)^2 \geq (2a^2)(2b^2c^2) - 4\lambda^2(abc)^2 =$$
$$= 4(abc)^2(1 - \lambda^2) > 0.$$

So for every choice of $(a, b, c) \in \Omega$ one has $r_{a,b,c} \cap S_\lambda = \emptyset$ if $0 \leq \lambda < 1$. Since the union of the lines $r_{a,b,c}$ as (a, b, c) varies in Ω is the whole of $\mathbb{R}^3$, it follows that S_λ has no real points for $0 \leq \lambda < 1$.

Summing up, we have proven that S_λ has real points if and only if $|\lambda| \geq 1$.

We describe an alternative method to prove that S_λ has no real points if $0 \leq \lambda < 1$. As above, note that, if one fixes on the coordinate hyperplane H_0 the homogeneous coordinates induced by the standard frame of $\mathbb{P}^3(\mathbb{C})$, for all $\lambda \in \mathbb{R}$ the curve $S_\lambda \cap H_0$ is defined by $x_1^4 + x_2^4 + x_3^4 = 0$, and so it does not contain any real point. Similarly one shows that the homogeneous coordinates of any real point of S_λ are all non-zero. So let x_0, x_1, x_2, x_3 be non-zero real numbers. From the inequality $(x_0^2 - x_1^2)^2 \geq 0$ one deduces $x_0^4 + x_1^4 \geq 2x_0^2x_1^2$, and analogously one has $x_2^4 + x_3^4 \geq 2x_2^2x_3^2$. Adding these inequalities one gets $x_0^4 + x_1^4 + x_2^4 + x_3^4 \geq 2(x_0^2x_1^2 + x_2^2x_3^2)$. Moreover one has

$$x_0^2x_1^2 + x_2^2x_3^2 = (|x_0x_1| - |x_2x_3|)^2 + 2|x_0x_1x_2x_3| \geq 2|x_0x_1x_2x_3|,$$

so that finally one has $x_0^4 + x_1^4 + x_2^4 + x_3^4 \geq 4|x_0x_1x_2x_3|$. Since $x_i \neq 0$ for $i = 0, \ldots, 3$, if $0 \leq \lambda < 1$ then

$$x_0^4 + x_1^4 + x_2^4 + x_3^4 \geq 4|x_0x_1x_2x_3| > 4\lambda x_0x_1x_2x_3.$$

It follows that $F_\lambda(x_0, x_1, x_2, x_3)$ is positive for every $[x_0, x_1, x_2, x_3] \in \mathbb{P}^3(\mathbb{R})$, so that S_λ contains no real points.

Exercise 119. Consider a plane H of $\mathbb{P}^3(\mathbb{C})$, a non-degenerate conic $C \subset H$, a point $P \in C$ and let $t \subset H$ be the tangent line to C at P. Let r and s be lines of $\mathbb{P}^3(\mathbb{C})$ such that:

(i) r and s are not contained in H;
(ii) $r \cap (s \cup C) = \emptyset$;
(iii) $s \cap C = P$.

Let $\pi \colon C \to r$ be the map defined by $\pi(Q) = L(s, Q) \cap r$ if $Q \neq P$ and $\pi(P) = L(s, t) \cap r$.

(a) Check that π is well defined and bijective.
(b) Prove that

$$X = \bigcup_{Q \in C} L(Q, \pi(Q))$$

is the support of a cubic surface S.
(c) Determine the singular points of S and the corresponding multiplicities, and say whether S is reducible and whether it is a cone (Fig. 3.1).

Solution. (a) The restriction of π to $C \setminus \{P\}$ coincides with the restriction of the projection onto r with centre s, and so it is well defined. Since s and r are skew by assumption, the plane $L(t, s)$ does not contain r and so the point $\pi(P)$ is well defined, too.

Let Q, Q' be points of C such that $\pi(Q) = \pi(Q')$. If $Q \neq P$ and $Q' \neq P$, one has $L(s, Q) = L(s, \pi(Q)) = L(s, \pi(Q')) = L(s, Q')$, so that the line $H \cap L(s, Q)$ meets C at P, Q, Q'. Since C is irreducible, we must have $Q = Q'$. Furthermore, if $Q = P$ and $Q' \neq P$, then the condition $\pi(P) = \pi(Q')$ would imply $L(s, t) = L(s, \pi(P)) = L(s, \pi(Q')) = L(s, Q')$. The line $t = H \cap L(s, t) = H \cap L(s, Q')$, which is tangent to C, would then meet C also at the point $Q' \neq P$, contradicting again the irreducibility of C. It follows that π is injective. The surjectivity of π is a consequence of the fact that, for any given point $A \in r$, the line $H \cap L(s, A)$ either is tangent to C at P, and in that case $A = \pi(P)$, or intersects C at a point $Q \neq P$, and in that case $A = \pi(Q)$.

(b) It is possible to choose a projective frame $\{P_0, \ldots, P_4\}$ of $\mathbb{P}^3(\mathbb{C})$ such that $P_0 = P$, $P_2 \in C$, P_1 is the intersection point of t with the tangent line to C at P_2, $P_3 \in s$ and $P_4 \in r$. In the coordinate system induced by this frame the plane H is defined by the equation $x_3 = 0$, P has coordinates $[1, 0, 0, 0]$, t is defined by $x_3 = x_2 = 0$, C is defined by the equation $x_0 x_2 - \lambda x_1^2 = 0$ for some $\lambda \neq 0$, and s

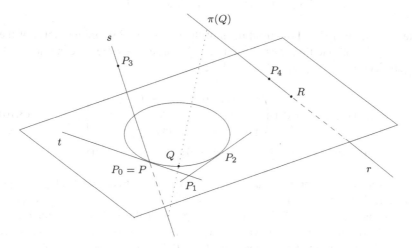

Fig. 3.1 The configuration described in Exercise 119

is given by $x_1 = x_2 = 0$. Let $[a, b, c, 0]$ be the coordinates of the point $R = r \cap H$, and let us determine first the conditions imposed on a, b, c by assumption (ii). Since $r = L(P_4, R)$, the lines r and s are skew if and only if there exists no non-zero linear combination of $(a, b, c, 0)$ and $(1, 1, 1, 1)$ that satisfies the conditions $x_1 = x_2 = 0$, i.e. if and only if $b - c \neq 0$. In addition, since $r \cap C = \emptyset$, one has $R \notin C$, so that $\lambda b^2 - ac \neq 0$.

Let $Y \in \mathbb{P}^3(\mathbb{C}) \backslash s$. The plane $L(Y, s)$ intersects H in a line, and therefore it intersects C, besides at P, at a second point Y', possibly not distinct from P. The points Y' and $\pi(Y') \in r$ are distinct because $r \cap C = \emptyset$, and the point Y is on X if and only if Y, Y' and $\pi(Y')$ are collinear.

If Y has coordinates $[y_0, y_1, y_2, y_3]$, then the plane $L(Y, s)$ has equation $y_2 x_1 - y_1 x_2 = 0$. Combining this equation and the equation of C, one sees that Y' has coordinates $[\lambda y_1^2, y_1 y_2, y_2^2, 0]$. In addition, if $Y' \neq P$ then one has $\pi(Y') = L(s, Y') \cap r = L(s, Y) \cap r$ by definition, so that $\pi(Y')$ is represented by the (unique up to multiples) non-zero linear combination of $(a, b, c, 0)$ and $(1, 1, 1, 1)$ that satisfies the equation $y_2 x_1 - y_1 x_2 = 0$. Then a simple computation shows that

$$\pi(Y') = [(a - c)y_1 + (b - a)y_2, (b - c)y_1, (b - c)y_2, -cy_1 + by_2]. \quad (3.8)$$

If instead $Y' = P$, since $L(s, t)$ has equation $x_2 = 0$, one easily gets $\pi(Y') = L(s, t) \cap r = [a - c, b - c, 0, -c]$. Note also that in this case, since $Y' = [\lambda y_1^2, y_1 y_2, y_2^2, 0] = [1, 0, 0, 0] = P$, one must necessarily have $y_2 = 0$, implying $y_1 \neq 0$ since $Y \notin s$. One deduces that the formula (3.8), obtained in the case when $Y' \neq P$, actually holds also when $Y' = P$. So Y belongs to X if and only if

$$\mathrm{rk} \begin{pmatrix} y_0 & \lambda y_1^2 & (a - c)y_1 + (b - a)y_2 \\ y_1 & y_1 y_2 & (b - c)y_1 \\ y_2 & y_2^2 & (b - c)y_2 \\ y_3 & 0 & -cy_1 + by_2 \end{pmatrix} \leq 2. \quad (3.9)$$

Note that the second and the third rows of the matrix in (3.9) are multiples of the row $(1, y_2, b - c)$ and they are both non-zero, because $Y \notin s$. Hence condition (3.9) is equivalent to $F(y_0, y_1, y_2, y_3) = 0$, where

$$F(y_0, y_1, y_2, y_3) = \det \begin{pmatrix} y_0 & \lambda y_1^2 & (a - c)y_1 + (b - a)y_2 \\ 1 & y_2 & (b - c) \\ y_3 & 0 & -cy_1 + by_2 \end{pmatrix} = \quad (3.10)$$

$$= y_3[(b - c)\lambda y_1^2 + (c - a)y_1 y_2 + (a - b)y_2^2] + (by_2 - cy_1)(y_0 y_2 - \lambda y_1^2).$$

Therefore we have proven that a point $Y \in \mathbb{P}^3(\mathbb{C}) \backslash s$ of coordinates $[y_0, \ldots, y_3]$ belongs to X if and only if $F(y_0, \ldots, y_3) = 0$, where F is a homogeneous polynomial of degree 3. So it follows that the support of the surface S defined by F (regarded from now on as a polynomial in x_0, x_1, x_2, x_3) coincides with the set X on $\mathbb{P}^3(\mathbb{C}) \backslash s$. Moreover it is evident that $F(a_0, 0, 0, a_3) = 0$ for all $(a_0, a_3) \in \mathbb{C}^2$, so that s is contained in the support of $\mathcal{S}$. Let us now show that $s \subseteq X$. One has that $P \in L(P, \pi(P)) \subseteq X$ and, since $Q \in s \setminus \{P\}$, the plane $L(Q, r)$ intersects H in a line different from t, and therefore intersects $\mathcal{C}$ in at least one point $Q_1 \neq P$. So $L(Q_1, \pi(Q_1)) = L(s, Q_1) \cap L(r, Q_1) = L(s, Q_1) \cap L(r, Q)$ intersects s at Q. This proves that s is contained in X, and therefore that X coincides with the support of $\mathcal{S}$.

(c) Using the equation F given in (3.10), it is immediate to check that all the points of the line s are singular for $\mathcal{S}$. In order to compute their multiplicities, note that $F_{x_1 x_1}(x_0, 0, 0, x_3) = 2\lambda(b - c)x_3$. So the points $Q \in s$, $Q \neq P$, verify $F_{x_1 x_1}(Q) \neq 0$ and so have multiplicity 2. Moreover $F_{x_1 x_2}(P) = -c$ and $F_{x_2 x_2}(P) = 2b$ are not both equal to zero because, as noted in the solution of (b), one has $\lambda b^2 - ac \neq 0$, hence also P has multiplicity 2.

Let us now show that $\mathcal{S}$ has no additional singular points. This verification can be carried out analytically, by computing ∇F and showing that all the solutions of $\nabla F = 0$ satisfy $x_1 = x_2 = 0$. Otherwise one can argue synthetically, as follows.

Let A be a point of $\mathcal{S} \setminus s$, let $K = L(s, A)$, and consider the plane cubic $\mathcal{S} \cap K$. Since the support of $\mathcal{S} \cap K$ contains s, by Bézout's Theorem one has $\mathcal{S} \cap K = s + \mathcal{Q}$, where $\mathcal{Q}$ is a conic in K. By Exercise 58 (a), the points of s are singular for $\mathcal{S} \cap K$ since $s \subseteq \operatorname{Sing}(\mathcal{S})$, and this implies easily that $s \subseteq \mathcal{Q}$. Again by Bézout's Theorem one has $\mathcal{Q} = s + s'$, where $s' \subseteq K$ is a line. More precisely, since r is contained in X and is disjoint from s, the support of the cubic $\mathcal{S} \cap K = 2s + s'$ cannot be contained in s, so that $s' \neq s$. In particular, the only singular points of $\mathcal{S} \cap K$ are given by the points of s, hence the curve $\mathcal{S} \cap K$ is smooth at A and, again by Exercise 58, $\mathcal{S}$ is smooth at A. So we have shown that $\operatorname{Sing}(\mathcal{S}) = s$.

Because all the points of $\operatorname{Sing}(\mathcal{S})$ have multiplicity 2, $\mathcal{S}$ is not a cone by Exercise 112.

Now let us show that $\mathcal{S}$ is not reducible. Assume by contradiction $\mathcal{S} = K + \mathcal{Q}$, where K is a plane and $\mathcal{Q}$ is a quadric. By Exercise 54, all the points of $K \cap \mathcal{Q}$ are singular for $\mathcal{S}$, therefore the support of $K \cap \mathcal{Q}$ is contained in s. Then the curve $K \cap \mathcal{Q}$, being a conic of K, coincides with the double line $2s$. In particular, s is contained in K. Since we have shown above that the support of the intersection of $\mathcal{S}$ with any plane containing s is the union of two distinct lines, we have reached a contradiction.

Note. It is natural to wonder how the situation described in Exercise 119 degenerates when the point $H \cap r$ belongs to $\mathcal{C}$. In that case it is easy to check that the polynomial F given in (3.10) turns out to be divisible by $by_2 - cy_1$, namely F vanishes on the plane $L(s, H \cap r)$. Then the cubic surface $\mathcal{S}$ defined by F splits as the union of $L(s, H \cap r)$ and of a quadric $\mathcal{Q}$, whose support coincides with the analogue of the set X defined here (cf. Exercise 186).

Chapter 4
Exercises on Conics and Quadrics

Affine and projective conics and quadrics. Polarity, centres and diametral hyperplanes. Conics and quadrics in Euclidean space. Principal hyperplanes, axes and vertices. Pencils of conics.

Abstract Solved problems on affine and projective conics and quadrics, polarity, conics and quadrics in Euclidean space, pencils of conics.

Notation: Throughout the whole chapter, the symbol $\mathbb{K}$ denotes $\mathbb{R}$ or $\mathbb{C}$.

Exercise 120. (a) Find the equation of a conic $\mathcal{C}$ of $\mathbb{P}^2(\mathbb{C})$ passing through the points
$P_1 = [1, 0, 1]$, $P_2 = [-1, 0, 0]$, $P_3 = [0, 1, 1]$, $P_4 = [0, -1, 0]$, $P_5 = [1, 3, 2]$.
(b) Check that $\mathcal{C}$ is non-degenerate and determine the pole of the line $5x_0 + x_1 - 3x_2 = 0$ with respect to the conic $\mathcal{C}$.

Solution. (a) The points P_1, P_2, P_3, P_4 are not collinear and hence the conics passing through these points form a pencil $\mathcal{F}$ (cf. Sect. 1.9.7) generated by the degenerate conics $L(P_1, P_2) + L(P_3, P_4)$ and $L(P_1, P_3) + L(P_2, P_4)$. Since

$$L(P_1, P_2) = \{x_1 = 0\}, \quad L(P_3, P_4) = \{x_0 = 0\},$$

$$L(P_1, P_3) = \{x_0 + x_1 - x_2 = 0\}, \quad L(P_2, P_4) = \{x_2 = 0\},$$

the generic conic $\mathcal{C}_{\lambda,\mu}$ of $\mathcal{F}$ has equation

$$\lambda x_0 x_1 + \mu x_2 (x_0 + x_1 - x_2) = 0.$$

This conic passes through P_5 if and only if $3\lambda + 4\mu = 0$; choosing $[\lambda, \mu] = [4, -3]$, we get the conic of equation $4x_0 x_1 - 3x_2(x_0 + x_1 - x_2) = 0$.

(b) The conic $\mathcal{C}$ is represented by the matrix $A = \begin{pmatrix} 0 & 4 & -3 \\ 4 & 0 & -3 \\ -3 & -3 & 6 \end{pmatrix}$. As $\det A \neq 0$, the conic is non-degenerate.

© Springer International Publishing Switzerland 2016
E. Fortuna et al., *Projective Geometry*, UNITEXT - La Matematica per il 3+2 104,
DOI 10.1007/978-3-319-42824-6_4

Since the map $\text{pol}_{\mathcal{C}}$ is a projective isomorphism (cf. Sect. 1.8.2), there exists exactly one point $R = [a, b, c]$ such that $\text{pol}_{\mathcal{C}}(R)$ is the line r of equation $5x_0 + x_1 - 3x_2 = 0$. As $\text{pol}_{\mathcal{C}}(R)$ has equation ${}^tRAX = 0$, that is

$$(4b - 3c)x_0 + (4a - 3c)x_1 + (-3a - 3b + 6c)x_2 = 0,$$

the values a, b, c we are looking for are those for which there exists $h \neq 0$ such that

$$(4b - 3c, 4a - 3c, -3a - 3b + 6c) = h(5, 1, -3).$$

The solutions of this linear system form the set $\{(t, 2t, t) \mid t \in \mathbb{C}\}$; therefore the point $R = [1, 2, 1]$ is the requested pole.

Observe that an alternative way to determine R consists in choosing two distinct points M, N on r and taking $R = \text{pol}_{\mathcal{C}}(M) \cap \text{pol}_{\mathcal{C}}(N)$: by reciprocity $\text{pol}_{\mathcal{C}}(R) = L(M, N) = r$.

Exercise 121. Assume that P_1, P_2, P_3 are non-collinear points in $\mathbb{P}^2(\mathbb{K})$ and let r be a line passing through P_1, which passes neither through P_2 nor through P_3. Consider the subset of the space Λ_2 of projective conics of $\mathbb{P}^2(\mathbb{K})$

$$\mathcal{F} = \{\mathcal{C} \in \Lambda_2 \mid \mathcal{C} \text{ passes through } P_1, P_2, P_3 \text{ and is tangent to } r \text{ at } P_1\}.$$

Show that $\mathcal{F}$ is a linear system and compute its dimension.

Solution (1). Imposing that a conic $\mathcal{C}$ is tangent to r at P_1 corresponds to imposing two independent linear conditions (cf. Exercise 95 and the Note following it). The conics of $\mathcal{F}$ are obtained by imposing that they pass also through the points P_2 and P_3, which are two additional linear conditions, so $\mathcal{F}$ is a linear system of dimension ≥ 1.

As a matter of fact $\mathcal{F}$ has dimension 1, i.e. it is a pencil. To prove that, assume by contradiction that $\mathcal{F}$ has dimension at least 2 and choose a fourth point $P_4 \notin r$ in such a way that P_1, P_2, P_3, P_4 are in general position. Let $\mathcal{F}'$ be the linear system of the conics of $\mathcal{F}$ that pass also through P_4; since $\dim \mathcal{F}' \geq \dim \mathcal{F} - 1 \geq 1$, we can choose two distinct conics $\mathcal{C}_1, \mathcal{C}_2$ in $\mathcal{F}'$. These conics meet in at least 5 points counted with multiplicity (because $I(\mathcal{C}_1, \mathcal{C}_2, P_1) \geq 2$ since $\mathcal{C}_1, \mathcal{C}_2$ are both tangent to the line r at P_1, cf. Sect. 1.9.3). As a consequence by Bézout's Theorem $\mathcal{C}_1$ and $\mathcal{C}_2$ contain a common line l and are both degenerate, that is $\mathcal{C}_1 = l + r_1$ and $\mathcal{C}_2 = l + r_2$. There are only two possibilities in order that the two previous conics be tangent to r at P_1: either $l = r$ or l and r_i, $i = 1, 2$, meet at P_1. The first situation cannot occur, because r_i should pass through the points P_2, P_3, P_4 which are not collinear. Also the second situation cannot occur, because at least two points among P_2, P_3, P_4 should lie either on l or on r_i and hence they would be collinear with P_1. So we have found a contradiction, and therefore $\dim \mathcal{F} = 1$.

Solution (2). It is also possible to solve the exercise analytically. Let R be a point of r not collinear with P_2 and P_3; choosing P_1, P_2, P_3, R as a projective frame, we have $P_1 = [1, 0, 0]$, $P_2 = [0, 1, 0]$, $P_3 = [0, 0, 1]$, $R = [1, 1, 1]$ and

$r = L(P_1, R) = \{x_1 - x_2 = 0\}$. Let C be a conic of $\mathcal{F}$ and let M be a symmetric matrix that represents it. Since C passes through $[1, 0, 0], [0, 1, 0], [0, 0, 1]$, the entries on the principal diagonal of M are necessarily zero, i.e. $M = \begin{pmatrix} 0 & a & b \\ a & 0 & c \\ b & c & 0 \end{pmatrix}$ where at least one among $a, b, c \in \mathbb{K}$ is non-zero. Therefore C has an equation of the form $F(x_0, x_1, x_2) = ax_0x_1 + bx_0x_2 + cx_1x_2 = 0$.

Since $\nabla F = (ax_1 + bx_2, ax_0 + cx_2, bx_0 + cx_1)$ and $\nabla F(1, 0, 0) = (0, a, b)$, the line r is tangent to C at P_1 if and only if $b = -a$. Therefore $\mathcal{F}$ consists precisely of the conics represented by matrices of the form

$$\begin{pmatrix} 0 & a & -a \\ a & 0 & c \\ -a & c & 0 \end{pmatrix}, \quad [a, c] \in \mathbb{P}^1(\mathbb{K})$$

so $\mathcal{F}$ is a linear system of dimension 1, i.e. it is a pencil.

Note. The result holds also when the line r passes, for instance, through P_2. Namely in this case every conic $C \in \mathcal{F}$ meets r in at least 3 points counted with multiplicity and therefore by Bézout's Theorem the line r is a component of C. As a consequence the conics of $\mathcal{F}$ are exactly those of the form $r + s$ when s varies in the pencil of lines centred at P_3, so $\mathcal{F}$ is a pencil (cf. Exercise 96).

Exercise 122. In $\mathbb{P}^2(\mathbb{R})$ consider the points

$$P_1 = [0, 1, 2], \quad P_2 = [0, 0, 1], \quad P_3 = [2, 1, 2], \quad P_4 = [3, 0, 1].$$

Determine, if it exists, an equation of a conic passing through P_1, P_2, P_3, P_4 and tangent at P_3 to the line r of equation $x_0 - x_2 = 0$.

Solution. Observe that the line r passes through the point P_3 and that P_1, P_2, P_3 are not collinear. The conic we are looking for belongs to the pencil $\mathcal{F}$ of conics passing through P_1, P_2, P_3 and tangent to r at P_3 (cf. Exercise 121 and the following Note). It is generated by the degenerate conics $r + L(P_1, P_2)$ and $L(P_1, P_3) + L(P_2, P_3)$.

Since $L(P_1, P_2) = \{x_0 = 0\}$, $L(P_1, P_3) = \{2x_1 - x_2 = 0\}$ and $L(P_2, P_3) = \{x_0 - 2x_1 = 0\}$, the generic conic $C_{\lambda,\mu}$ of $\mathcal{F}$ has equation

$$\lambda x_0(x_0 - x_2) + \mu(2x_1 - x_2)(x_0 - 2x_1) = 0.$$

The conic $C_{\lambda,\mu}$ passes through P_4 if and only if $6\lambda - 3\mu = 0$; hence, choosing for instance $[\lambda, \mu] = [1, 2]$, we find that the conic of equation

$$x_0^2 - 8x_1^2 + 4x_0x_1 - 3x_0x_2 + 4x_1x_2 = 0$$

satisfies the required properties.

Exercise 123. Determine, if it exists, a non-degenerate conic C of $\mathbb{P}^2(\mathbb{C})$ such that

(i) C passes through the points $A = [0, 0, 1]$ and $B = [0, 1, 1]$;
(ii) C is tangent to the line $x_0 - x_2 = 0$ at the point $P = [1, 1, 1]$;
(iii) the polar of the point $[2, 4, 3]$ with respect to C is the line $3x_1 - 4x_2 = 0$.

Solution. The points A, B and P are not collinear. If r is the line of equation $x_0 - x_2 = 0$, a conic satisfying conditions (i) and (ii) belongs to the pencil generated by the degenerate conics $L(A, B) + r$ and $L(A, P) + L(B, P)$ (cf. Exercise 121). It is easy to compute that $L(A, B) = \{x_0 = 0\}$, $L(A, P) = \{x_0 - x_1 = 0\}$, $L(B, P) = \{x_1 - x_2 = 0\}$, so the generic conic $C_{\lambda,\mu}$ of the pencil has equation

$$\lambda x_0(x_0 - x_2) + \mu(x_0 - x_1)(x_1 - x_2) = 0,$$

and is therefore represented by the matrix

$$A_{\lambda,\mu} = \begin{pmatrix} 2\lambda & \mu & -\lambda - \mu \\ \mu & -2\mu & \mu \\ -\lambda - \mu & \mu & 0 \end{pmatrix}.$$

Therefore, if $C_{\lambda,\mu}$ is non-degenerate, the polar of the point $[2, 4, 3]$ with respect to $C_{\lambda,\mu}$ has equation $(\lambda + \mu)x_0 - 3\mu x_1 + (2\mu - 2\lambda)x_2 = 0$ and this line coincides with the line of equation $3x_1 - 4x_2 = 0$ if and only if $\lambda + \mu = 0$. So, if we choose the homogeneous pair $[\lambda, \mu] = [1, -1]$, we find the conic of equation

$$x_0^2 + x_1^2 - x_0x_1 - x_1x_2 = 0$$

which, being non-degenerate, satisfies the required properties.

Exercise 124. Consider the curve D of $\mathbb{P}^2(\mathbb{C})$ of equation

$$F(x_0, x_1, x_2) = x_2^2(x_1 - x_2)^2 - x_1(x_1 + x_2)x_0^2 = 0.$$

(a) Find the singular points of D and compute their multiplicities and principal tangents.
(b) Determine the dimension of the linear system $\mathcal{F}$ of the conics of $\mathbb{P}^2(\mathbb{C})$ passing through the singular points of D and tangent at $Q = [1, 0, 0]$ to the line τ of equation $x_1 = 0$.
(c) Compute the points of intersection between D and C when C varies in $\mathcal{F}$.

Solution. (a) The singular points D are the points whose homogeneous coordinates annihilate the gradient of F. By solving the system

$$\begin{cases} F_{x_0} = -2x_0x_1(x_1 + x_2) = 0 \\ F_{x_1} = 2x_2^2(x_1 - x_2) - x_0^2(x_1 + x_2) - x_1x_0^2 = 0 \\ F_{x_2} = 2x_2(x_1 - x_2)^2 - 2x_2^2(x_1 - x_2) - x_1x_0^2 = 0 \end{cases}$$

we find that the singular points of $\mathcal{D}$ are $A = [0, 1, 0]$, $B = [0, 1, 1]$ and $Q = [1, 0, 0]$.

If we work in the affine chart U_0 using the coordinates $x = \frac{x_1}{x_0}$, $y = \frac{x_2}{x_0}$, the point Q has coordinates $(0, 0)$ and $\mathcal{D} \cap U_0$ has equation $y^2(x - y)^2 - x(x + y) = 0$. Thus we see that Q is an ordinary double point for $\mathcal{D}$ where the principal tangents are the projective closures of the lines $x = 0$ and $x + y = 0$, that is the lines $x_1 = 0$ and $x_1 + x_2 = 0$.

If instead we work in the affine chart U_1 using the coordinates $u = \frac{x_0}{x_1}$, $v = \frac{x_2}{x_1}$, we have $A = (0, 0)$, $B = (0, 1)$ and the affine part $\mathcal{D} \cap U_1$ of $\mathcal{D}$ has equation

$$v^2(1 - v)^2 - (1 + v)u^2 = 0.$$

Thus we immediately see that A is an ordinary double point with principal tangents the projective closures of the lines $v - u = 0$ and $v + u = 0$, that is the lines $x_2 - x_0 = 0$ and $x_2 + x_0 = 0$.

In order to study the multiplicity of B, if we perform the translation $(w, z) \to (u, v) = (w, z + 1)$ that maps $(0, 0)$ to B, we obtain the equation

$$(z + 1)^2 z^2 - (z + 2)w^2 = 0.$$

Therefore, B is an ordinary double point with principal tangents the projective closures of the lines $z - \sqrt{2}w = 0$ and $z + \sqrt{2}w = 0$, that is the lines $x_2 - x_1 - \sqrt{2}x_0 = 0$ and $x_2 - x_1 + \sqrt{2}x_0 = 0$.

(b) The points A, B, Q are not collinear, $A \notin \tau$ and $B \notin \tau$; hence the set of conics passing through A, B, Q and tangent to τ at Q is a linear system of dimension 1, that is a pencil (cf. Exercise 121).

(c) Since the conics $\tau + L(A, B)$ and $L(Q, A) + L(Q, B)$ generate the pencil $\mathcal{F}$ and have equations $x_1 x_0 = 0$ and $x_2(x_1 - x_2) = 0$, respectively, the generic conic $C_{\lambda,\mu}$ of $\mathcal{F}$ has equation $\lambda x_1 x_0 + \mu x_2(x_1 - x_2) = 0$.

If $\mu = 0$, then $C_{1,0} \cap \mathcal{D} = \{Q, A, B\}$. Thus, if we suppose $\mu \neq 0$, we can set $t = \frac{\lambda}{\mu}$ and find the solutions of the system given by the equations of $\mathcal{D}$ and of $C_{t,1} = C_{\lambda,\mu}$. From the equation of $C_{t,1}$ we obtain the equality $x_2(x_1 - x_2) = -t x_0 x_1$. By squaring and substituting into the equation of $\mathcal{D}$, one gets

$$x_0^2 x_1(t^2 x_1 - (x_1 + x_2)) = 0.$$

In correspondence with the roots of the factors x_0 and x_1 we find the points Q, A, B. If instead $t^2 x_1 - (x_1 + x_2) = 0$, if we substitute $x_2 = (t^2 - 1)x_1$ into the equation of $C_{t,1}$ and if we divide by x_1, we obtain the equation

$$t x_0 + (t^2 - 1)(2 - t^2)x_1 = 0,$$

having as unique homogeneous solution the pair $[x_0, x_1] = [(1 - t^2)(2 - t^2), t]$. Thus it turns out that $C_{t,1}$ intersects $\mathcal{D}$, besides at Q, A, B, only at the point of homogeneous coordinates $[(1 - t^2)(2 - t^2), t, t(t^2 - 1)]$. This point coincides with Q if $t = 0$, with A if $t = \pm 1$, with B if $t = \pm\sqrt{2}$, otherwise it is distinct from Q, A and B.

Note. Let $f : \mathbb{P}^1(\mathbb{C}) \to \mathbb{P}^2(\mathbb{C})$ be the map defined by

$$f([\lambda, \mu]) = [(\mu^2 - \lambda^2)(2\mu^2 - \lambda^2), \lambda\mu^3, \lambda\mu(\lambda^2 - \mu^2)].$$

The argument in the proof of part (c) of the previous exercise shows that the image of f is contained in $\mathcal{D}$. Since for every $P \in \mathcal{D}$ there exists a conic of $\mathcal{F}$ passing through P, it easily follows that $f(\mathbb{P}^1(\mathbb{C})) = \mathcal{D}$. Therefore, the map f gives a parametrization of $\mathcal{D}$. Observe that such a parametrization is not injective, because $f([0, 1]) = f([1, 0]) = Q$, $f([1, 1]) = f([1, -1]) = A$, $f([\sqrt{2}, 1]) = f([\sqrt{2}, -1]) = B$.

Exercise 125. Let $P_1, P_2 \in \mathbb{P}^2(\mathbb{K})$ be distinct points and let r_i be a projective line passing through P_i, $i = 1, 2$. Assume that $P_2 \notin r_1$, $P_1 \notin r_2$. Show that the set

$$\mathcal{F} = \{\mathcal{C} \in \Lambda_2 \mid r_i \text{ is tangent to } \mathcal{C} \text{ at } P_i, \ i = 1, 2\}$$

is a linear system and compute its dimension.

Solution (1). The conics of $\mathcal{F}$ can be obtained by imposing tangency conditions to two given lines at the points P_1 and P_2; this corresponds to imposing four linear conditions, so $\mathcal{F}$ is a linear system of dimension ≥ 1.

If $\mathcal{C}$ is a reducible conic in $\mathcal{F}$, then $\mathcal{C}$ is tangent to r_1 at P_1 if and only if either r_1 is an irreducible component of $\mathcal{C}$ or both irreducible components of the conic pass through P_1. It follows that the only reducible conics in $\mathcal{F}$, which must fulfil the tangency conditions both at P_1 and at P_2, are the conics $r_1 + r_2$ and $2L(P_1, P_2)$.

Let us now see that $\mathcal{F}$ has dimension 1. Assume by contradiction that $\mathcal{F}$ has dimension at least 2 and choose a point $P_3 \notin r_1 \cup r_2$ not collinear with P_1 and P_2. Then the set $\mathcal{F}'$ consisting of the conics of $\mathcal{F}$ passing also through P_3 is a linear system of dimension ≥ 1, so that we can choose two distinct conics $\mathcal{C}_1, \mathcal{C}_2$ in it. These conics meet in at least 5 points counted with multiplicity (because $I(\mathcal{C}_1, \mathcal{C}_2, P_i) \geq 2$ for $i = 1, 2$, cf. Sect. 1.9.3). As a consequence, by Bézout's Theorem $\mathcal{C}_1$ and $\mathcal{C}_2$ are reducible and share a common line. Then, by the previous considerations, for $i = 1, 2$ either $\mathcal{C}_i = r_1 + r_2$ or $\mathcal{C}_i = 2L(P_1, P_2)$, but both these conics pass through P_3. Therefore the hypothesis that $\mathcal{F}$ has dimension at least 2 leads to a contradiction, and hence $\dim \mathcal{F} = 1$.

Solution (2). As for Exercise 121, we can give an analytic solution of the problem.

Set $Q = r_1 \cap r_2$. By hypothesis Q, P_1, P_2 are in general position, so we can choose coordinates in such a way that $Q = [1, 0, 0]$, $P_1 = [0, 1, 0]$, $P_2 = [0, 0, 1]$; consequently $r_1 = \{x_2 = 0\}, r_2 = \{x_1 = 0\}$. Let $\mathcal{C}$ be a conic of $\mathcal{F}$; since $\mathcal{C}$ passes both through P_1 and through P_2, it is represented by a matrix $M = \begin{pmatrix} a & b & c \\ b & 0 & d \\ c & d & 0 \end{pmatrix}$ with at least one among $a, b, c, d \in \mathbb{K}$ non-zero. Since $\mathrm{pol}(P_1)$ has equation $bx_0 + dx_2 = 0$, the line r_1 is tangent to $\mathcal{C}$ at P_1 if and only if $b = 0$. Similarly, since $\mathrm{pol}(P_2)$ has equation $cx_0 + dx_1 = 0$, the line r_2 is tangent to $\mathcal{C}$ at P_2 if and only if $c = 0$. Therefore, $\mathcal{F}$ consists precisely of the conics represented by matrices of the form

$$\begin{pmatrix} a & 0 & 0 \\ 0 & 0 & d \\ 0 & d & 0 \end{pmatrix}, \quad [a, d] \in \mathbb{P}^1(\mathbb{K})$$

and then $\mathcal{F}$ is a linear system of dimension 1, i.e. it is a pencil.

Exercise 126. Compute, if it exists, an equation of a non-degenerate conic $\mathcal{C}$ of $\mathbb{P}^2(\mathbb{C})$ such that:

(i) $\mathcal{C}$ is tangent to the line $x_0 + x_2 = 0$ at the point $A = [1, 1, -1]$;
(ii) $\mathcal{C}$ is tangent to the line $x_0 - x_1 - x_2 = 0$ at the point $B = [1, 1, 0]$;
(iii) the polar of the point $[2, 0, 1]$ with respect to $\mathcal{C}$ passes through the point $[4, 2, 3]$.

Solution. A conic fulfilling conditions (i) and (ii) belongs to the pencil generated by the degenerate conics $(x_0 + x_2)(x_0 - x_1 - x_2) = 0$ and $2L(A, B)$ (cf. Exercise 125). Since $L(A, B)$ has equation $x_0 - x_1 = 0$, the generic conic $C_{\lambda,\mu}$ of the pencil has equation

$$\lambda(x_0 + x_2)(x_0 - x_1 - x_2) + \mu(x_0 - x_1)^2 = 0$$

and is thus represented by the matrix

$$A_{\lambda,\mu} = \begin{pmatrix} 2\lambda + 2\mu & -\lambda - 2\mu & 0 \\ -\lambda - 2\mu & 2\mu & -\lambda \\ 0 & -\lambda & -2\lambda \end{pmatrix}.$$

Since $A_{\lambda,\mu} \begin{pmatrix} 2 \\ 0 \\ 1 \end{pmatrix} = \begin{pmatrix} 4\lambda + 4\mu \\ -3\lambda - 4\mu \\ -2\lambda \end{pmatrix}$, the point $[2, 0, 1]$ is non-singular for every $C_{\lambda,\mu}$ and the polar of $[2, 0, 1]$ with respect to $\mathcal{C}$ is the line of equation $(4\lambda + 4\mu)x_0 + (-3\lambda - 4\mu)x_1 - 2\lambda x_2 = 0$. It contains the point $[4, 2, 3]$ if and only if $4\lambda + 8\mu = 0$; therefore, in correspondence with the homogeneous pair $[\lambda, \mu] = [2, -1]$ we find the conic of equation

$$x_0^2 - x_1^2 - 2x_2^2 - 2x_1 x_2 = 0$$

which, being non-degenerate, satisfies the required properties.

Exercise 127. In $\mathbb{P}^2(\mathbb{C})$ consider the lines

$$r = \{x_2 = 0\} \quad s = \{x_1 - 2x_2 = 0\} \quad t = \{x_1 = 0\} \quad l = \{x_0 - x_2 = 0\}$$

and the points $P = r \cap t$, $Q = t \cap l$, $A = l \cap r$, $R = [2, -1, 1]$.

(a) Find an explicit formula for a projectivity f of $\mathbb{P}^2(\mathbb{C})$ such that $f(r) = s$, $f(l) = l$, $f(t) = t$, $f(R)$ belongs to the line of equation $x_0 + x_1 = 0$ and t is the only line in the pencil of lines centred at P that is invariant under f.

(b) Find, if it exists, a conic $\mathcal{C}$ of $\mathbb{P}^2(\mathbb{C})$ tangent at P to t, tangent at A to l and such that $f(\mathcal{C})$ passes through the point $[2, -2, 1]$.

Solution. (a) It turns out that $P = [1, 0, 0]$, $Q = [1, 0, 1]$, $A = [0, 1, 0]$ and $B = l \cap s = [1, 2, 1]$. Moreover, we observe that the lines r, s, t pass through P while $P \notin l$.

If f is a projectivity such that $f(r) = s$, $f(l) = l$, $f(t) = t$, then $f(A) = B$, $f(P) = P$, $f(Q) = Q$. Imposing the first two conditions, we see that f is represented by a matrix of the form $M = \begin{pmatrix} 1 & b & c \\ 0 & 2b & d \\ 0 & b & e \end{pmatrix}$ with $b(2e - d) \neq 0$.

Also, we have that $f(Q) = Q$ if and only if $d = 0$ and $e = 1 + c$, so that $M = \begin{pmatrix} 1 & b & c \\ 0 & 2b & 0 \\ 0 & b & 1+c \end{pmatrix}$. Since $f(R) = [2 - b + c, -2b, -b + 1 + c]$, the point $f(R)$ belongs to the line of equation $x_0 + x_1 = 0$ if and only if $c = 3b - 2$; hence necessarily $M = \begin{pmatrix} 1 & b & 3b - 2 \\ 0 & 2b & 0 \\ 0 & b & 3b - 1 \end{pmatrix}$.

Let $\mathcal{F}_P$ be the pencil of lines centred at P; the lines of this pencil have equations $\alpha x_1 + \beta x_2 = 0$, with $[\alpha, \beta] \in \mathbb{P}^1(\mathbb{C})$. The pencil $\mathcal{F}_P$ is a line of $\mathbb{P}^2(\mathbb{C})^*$ that is invariant under the dual projectivity $f_* : \mathbb{P}^2(\mathbb{C})^* \to \mathbb{P}^2(\mathbb{C})^*$, which is represented by the matrix ${}^t M^{-1}$ in the system of dual homogeneous coordinates on $\mathbb{P}^2(\mathbb{C})^*$ (cf. Sect. 1.4.5). In the system of coordinates induced on $\mathcal{F}_P$ the line $\alpha x_1 + \beta x_2 = 0$ has coordinates $[\alpha, \beta]$ and the restriction of f_* to $\mathcal{F}_P$ is represented by the matrix $\begin{pmatrix} 3b - 1 & -b \\ 0 & 2b \end{pmatrix}$. Then t, which has coordinates $[1, 0]$, is the only line of $\mathcal{F}_P$ invariant under f if and only if the eigenvectors of the latter matrix are precisely the non-zero multiples of $(1, 0)$, which happens if and only if $b = 1$.

So the only projectivity having the required properties is the one represented by the matrix $M = \begin{pmatrix} 1 & 1 & 1 \\ 0 & 2 & 0 \\ 0 & 1 & 2 \end{pmatrix}$.

(b) The pencil of conics tangent to t at P and tangent to l at A is generated by the degenerate conics $l + t$ and $2r$. Therefore, the generic conic $\mathcal{C}_{\lambda,\mu}$ of this pencil has equation

$$\lambda x_1(x_0 - x_2) + \mu x_2^2 = 0.$$

Since $f([2, -1, 1]) = [2, -2, 1]$, $f(\mathcal{C}_{\lambda,\mu})$ passes through the point $[2, -2, 1]$ if and only if $\mathcal{C}_{\lambda,\mu}$ passes through the point $[2, -1, 1]$; this occurs if and only if $\mu - \lambda = 0$. Thus, if we choose the homogeneous pair $[\lambda, \mu] = [1, 1]$, we get the conic $x_1(x_0 - x_2) + x_2^2 = 0$ which satisfies all the required properties.

Exercise 128. Given four distinct lines r_1, r_2, r_3, r_4 in $\mathbb{P}^2(\mathbb{K})$, prove that there exists at least one non-degenerate conic which is tangent to the four lines if and only if r_1, r_2, r_3, r_4 are in general position.

Solution (1). Assume that there exists a non-degenerate conic which is tangent to r_1, r_2, r_3, r_4 and, for any $i \neq j$, let $P_{ij} = r_i \cap r_j$. Since for every point of $\mathbb{P}^2(\mathbb{K})$ there exist at most two lines passing through that point and tangent to the conic (cf. Sect. 1.8.2), the line r_h cannot pass through P_{ij} if $h \notin \{i, j\}$. Therefore, any three lines chosen among r_1, r_2, r_3, r_4 are not concurrent, that is the four lines are in general position.

Conversely, let $\mathcal{D}$ be a non-degenerate and non-empty conic and denote by l_1, l_2, l_3, l_4 the lines tangent to $\mathcal{D}$ at four distinct points of $\mathcal{D}$. As seen here above, the four lines are in general position. For any $i \neq j$ let $Q_{ij} = l_i \cap l_j$. One can easily check that each of the quadruples $P_{12}, P_{23}, P_{34}, P_{41}$ and $Q_{12}, Q_{23}, Q_{34}, Q_{41}$ consists of points in general position. Therefore, there exists a projectivity T of $\mathbb{P}^2(\mathbb{K})$ such that $T(Q_{12}) = P_{12}$, $T(Q_{23}) = P_{23}$, $T(Q_{34}) = P_{34}$ and $T(Q_{41}) = P_{41}$. This immediately implies that $T(l_i) = r_i$ for $i = 1, 2, 3, 4$. Then the conic $\mathcal{C} = T(\mathcal{D})$ is tangent to r_1, r_2, r_3, r_4.

Solution (2). Arguing by duality, we can regard the lines r_1, r_2, r_3, r_4 as points R_1, R_2, R_3, R_4 of $\mathbb{P}^2(\mathbb{K})^*$. The lines r_1, r_2, r_3, r_4 are in general position in $\mathbb{P}^2(\mathbb{K})$ if and only if the corresponding points in $\mathbb{P}^2(\mathbb{K})^*$ are in general position; in this case there exists a non-degenerate conic passing through R_1, R_2, R_3, R_4 and then the dual conic is tangent to r_1, r_2, r_3, r_4 (cf. Sect. 1.8.2).

Note. The same argument used in Solution (2) proves that, if r_1, r_2, r_3, r_4, r_5 are five lines of $\mathbb{P}^2(\mathbb{K})$ in general position (that is the corresponding points R_1, R_2, R_3, R_4, R_5 of $\mathbb{P}^2(\mathbb{K})^*$ are in general position), then there exists exactly one non-degenerate conic tangent to the five lines.

Exercise 129. Let A, B, C, D be points of $\mathbb{P}^2(\mathbb{K})$ in general position and assume that r is a line of $\mathbb{P}^2(\mathbb{K})$ passing through D and such that $A \notin r$, $B \notin r$ and $C \notin r$. Prove that there exists only one non-degenerate conic $\mathcal{Q}$ of $\mathbb{P}^2(\mathbb{K})$ fulfilling the following conditions:

(i) $\mathrm{pol}_{\mathcal{Q}}(A) = L(B, C)$;
(ii) $\mathrm{pol}_{\mathcal{Q}}(B) = L(A, C)$;
(iii) $D \in \mathcal{Q}$ and the line r is tangent to $\mathcal{Q}$ at D.

Solution. Observe that, if $\mathcal{Q}$ is a non-degenerate conic satisfying conditions (i) and (ii), then the point C lies both on $\mathrm{pol}_{\mathcal{Q}}(A)$ and on $\mathrm{pol}_{\mathcal{Q}}(B)$; hence, by reciprocity, necessarily $\mathrm{pol}_{\mathcal{Q}}(C) = L(A, B)$. As a consequence the points A, B, C are the vertices of a self-polar triangle for the conic.

In a suitable system of homogeneous coordinates in $\mathbb{P}^2(\mathbb{K})$ we have that $A = [1, 0, 0]$, $B = [0, 1, 0]$, $C = [0, 0, 1]$, $D = [1, 1, 1]$. Then the non-degenerate conics $\mathcal{Q}$ having A, B, C as vertices of a self-polar triangle have equations of the form $\alpha x_0^2 + \beta x_1^2 + \gamma x_2^2 = 0$, with at least one among α, β, γ non-zero (cf. Sect. 1.8.4).

In order for $\mathcal{Q}$ to satisfy property (iii) too, the line r must be the polar of D with respect to $\mathcal{Q}$. This polar has equation $ax_0 + bx_1 + cx_2 = 0$ with $a + b + c = 0$, as $D \in r$, and with $a \neq 0, b \neq 0, c \neq 0$, as r does not pass through any of the points A, B and C. Instead, the line $\text{pol}_{\mathcal{Q}}(D)$ has equation $\alpha x_0 + \beta x_1 + \gamma x_2 = 0$.

Therefore, we have $\text{pol}_{\mathcal{Q}}(D) = r$ if and only if the vectors (α, β, γ) and (a, b, c) are proportional. Hence the only conic having the required properties is the one defined by the equation $ax_0^2 + bx_1^2 + cx_2^2 = 0$.

Exercise 130. Denote by $\mathcal{F}$ the pencil of conics of $\mathbb{P}^2(\mathbb{C})$ of equation

$$\lambda(4x_0x_1 - x_2^2 - 4x_1^2) + \mu(x_0x_1 + x_2^2 + 4x_1^2 - 5x_1x_2) = 0 \quad [\lambda, \mu] \in \mathbb{P}^1(\mathbb{C}).$$

(a) Find the degenerate conics and the base points of the pencil $\mathcal{F}$.
(b) Describe the projectivities f of $\mathbb{P}^2(\mathbb{C})$ such that $f([1, 1, 0]) = [1, 1, 0]$ and $f(\mathcal{C}) \in \mathcal{F}$ for every $\mathcal{C} \in \mathcal{F}$.

Solution. (a) The generic conic $\mathcal{C}_{\lambda,\mu}$ of the pencil is represented by the matrix

$$A_{\lambda,\mu} = \begin{pmatrix} 0 & 2\lambda + \frac{\mu}{2} & 0 \\ 2\lambda + \frac{\mu}{2} & -4\lambda + 4\mu & -\frac{5}{2}\mu \\ 0 & -\frac{5}{2}\mu & -\lambda + \mu \end{pmatrix}.$$

Since $\det A_{\lambda,\mu} = \left(2\lambda + \frac{\mu}{2}\right)^2 (\mu - \lambda)$, the only degenerate conics of the pencil are the ones corresponding to the homogeneous pairs $[\lambda, \mu] = [1, 1]$ and $[\lambda, \mu] = [-1, 4]$, i.e. the conics $\mathcal{D}_1$ and $\mathcal{D}_2$ of equations

$$x_1(x_0 - x_2) = 0 \quad \text{and} \quad (2x_1 - x_2)^2 = 0,$$

respectively. If we intersect $\mathcal{D}_1$ and $\mathcal{D}_2$, we get that the base points of the pencil are $P = [1, 0, 0]$ and $Q = [2, 1, 2]$.

(b) The conic $\mathcal{D}_1$ is simply degenerate and its irreducible components are the lines $r_1 = \{x_1 = 0\}$ and $r_2 = \{x_0 - x_2 = 0\}$, which meet at $R = [1, 0, 1]$. The conic $\mathcal{D}_2$ has the line $L(P, Q) = \{2x_1 - x_2 = 0\}$ as an irreducible component of multiplicity 2 and so it is doubly degenerate.

Recall (cf. Sect. 1.8.1) that every projectivity of $\mathbb{P}^2(\mathbb{C})$ transforms a non-degenerate (respectively, simply degenerate, doubly degenerate) conic into a non-degenerate (respectively, simply degenerate, doubly degenerate) conic.

Therefore, if f is a projectivity that transforms the conics of the pencil $\mathcal{F}$ into conics of the same pencil, necessarily $f(\mathcal{D}_1) = \mathcal{D}_1$ and $f(\mathcal{D}_2) = \mathcal{D}_2$. In particular f must leave the line $L(P, Q)$ invariant and must either leave the components r_1, r_2 of $\mathcal{D}_1$ invariant or exchange them. In any case the point $R = [1, 0, 1]$ where r_1 and r_2 meet must be invariant under f, while the points P and Q may be fixed or may be exchanged by f.

If we set $S = [1, 1, 0]$, the points R, P, Q, S are in general position, so their images completely determine the projectivity f.

If f fixes the 4 points, then f is the identity map. The only other possibility is the projectivity that fixes R, S and exchanges P and Q. One can easily check that this latter projectivity is represented by the matrix

$$\begin{pmatrix} 2 & -1 & -1 \\ 1 & 0 & -1 \\ 2 & -2 & -1 \end{pmatrix}.$$

Finally note that, besides the identity map, also this second projectivity satisfies the properties required by the exercise: namely by construction $f(\mathcal{D}_1) = \mathcal{D}_1$ and $f(\mathcal{D}_2) = \mathcal{D}_2$, so that all the conics of the pencil generated by these two conics, i.e. of the pencil $\mathcal{F}$, are transformed into conics of the same pencil.

Exercise 131. Assume that the points R, P_1, P_2 of $\mathbb{P}^2(\mathbb{C})$ are not collinear and that f is a projectivity of $\mathbb{P}^2(\mathbb{C})$ such that $f(R) = R$, $f(P_1) = P_2$ and $f(P_2) = P_1$. Let $\mathcal{F}$ be the pencil of conics tangent at P_1 to the line $r_1 = L(R, P_1)$ and tangent at P_2 to the line $r_2 = L(R, P_2)$. Prove that $f^2 = \mathrm{Id}$ if and only if $f(\mathcal{C}) = \mathcal{C}$ for every conic $\mathcal{C} \in \mathcal{F}$.

Solution. The pencil $\mathcal{F}$ is generated by the degenerate conics $\mathcal{C}_1 = r_1 + r_2$ and $\mathcal{C}_2 = 2L(P_1, P_2)$ and its only base points are P_1 and P_2 (cf. Exercise 125). The hypotheses immediately imply that $f(r_1) = r_2$, $f(r_2) = r_1$ and that the line $L(P_1, P_2)$ is f-invariant; therefore, the degenerate conics $\mathcal{C}_1$ and $\mathcal{C}_2$ are f-invariant. Since f induces a projectivity of the space Λ_2 of conics of $\mathbb{P}^2(\mathbb{C})$ (cf. Sect. 1.9.5), in particular we obtain that f transforms the conics of the pencil $\mathcal{F}$ into conics of the same pencil. Moreover, by the Fundamental theorem of projective transformations we have that $f(\mathcal{C}) = \mathcal{C}$ for every $\mathcal{C} \in \mathcal{F}$ if and only if there exists another conic $\mathcal{C}_3 \in \mathcal{F}$ distinct from $\mathcal{C}_1$ and from $\mathcal{C}_2$ such that $f(\mathcal{C}_3) = \mathcal{C}_3$.

If $f^2 = \mathrm{Id}$, the fixed-point set of f is the union of a line r and a point $P \notin r$ (cf. Exercise 49 and the following Note). If we had $P = R$, then r would intersect the lines r_1 and r_2 in two distinct points (each of them distinct from R) which would be fixed under f, while the only fixed point lying on the two lines is R. Therefore, the line r of fixed points necessarily passes through R and, in particular, $P \neq R$.

Let Q be a point of $r \setminus L(P_1, P_2)$, $Q \neq R$ (cf. Fig. 4.1). Since Q is not a base point of the pencil, there exists only one conic $\mathcal{C}_3 \in \mathcal{F}$ passing through Q; it is distinct both from $\mathcal{C}_1$ and from $\mathcal{C}_2$ because $Q \notin \mathcal{C}_1 \cup \mathcal{C}_2$. The conic $f(\mathcal{C}_3)$ belongs to $\mathcal{F}$ and passes through $f(Q) = Q$; then necessarily $f(\mathcal{C}_3) = \mathcal{C}_3$ and so $f(\mathcal{C}) = \mathcal{C}$ for any $\mathcal{C} \in \mathcal{F}$.

Conversely suppose that $f(\mathcal{C}) = \mathcal{C}$ for any $\mathcal{C} \in \mathcal{F}$. Since $f(R) = R$, f acts as a projectivity on the pencil of lines centred at R (which is isomorphic to $\mathbb{P}^1(\mathbb{C})$). Since every projectivity of a complex projective line has at least one fixed point, there exists a line s passing through R which is invariant under f. By the hypotheses neither r_1 nor r_2 are invariant under f, so necessarily $s \neq r_1$ and $s \neq r_2$. Let $\mathcal{C}_3$ be a non-degenerate conic of $\mathcal{F}$ (hence distinct from $\mathcal{C}_1$ and from $\mathcal{C}_2$) and denote

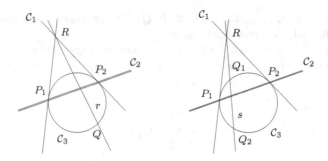

Fig. 4.1 *Left*, if $f^2 = \mathrm{Id}$, then $f(\mathcal{C}) = \mathcal{C}$ for every $\mathcal{C} \in \mathcal{F}$; *right*, if $f(\mathcal{C}) = \mathcal{C}$ for every $\mathcal{C} \in \mathcal{F}$, then $f^2 = \mathrm{Id}$

by Q_1, Q_2 the points where C_3 meets the invariant line s (cf. Fig. 4.1); one easily checks that $Q_1 \neq Q_2$. Observe that the points P_1, P_2, Q_1, Q_2 are in general position. Evidently $f(\{Q_1, Q_2\}) = \{Q_1, Q_2\}$ and hence either $f(Q_1) = Q_1$ and $f(Q_2) = Q_2$ or $f(Q_1) = Q_2$ and $f(Q_2) = Q_1$. In both cases f^2 has P_1, P_2, Q_1, Q_2 as fixed points and thus $f^2 = \mathrm{Id}$.

Exercise 132. Let $\mathcal{C}$ be a non-degenerate conic of $\mathbb{P}^2(\mathbb{C})$. Let P be a point of $\mathcal{C}$ and r the tangent line to $\mathcal{C}$ at P. Assume that R is a point of $\mathbb{P}^2(\mathbb{C})$ such that $R \notin \mathcal{C} \cup r$. Prove that there exist exactly two projectivities f of $\mathbb{P}^2(\mathbb{C})$ such that

$$f(\mathcal{C}) = \mathcal{C}, \quad f(R) = R, \quad f(P) = P.$$

Solution. Assume that f is a projectivity of $\mathbb{P}^2(\mathbb{C})$ under which $\mathcal{C}, P$ and R are invariant. Then each of the two lines passing through R and tangent to $\mathcal{C}$ is transformed into a line passing through R and tangent to $\mathcal{C}$. If A and B are the points where the tangent lines to $\mathcal{C}$ passing through R meet the conic, then there are only two possibilities: either $f(A) = A$ and $f(B) = B$, or $f(A) = B$ and $f(B) = A$.

In the first case f fixes four points in general position (that is A, B, P, R) and therefore it is the identity map, which of course fixes the conic too (Fig. 4.2).

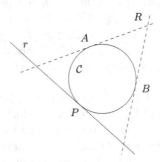

Fig. 4.2 The configuration described in Exercise 132

If $f(A) = B$ and $f(B) = A$, the additional conditions $f(P) = P$ and $f(R) = R$ uniquely determine a projectivity f. In that case $f^2 = \mathrm{Id}$, because f^2 fixes the points in general position A, B, P, R. Since C belongs to the pencil of conics tangent at A to $L(A, R)$ and tangent at B to $L(B, R)$, Exercise 131 implies that $f(C) = C$.

Thus we have found exactly two projectivities with the required properties.

Exercise 133. If C is a non-degenerate conic of $\mathbb{P}^2(\mathbb{K})$, let f be a projectivity of $\mathbb{P}^2(\mathbb{K})$ such that $f \neq \mathrm{Id}$ and $f(C) = C$. Prove that:

(a) If $f^2 = \mathrm{Id}$, then the fixed-point set of f is the union of a line r which is not tangent to C and its pole R with respect to C.

(b) If there exists a line of fixed points for f, then $f^2 = \mathrm{Id}$.

Solution. (a) If $f^2 = \mathrm{Id}$, the fixed-point set of f is the union of a line r and a point $P \notin r$ (for a proof of this fact see Exercise 49 and the following Note).

Assume by contradiction that r is tangent to C, and let $Q = r \cap C$ (cf. Fig. 4.3). For any $B \in C \setminus \{Q\}$, denote by r_B the tangent to C at B. Clearly $r_B \neq r$, and so $r_B \cap r$ consists of only one point S. By construction, $\mathrm{pol}_C(S) = L(Q, B)$, so $f(\{Q, B\}) = \{Q, B\}$, since $f(C) = C$ and $f(S) = S$. On the other hand, $f(Q) = Q$ and f is injective, so $f(B) = B$. Thus we have shown that f fixes all points of C, and hence $f = \mathrm{Id}$, which contradicts the hypotheses. Therefore, r is not tangent to C.

Set $R = \mathrm{pol}_C^{-1}(r)$. Since $f(r) = r$ and $f(C) = C$, it follows that $f(R) = R$. Moreover, since r is not tangent to C, we obtain that $R \notin r$, and so necessarily $P = R$.

(b) It is sufficient to consider the case $\mathbb{K} = \mathbb{C}$. Namely, if $\mathbb{K} = \mathbb{R}$, denote by $\mathcal{D}$ the complexified conic of C and by g the projectivity of $\mathbb{P}^2(\mathbb{C})$ induced by f. Clearly $\mathcal{D}$ is non-degenerate, $g(\mathcal{D}) = \mathcal{D}$ and, if r is a line consisting of fixed points of f, the complex line $r_\mathbb{C}$ is a line of fixed points of g; moreover, the property $g^2 = \mathrm{Id}_{\mathbb{P}^2(\mathbb{C})}$ implies that $f^2 = \mathrm{Id}_{\mathbb{P}^2(\mathbb{R})}$.

Thus assume that $\mathbb{K} = \mathbb{C}$ and that $r \subset \mathbb{P}^2(\mathbb{C})$ is a line of fixed points of f. Arguing as in the proof of (a), one shows that r is not tangent to C and that the pole R of

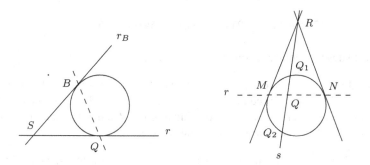

Fig. 4.3 The configuration described in Exercise 133: on the *left*, the hypothesis that r is tangent to C leads to a contradiction; on the *right*, the solution of part (b)

r with respect to C is fixed under f. Denote by M, N the points where r intersects C and let s be any line passing through R and not tangent to C (cf. Fig. 4.3). The line s is distinct from r, since $R \notin r$, and it intersects r in a point Q which is fixed under f. As a consequence $s = L(R, Q)$ is f-invariant. If Q_1, Q_2 are the two points where s intersects C, since $f(C) = C$ we have that $f(\{Q_1, Q_2\}) = \{Q_1, Q_2\}$. So f may either fix the points Q_1, Q_2 or exchange them: in both cases Q_1, Q_2 are fixed under f^2. Since f^2 fixes the points M, N, Q_1, Q_2 which are in general position, then $f^2 = \mathrm{Id}$.

Exercise 134. Consider the curve $\mathcal{D}$ of $\mathbb{P}^2(\mathbb{C})$ of equation

$$F(x_0, x_1, x_2) = x_0^3 + x_1^3 + x_2^3 - 5x_0x_1x_2 = 0.$$

Compute, if it exists, the equation of a conic tangent to the curve $\mathcal{D}$ at the points $P = [1, -1, 0]$ and $Q = [1, 2, 1]$ and passing through $R = [1, 2, -3]$.

Solution. First of all we can compute the tangents to $\mathcal{D}$ at the points P and Q. Since

$$\nabla F = (3x_0^2 - 5x_1x_2, 3x_1^2 - 5x_0x_2, 3x_2^2 - 5x_0x_1),$$

then $\nabla F(1, -1, 0) = (3, 3, 5)$ and $\nabla F(1, 2, 1) = (-7, 7, -7)$. Therefore the curve $\mathcal{D}$ is tangent at P to the line $r_P = \{3x_0 + 3x_1 + 5x_2 = 0\}$ and at Q to the line $r_Q = \{x_0 - x_1 + x_2 = 0\}$. In order to find a conic which is tangent to $\mathcal{D}$ at the points P and Q, so it is enough to find a conic which is tangent to r_P at P and to r_Q at Q (cf. Sect. 1.9.3).

The pencil of conics tangent at P to r_P and at Q to r_Q is generated by the conics $r_P + r_Q$ and $2L(P, Q)$; the generic conic of this pencil has therefore equation

$$\lambda(3x_0 + 3x_1 + 5x_2)(x_0 - x_1 + x_2) + \mu(x_0 + x_1 - 3x_2)^2 = 0$$

as $[\lambda, \mu]$ varies in $\mathbb{P}^1(\mathbb{C})$. The conic passes through R if and only if $\lambda + 6\mu = 0$; choosing the homogeneous pair $[\lambda, \mu] = [6, -1]$ we obtain the conic

$$17x_0^2 - 19x_1^2 + 21x_2^2 - 2x_0x_1 + 54x_0x_2 - 6x_1x_2 = 0$$

which satisfies the required properties.

Exercise 135. In $\mathbb{P}^2(\mathbb{C})$ consider the points $P = [1, -1, 0]$, $R = [1, 0, 0]$, $S = [0, 2, 1]$ and the conic C of equation $F(x_0, x_1, x_2) = 2x_0^2 + 2x_0x_1 + 3x_2^2 = 0$. Find, if it exists, a non-degenerate conic $\mathcal{D}$ passing through R and S, tangent to C at P and such that $\mathrm{pol}_{\mathcal{D}}(R)$ passes through the point $[7, 3, 3]$.

Solution. It is immediate to see that the point P belongs to the conic C; since $\nabla F(1, -1, 0) = (2, 2, 0)$, the tangent to the conic C at P is the line r of equation $x_0 + x_1 = 0$.

The conics passing through R and S and tangent to r at P form a pencil $\mathcal{F}$ generated by the degenerate conics $L(P, R) + L(P, S)$ and $L(R, S) + r$ (cf. Exercise 121). By easy computations we obtain that the pencil $\mathcal{F}$ consists of the conics of equations

$$\lambda x_2(x_0 + x_1 - 2x_2) + \mu(x_0 + x_1)(x_1 - 2x_2) = 0, \quad [\lambda, \mu] \in \mathbb{P}^1(\mathbb{C}).$$

The generic conic $\mathcal{C}_{\lambda,\mu}$ of $\mathcal{F}$ is thus represented by the matrix

$$A_{\lambda,\mu} = \begin{pmatrix} 0 & \mu & \lambda - 2\mu \\ \mu & 2\mu & \lambda - 2\mu \\ \lambda - 2\mu & \lambda - 2\mu & -4\lambda \end{pmatrix}.$$

Since $(1, 0, 0) \notin \ker A_{\lambda,\mu}$ for any $[\lambda, \mu] \in \mathbb{P}^1(\mathbb{C})$, the polar of R with respect to $\mathcal{C}_{\lambda,\mu}$ is the line of equation $\mu x_1 + (\lambda - 2\mu)x_2 = 0$; it passes through the point $[7, 3, 3]$ if and only if $\lambda = \mu$. Thus we find the conic

$$x_1^2 - 2x_2^2 + x_0 x_1 - x_0 x_2 - x_1 x_2 = 0$$

which is non-degenerate and therefore satisfies the requested properties.

Exercise 136. Assume that $\mathcal{C}$ is a non-degenerate conic of $\mathbb{P}^2(\mathbb{C})$. Let P be a point not lying on $\mathcal{C}$ and denote by r the polar of P with respect to $\mathcal{C}$. Let s be a line passing through P and not tangent to $\mathcal{C}$; denote by Q and R the points where s intersects $\mathcal{C}$.

Show that r and s meet only at one point D, and that $\beta(P, D, Q, R) = -1$.

Solution (1). Since $P \notin \mathcal{C}$, the polar r does not pass through P; hence the lines r and s are necessarily distinct, so they meet only at one point D.

The point D cannot lie on the conic $\mathcal{C}$: otherwise, since D is a point of the polar of P, the line $L(D, P) = s$ would be tangent to the conic (cf. Sect. 1.8.2), which contradicts the hypotheses. Therefore, in particular D is distinct both from R and from Q (Fig. 4.4).

The polar $\mathrm{pol}_\mathcal{C}(D)$ passes through P and intersects r at a point M such that P, D, M are the vertices of a self-polar triangle for $\mathcal{C}$. Then there exists a system of homogeneous coordinates in $\mathbb{P}^2(\mathbb{C})$ in which we have $P = [1, 0, 0]$, $D = [0, 1, 0]$, $M = [0, 0, 1]$. In these coordinates s has equation $x_2 = 0$ and a matrix associated to the conic is of the form $\begin{pmatrix} 1 & 0 & 0 \\ 0 & a & 0 \\ 0 & 0 & b \end{pmatrix}$ with a and b non-zero, i.e. $x_0^2 + ax_1^2 + bx_2^2 = 0$ is an equation of the conic. If we compute the intersections between $\mathcal{C}$ and s, we find that $Q = [\alpha, 1, 0]$ and $R = [-\alpha, 1, 0]$ with $\alpha^2 = -a$.

In the system of coordinates x_0, x_1 induced on s we have $P = [1, 0]$, $D = [0, 1]$, $Q = [\alpha, 1]$, $R = [-\alpha, 1]$. Therefore $\beta(P, D, Q, R) = -1$.

Solution (2). Since $P \notin \mathcal{C}$, the line $\mathrm{pol}_\mathcal{C}(P)$ intersects $\mathcal{C}$ at two distinct points, say A and B. It is easy to check that A, B, P, Q are in general position, so that

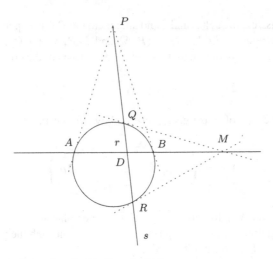

Fig. 4.4 The configuration described in Exercise 136

there exists a system of coordinates of $\mathbb{P}^2(\mathbb{C})$ in which we have $P = [1, 0, 0]$, $A = [0, 1, 0]$, $B = [0, 0, 1]$ and $Q = [1, 1, 1]$. If we impose that $\mathrm{pol}_{\mathcal{C}}(A) = L(A, P) = \{x_2 = 0\}$, $\mathrm{pol}_{\mathcal{C}}(B) = L(B, P) = \{x_1 = 0\}$ and $Q \in \mathcal{C}$, we easily obtain that $\mathcal{C}$ has equation $x_0^2 - x_1 x_2 = 0$. Since $s = L(P, Q)$ has equation $x_1 = x_2$, it follows that $R = [-1, 1, 1]$. Moreover, since $r = L(A, B)$ has equation $x_0 = 0$, we get that $D = r \cap s = [0, 1, 1]$. Hence

$$\beta(P, D, Q, R) = \beta([1, 0, 0], [0, 1, 1], [1, 1, 1], [-1, 1, 1]) = -1.$$

Solution (3). Arguing as in Solution (1), we show that there is exactly one point D where r and s meet. Assume that A, B are defined as in Solution (2). The set $\mathcal{F}$ of the conics that are tangent at A to the line $L(P, A)$ and tangent at B to the line $L(P, B)$ is a pencil having A and B as base points. Since R is distinct both from A and from B, in the pencil there is exactly one conic passing through R: that conic is $\mathcal{C}$.

Since the points P, Q, A, B and the points P, R, A, B are in general position, there exists exactly one projectivity $f : \mathbb{P}^2(\mathbb{C}) \to \mathbb{P}^2(\mathbb{C})$ such that $f(P) = P$, $f(Q) = R$, $f(A) = A$ and $f(B) = B$. The map f leaves the lines $L(P, A)$ and $L(P, B)$ invariant and transforms $\mathcal{C}$ into a conic which is tangent to $L(P, A)$ at A, tangent to $L(P, B)$ at B and passes through R. By the uniqueness property recalled above, $f(\mathcal{C}) = \mathcal{C}$. Moreover, $f(s) = s$, so that $f(s \cap \mathcal{C}) = s \cap \mathcal{C}$, that is $f(\{Q, R\}) = \{Q, R\}$. As $f(Q) = R$, it follows that $f(R) = Q$. Finally, since $f(r) = r$, we also have that $f(D) = D$. Hence

$$\beta(P, D, Q, R) = \beta(f(P), f(D), f(Q), f(R)) = \beta(P, D, R, Q) = \beta(P, D, Q, R)^{-1}$$

(cf. Sect. 1.5.2). Then $\beta(P, D, Q, R)^2 = 1$; since $Q \neq R$, it follows that $\beta(P, D, Q, R) = -1$.

Exercise 137. Consider the following pencil of conics of $\mathbb{P}^2(\mathbb{R})$:

$$2\lambda x_0^2 - (\mu + \lambda)x_1^2 + (\mu - \lambda)x_2^2 - 2\mu x_0 x_1 - 2\mu x_0 x_2 = 0, \qquad [\lambda, \mu] \in \mathbb{P}^1(\mathbb{R}).$$

(a) Find the degenerate conics and the base points of the pencil.
(b) Show that there is exactly one line which is tangent to all conics of the pencil.
(c) Determine all conics of the pencil that are parabolas in the affine chart U_0.

Solution. (a) The generic conic $\mathcal{C}_{\lambda,\mu}$ of the pencil is represented by the matrix

$$A_{\lambda,\mu} = \begin{pmatrix} 2\lambda & -\mu & -\mu \\ -\mu & -\lambda - \mu & 0 \\ -\mu & 0 & \mu - \lambda \end{pmatrix}.$$

Since $\det A_{\lambda,\mu} = 2\lambda^3$, the pencil contains exactly one degenerate conic $\mathcal{D}_1$ of equation $-x_1^2 + x_2^2 - 2x_0 x_1 - 2x_0 x_2 = 0$, that is $(2x_0 + x_1 - x_2)(x_1 + x_2) = 0$. Denote by l_1 and l_2 the irreducible components of $\mathcal{D}_1$ having equations $2x_0 + x_1 - x_2 = 0$ and $x_1 + x_2 = 0$ respectively; they meet at the point $P = [1, -1, 1]$.

If we choose for instance the homogeneous pair $[\lambda, \mu] = [1, 0]$, we find the conic $\mathcal{D}_2$ of the pencil having equation $2x_0^2 - x_1^2 - x_2^2 = 0$. By intersecting the conics $\mathcal{D}_1$ and $\mathcal{D}_2$ spanning the pencil, we obtain that the base points are $P = [1, -1, 1]$ and $Q = [1, 1, -1]$.

(b) The irreducible conic $\mathcal{D}_2$ is tangent at the point P to the component l_1 of $\mathcal{D}_1$; as a consequence, all conics of the pencil are tangent to the line l_1 at P (cf. Sect. 1.9.5).

If r is a line which is tangent to all conics of the pencil, in particular r must be tangent to the degenerate conic $\mathcal{D}_1 = l_1 + l_2$. Hence r must pass through P. On the other hand, every line passing through P and distinct from l_1 is not tangent to $\mathcal{D}_2$. Therefore, the only line which is tangent to all conics of the pencil is l_1.

(c) In the affine chart U_0, using the affine coordinates $x = \frac{x_1}{x_0}, y = \frac{x_2}{x_0}$, the conic $\mathcal{C}_{\lambda,\mu} \cap U_0$ has equation

$$2\lambda - (\mu + \lambda)x^2 + (\mu - \lambda)y^2 - 2\mu x - 2\mu y = 0.$$

It is a parabola if it is non-degenerate (that is $\lambda \neq 0$) and its projective closure intersects the improper line only at one point. That occurs when $\lambda^2 - \mu^2 = 0$ and therefore for $[\lambda, \mu] = [1, 1]$ and $[\lambda, \mu] = [1, -1]$.

Exercise 138. Consider the pencil of conics $\mathcal{C}_{\lambda,\mu}$ of $\mathbb{P}^2(\mathbb{R})$ of equation

$$(\lambda + \mu)x_0^2 - \mu x_1^2 - (\lambda + \mu)x_0 x_2 + \mu x_1 x_2 = 0, \qquad [\lambda, \mu] \in \mathbb{P}^1(\mathbb{R}).$$

(a) Find the degenerate conics and the base points of the pencil.
(b) Let r be the line of equation $x_2 = 0$. Check that, for every $P \in r$, the pencil contains exactly one conic $\mathcal{C}_{\lambda,\mu}(P)$ passing through P.
(c) Consider the map $f: r \to r$ defined by

$$f(P) = \begin{cases} (\mathcal{C}_{\lambda,\mu}(P) \cap r) \setminus \{P\} & \text{if the conic } \mathcal{C}_{\lambda,\mu}(P) \text{ is not tangent to } r \\ P & \text{otherwise} \end{cases}$$

Check that f is a projectivity of r such that $f^2 = \text{Id}$.

(d) Determine the homogeneous pairs $[\lambda, \mu] \in \mathbb{P}^1(\mathbb{R})$ such that the affine part of the conic $\mathcal{C}_{\lambda,\mu}$ in the affine chart $U = \mathbb{P}^2(\mathbb{R}) \setminus \{x_0 + x_2 = 0\}$ is a parabola.

Solution. (a) The generic conic of the pencil is represented by the matrix

$$A_{\lambda,\mu} = \begin{pmatrix} 2\lambda + 2\mu & 0 & -\lambda - \mu \\ 0 & -2\mu & \mu \\ -\lambda - \mu & \mu & 0 \end{pmatrix}.$$

Since $\det A_{\lambda,\mu} = 2\lambda\mu(\lambda + \mu)$, the degenerate conics of the pencil correspond to the homogeneous pairs $[0, 1]$, $[1, 0]$ and $[1, -1]$. Therefore, they have equations

$$(x_0 - x_1)(x_0 + x_1 - x_2) = 0, \quad x_0(x_0 - x_2) = 0, \quad x_1(x_1 - x_2) = 0.$$

Intersecting two of these degenerate conics, we find that the base points of the pencil are $[0, 0, 1]$, $[0, 1, 1]$, $[1, 1, 1]$ and $[1, 0, 1]$.

(b) As the line r does not contain any base point of the pencil, every point $P \in r$ imposes a non-trivial linear condition on the conics of the pencil; hence there is exactly one conic of the pencil passing through P. Explicitly, if $P = [y_0, y_1, 0]$, the only conic of the pencil passing through P is the conic $\mathcal{C}_{\lambda,\mu}(P)$ defined by the equation

$$y_1^2 x_0^2 - y_0^2 x_1^2 - y_1^2 x_0 x_2 + y_0^2 x_1 x_2 = 0.$$

(c) The points of intersection between $\mathcal{C}_{\lambda,\mu}(P)$ and r are the points $[x_0, x_1, 0]$ such that $y_1^2 x_0^2 - y_0^2 x_1^2 = 0$. We obtain only one intersection point if and only if either $y_1 = 0$ or $y_0 = 0$; hence $\mathcal{C}_{\lambda,\mu}(P)$ is not tangent to r provided that $P \neq P_1 = [1, 0, 0]$ and $P \neq P_2 = [0, 1, 0]$. In that case the point of $\mathcal{C}_{\lambda,\mu}(P) \cap r$ distinct from P is the point $[y_0, -y_1, 0]$.

So in the system of coordinates induced on r we have that $f([y_0, y_1]) = [y_0, -y_1]$ when $[y_0, y_1] \neq [1, 0]$ and $[y_0, y_1] \neq [0, 1]$. On the other hand, by definition $f(P_1) = P_1$ and $f(P_2) = P_2$, so that f is represented in coordinates by $[y_0, y_1] \mapsto [y_0, -y_1]$ on the whole line r. Thus it is evident that f is a projectivity such that $f^2 = \text{Id}$.

(d) The affine conic $\mathcal{C}_{\lambda,\mu} \cap U$ is a parabola when $\mathcal{C}_{\lambda,\mu}$ is non-degenerate and $\mathcal{C}_{\lambda,\mu} \cap \{x_0 + x_2 = 0\}$ consists of only one point (cf. Sect. 1.8.7). By intersecting $\mathcal{C}_{\lambda,\mu}$ with the line $x_0 + x_2 = 0$, we find the equation $2(\lambda + \mu)x_0^2 - \mu x_0 x_1 - \mu x_1^2 = 0$, which admits only one solution if and only if $\mu(8\lambda + 9\mu) = 0$. Since the conic corresponding to $[\lambda, \mu] = [1, 0]$ is degenerate, the only conic whose affine part is a parabola is the one obtained in correspondence with $[\lambda, \mu] = [9, -8]$, which has equation

$$x_0^2 + 8x_1^2 - x_0 x_2 - 8x_1 x_2 = 0.$$

Exercise 139. Denote by C the conic of $\mathbb{P}^2(\mathbb{R})$ of equation

$$x_0^2 + 2x_1^2 + 2x_0x_2 - 6x_1x_2 + x_2^2 = 0.$$

Check that C is non-degenerate and find the vertices of a self-polar triangle with respect to C containing the line $r = \{x_0 + x_1 + x_2 = 0\}$.

Solution. The conic C is represented by the matrix

$$A = \begin{pmatrix} 1 & 0 & 1 \\ 0 & 2 & -3 \\ 1 & -3 & 1 \end{pmatrix},$$

whose determinant is equal to -9; hence C is non-degenerate.

Let $P = [1, -1, 0] \in r \setminus C$. Since $\begin{pmatrix} 1 & -1 & 0 \end{pmatrix} A = \begin{pmatrix} 1 & -2 & 4 \end{pmatrix}$, the line $\mathrm{pol}_C(P)$ has equation $x_0 - 2x_1 + 4x_2 = 0$. Therefore, setting $Q = \mathrm{pol}_C(P) \cap r$, we have $Q = [-2, 1, 1]$, and so $Q \notin C$. Finally, $\mathrm{pol}_C(Q)$ has equation $x_0 + x_1 + 4x_2 = 0$, which implies that $R = \mathrm{pol}_C(P) \cap \mathrm{pol}_C(Q) = [4, 0, -1]$. By construction, the triangle with vertices P, Q, R satisfies the requested property.

Exercise 140. Let $\mathcal{F}$ be a pencil of conics of $\mathbb{P}^2(\mathbb{R})$. Prove that:

(a) $\mathcal{F}$ contains at least one degenerate conic.
(b) $\mathcal{F}$ contains infinitely many non-empty conics.

Solution. (a) If C_1 and C_2 are distinct conics of $\mathcal{F}$ of equations ${}^t X A X = 0$ and ${}^t X B X = 0$ respectively, the generic conic of the pencil $\mathcal{F}$ has equation ${}^t X (\lambda A + \mu B) X = 0$ as $[\lambda, \mu]$ varies in $\mathbb{P}^1(\mathbb{R})$. Let $G(\lambda, \mu) = \det(\lambda A + \mu B)$. If $G = 0$, all conics of the pencil are degenerate. If $G \neq 0$, then G is a real homogeneous polynomial of degree 3, and so there exists at least one homogeneous pair $[\lambda_0, \mu_0]$ in $\mathbb{P}^1(\mathbb{R})$ such that $\det(\lambda_0 A + \mu_0 B) = 0$; so the corresponding conic of the pencil is degenerate.

(b) If all conics of $\mathcal{F}$ are non-empty, the statement is trivially true. So assume that there exists at least an empty conic C_1 in $\mathcal{F}$ and let C_2 be another conic of the pencil. The conic C_1 can be represented by a symmetric positive definite matrix, so that by the Spectral theorem we may choose a system of coordinates in $\mathbb{P}^2(\mathbb{R})$ where C_1 is represented by the identity matrix and C_2 is represented by a diagonal matrix $\begin{pmatrix} a & 0 & 0 \\ 0 & b & 0 \\ 0 & 0 & c \end{pmatrix}$; without loss of generality we may assume $a \leq b \leq c$. As $C_1 \neq C_2$, it cannot occur that $a = b = c$.

At first consider the case $b < c$. The generic conic of the pencil is represented by the matrix $\begin{pmatrix} \lambda + \mu a & 0 & 0 \\ 0 & \lambda + \mu b & 0 \\ 0 & 0 & \lambda + \mu c \end{pmatrix}$; if we exclude the case $[\lambda, \mu] = [1, 0]$, corresponding to the empty conic C_1, the conics of $\mathcal{F} \setminus \{C_1\}$ are represented by the matrix $A_t = \begin{pmatrix} t+a & 0 & 0 \\ 0 & t+b & 0 \\ 0 & 0 & t+c \end{pmatrix}$ as t varies in $\mathbb{R}$. If $t \in (-c, -b)$ we have that

$t + b < 0$ and $t + c > 0$; therefore the symmetric matrix A_t is indefinite and hence every conic of $\mathcal{F}$ having equation ${}^t X A_t X = 0$ with $t \in (-c, -b)$ is non-empty.

Using a similar argument we can prove the thesis when $a < b$.

Exercise 141. Assume that $\mathcal{F}$ is a pencil of conics of $\mathbb{P}^2(\mathbb{K})$ containing a doubly degenerate conic C_1 and a non-degenerate conic C_2. Prove that $\mathcal{F}$ contains at most two degenerate conics.

Solution (1). First of all observe that it suffices to deal with the case $\mathbb{K} = \mathbb{C}$. Namely, if $\mathbb{K} = \mathbb{R}$ consider the pencil $\mathcal{G}$ of complex conics generated by $(C_1)_\mathbb{C}$ and $(C_2)_\mathbb{C}$. Since $\mathcal{G}$ contains the complexifications of all conics of $\mathcal{F}$, the number of degenerate conics of $\mathcal{F}$ is lower than or equal to the number of degenerate conics of $\mathcal{G}$.

So assume $\mathbb{K} = \mathbb{C}$. By hypothesis $C_1 = 2r$, where r is a line of $\mathbb{P}^2(\mathbb{C})$. Since C_2 is non-degenerate, r intersects C_2 at two (possibly coincident) points A and B.

If $A \neq B$, consider the line τ_A tangent to C_2 at A and the line τ_B tangent to C_2 at B. The lines τ_A and τ_B are distinct and they meet only at one point $M \notin C_2$. If we choose a point $P \in C_2 \setminus r$, since the points M, A, B, P are in general position, there exists a system of homogeneous coordinates in $\mathbb{P}^2(\mathbb{C})$ in which we have $M = [1, 0, 0]$, $A = [0, 1, 0]$, $B = [0, 0, 1]$, $P = [1, 1, 1]$; consequently $r = L(A, B)$ has equation $x_0 = 0$. If we impose that C_2 passes through A, B, P and that the polar of M with respect to C_2 is the line $r = \{x_0 = 0\}$, we easily see that C_2 has equation $x_0^2 - x_1 x_2 = 0$. Then the generic conic $C_{\lambda,\mu}$ of the pencil $\mathcal{F}$ generated by C_1 and C_2 has equation ${}^t X A_{\lambda,\mu} X = 0$ where

$$A_{\lambda,\mu} = \lambda \begin{pmatrix} 1 & 0 & 0 \\ 0 & 0 & 0 \\ 0 & 0 & 0 \end{pmatrix} + \mu \begin{pmatrix} 2 & 0 & 0 \\ 0 & 0 & -1 \\ 0 & -1 & 0 \end{pmatrix} = \begin{pmatrix} \lambda + 2\mu & 0 & 0 \\ 0 & 0 & -\mu \\ 0 & -\mu & 0 \end{pmatrix}.$$

Since $\det A_{\lambda,\mu} = \mu^2(\lambda + 2\mu)$, the reducible conics of the pencil correspond to $[\lambda, \mu] = [1, 0]$ (which recovers the degenerate conic C_1) and to $[\lambda, \mu] = [2, -1]$. Observe that the degenerate conic $C_{2,-1}$ has equation $x_1 x_2 = 0$ and hence it coincides with $\tau_A + \tau_B$.

If $A = B$, the line r is tangent to C_2 at A. Choose $P_1 \in r \setminus \{A\}$ and two distinct points P_2, P_3 of $C_2 \setminus r$ so that P_1, P_2, P_3 are not collinear. As the points A, P_1, P_2, P_3 are in general position, there exists a system of homogeneous coordinates in $\mathbb{P}^2(\mathbb{C})$ in which $A = [1, 0, 0]$, $P_1 = [0, 1, 0]$, $P_2 = [0, 0, 1]$, $P_3 = [1, 1, 1]$. As a consequence, r has equation $x_2 = 0$ and C_2 is represented by a matrix of the form $\begin{pmatrix} 0 & 0 & 1 \\ 0 & -2(1+c) & c \\ 1 & c & 0 \end{pmatrix}$ with $c \neq -1$. Then the generic conic $C_{\lambda,\mu}$ of the pencil $\mathcal{F}$ generated by C_1 and C_2 has equation ${}^t X A_{\lambda,\mu} X = 0$ where

$$A_{\lambda,\mu} = \lambda \begin{pmatrix} 0 & 0 & 0 \\ 0 & 0 & 0 \\ 0 & 0 & 1 \end{pmatrix} + \mu \begin{pmatrix} 0 & 0 & 1 \\ 0 & -2(1+c) & c \\ 1 & c & 0 \end{pmatrix} = \begin{pmatrix} 0 & 0 & \mu \\ 0 & -2\mu(1+c) & \mu c \\ \mu & \mu c & \lambda \end{pmatrix}.$$

Since $\det A_{\lambda,\mu} = 2\mu^3(1+c)$, the only reducible conic of the pencil is obtained for $[\lambda, \mu] = [1, 0]$, and so it is the conic $C_1 = 2r$.

Solution (2). As observed in Solution (1), it suffices to deal with the case $\mathbb{K} = \mathbb{C}$.

We have $C_1 = 2r$, where r is a line of $\mathbb{P}^2(\mathbb{C})$; denote by A and B the two (possibly coincident) points of intersection between C_1 and C_2.

If $A \neq B$ and if we denote by τ_A the line which is tangent to C_2 at A and by τ_B the line which is tangent to C_2 at B, for every conic C of $\mathcal{F}$ we have that $I(C, \tau_A, A) \geq 2$ and $I(C, \tau_B, B) \geq 2$. Therefore, if C is degenerate, the only possibilities are either $C = 2r$ or $C = \tau_A + \tau_B$.

If $A = B$, then the line r is tangent to C_2 at A. Moreover, the only base point of $\mathcal{F}$ is A, so that, if $C = l + l'$ is a degenerate conic of $\mathcal{F}$, necessarily $l \cap C = \{A\}$, $l' \cap C = \{A\}$. This latter fact can happen only if $l = l' = r$, and hence the only degenerate conic of the pencil is $C_1 = 2r$.

Note. If $\mathbb{K} = \mathbb{R}$ in both solutions of the Exercise we reduced to the complex case by considering the pencil generated by the complexified conics of $C_1 = 2r$ and C_2.

Alternatively, it is possible to adapt the two solutions to the case of a pencil of real conics arguing as follows. If the set $r \cap C_2$ is not empty, the considerations made in the previous solutions still hold. Therefore we need to deal only with the case when $r \cap C_2$ is empty, that is the case when the complexification of r intersects $(C_2)_{\mathbb{C}}$ at two distinct and conjugate points, A and $B = \sigma(A)$. In this case we showed in both solutions that the pencil of complex conics generated by $(C_1)_{\mathbb{C}}$ and $(C_2)_{\mathbb{C}}$ contains, besides $(C_1)_{\mathbb{C}}$, only another non-degenerate conic $\mathcal{Q} = \tau_A + \tau_B$, where τ_A (respectively τ_B) is the tangent to $(C_2)_{\mathbb{C}}$ at A (respectively at B). It follows that $\sigma(\tau_A) = \tau_{\sigma(A)} = \tau_B$, and hence $\sigma(\mathcal{Q}) = \mathcal{Q}$, so that, by Exercise 59, the conic $\mathcal{Q}$ is the complexification of a real conic C. It is not difficult to prove, arguing for instance as in the solution of Exercise 59, that C belongs to the pencil of real conics $\mathcal{F}$. Therefore, if the set $r \cap C_2$ is empty, then $\mathcal{F}$ contains exactly two degenerate conics.

Exercise 142. (*Cross-ratio of four points on a conic*)

(a) Assume that A and B are two distinct points of a non-degenerate conic C of $\mathbb{P}^2(\mathbb{K})$. Denote by t_A and t_B the lines which are tangent to C at A and B respectively, and by $L(A, B)$ the line joining A and B. Let $\mathcal{F}_A$ and $\mathcal{F}_B$ be the pencils of lines with centres A and B respectively. Consider the map $\psi: \mathcal{F}_A \to \mathcal{F}_B$ defined by

(i) $\psi(t_A) = L(A, B)$;

(ii) $\psi(L(A, B)) = t_B$;

(iii) $\psi(r) = L(B, Q)$ if $r \in \mathcal{F}_A$, $r \neq t_A$, $r \neq L(A, B)$, where Q is the intersection between r and C different from A (cf. Fig. 4.5). Prove that ψ is a projective isomorphism.

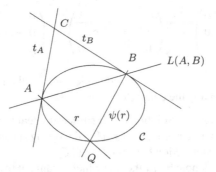

Fig. 4.5 If A and B are points of a conic $\mathcal{C}$, then the conic $\mathcal{C}$ establishes a projective isomorphism between the pencils of lines $\mathcal{F}_A$ and $\mathcal{F}_B$ (cf. Exercise 142)

(b) Let P_1, P_2, P_3, P_4 be distinct points on a non-degenerate conic $\mathcal{C}$ of $\mathbb{P}^2(\mathbb{K})$. Define the cross-ratio of the four points as follows:

$$\beta(P_1, P_2, P_3, P_4) = \beta(L(A, P_1), L(A, P_2), L(A, P_3), L(A, P_4))$$

where A is a point of $\mathcal{C}$ different from each P_i. Prove that $\beta(P_1, P_2, P_3, P_4)$ does not depend on the choice of A.

Solution. (a) Since the conic is non-degenerate, the lines t_A and t_B are distinct and hence they meet at a point C which does not lie on the conic. If R is a point on $\mathcal{C}$ different both from A and from B, it is easy to check that $\{A, B, C, R\}$ is a projective frame (a real non-degenerate conic which is not empty contains infinitely many points, so that R exists also in the real case). In the induced system of homogeneous coordinates we have $A = [1, 0, 0]$, $B = [0, 1, 0]$, $C = [0, 0, 1]$, $R = [1, 1, 1]$ and the lines t_A and t_B have equations $x_1 = 0$ and $x_0 = 0$ respectively. If we impose that $\mathcal{C}$ passes through $[1, 1, 1]$ and has the lines $x_1 = 0$ and $x_0 = 0$ as polars at $[1, 0, 0]$ and $[0, 1, 0]$ respectively, we get immediately that the conic is represented by a matrix of the form $M = \begin{pmatrix} 0 & b & 0 \\ b & 0 & 0 \\ 0 & 0 & -2b \end{pmatrix}$ with $b \neq 0$, so that $\mathcal{C}$ has equation $x_2^2 - x_0 x_1 = 0$.
Moreover, $\mathcal{F}_A$ consists of the lines of equations $a_1 x_1 + a_2 x_2 = 0$ with $[a_1, a_2] \in \mathbb{P}^1(\mathbb{K})$, so that a_1, a_2 is a system of homogeneous coordinates for $\mathcal{F}_A$. Similarly, $\mathcal{F}_B$ consists of the lines of equations $b_0 x_0 + b_2 x_2 = 0$ with $[b_0, b_2] \in \mathbb{P}^1(\mathbb{K})$, and b_0, b_2 is a system of homogeneous coordinates for $\mathcal{F}_B$.

Let r be a line in $\mathcal{F}_A$, with $r \neq t_A$, $r \neq L(A, B)$. Then r has equation $a_1 x_1 + a_2 x_2 = 0$ with $a_1 \neq 0$ and $a_2 \neq 0$. If we compute the intersections between r and $\mathcal{C}$ we find, besides the point A, the point $Q = [a_1^2, a_2^2, -a_1 a_2]$. Then the line $L(B, Q)$ has equation $a_1^2 x_2 + a_1 a_2 x_0 = 0$; since $a_1 \neq 0$, the line has equation $a_1 x_2 + a_2 x_0 = 0$ and hence it has coordinates $[a_2, a_1]$.

It follows that the restriction of the map ψ to $\mathcal{F}_A \setminus \{t_A, L(A, B)\}$ coincides with the restriction to $\mathcal{F}_A \setminus \{t_A, L(A, B)\}$ of the map $f : \mathcal{F}_A \to \mathcal{F}_B$ defined in coordinates by $f([a_1, a_2]) = [a_2, a_1]$, which is evidently a projective isomorphism.

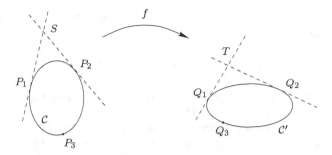

Fig. 4.6 The construction of the projectivity required in Exercise 143 (a)

Observe that, in the system of coordinates chosen in $\mathcal{F}_A$, the line t_A has coordinates $[1, 0]$ and $L(A, B)$ has coordinates $[0, 1]$. Similarly, in the system of coordinates chosen in $\mathcal{F}_B$, $L(A, B)$ has coordinates $[0, 1]$ and t_B has coordinates $[1, 0]$. Since $f(t_A) = f([1, 0]) = [0, 1] = L(A, B) = \psi(t_A)$ and $f(L(A, B)) = f([0, 1]) = [1, 0] = t_B = \psi(L(A, B))$, then ψ coincides with f and hence it is a projective isomorphism.

(b) Let B be a point of $\mathcal{C}$ distinct from A and from each P_i. Then the map $\psi \colon \mathcal{F}_A \to \mathcal{F}_B$ considered in (a) is a projective isomorphism such that $\psi(L(A, P_i)) = L(B, P_i)$ for $i = 1, \ldots, 4$. As the cross-ratio is invariant under projective isomorphisms, we have that $\beta(L(A, P_1), L(A, P_2), L(A, P_3), L(A, P_4)) = \beta(L(B, P_1), L(B, P_2), L(B, P_3), L(B, P_4))$ which implies the thesis.

Exercise 143. Consider two non-degenerate conics $\mathcal{C}$ and $\mathcal{C}'$ in $\mathbb{P}^2(\mathbb{K})$. Let P_1, P_2, P_3, P_4 be distinct points of $\mathcal{C}$ and Q_1, Q_2, Q_3, Q_4 distinct points of $\mathcal{C}'$.

(a) Show that there is exactly one projectivity $f \colon \mathbb{P}^2(\mathbb{K}) \to \mathbb{P}^2(\mathbb{K})$ such that $f(\mathcal{C}) = \mathcal{C}'$ and $f(P_i) = Q_i$ for $i = 1, 2, 3$.
(b) Show that there exists a projectivity f of $\mathbb{P}^2(\mathbb{K})$ such that $f(\mathcal{C}) = \mathcal{C}'$ and $f(P_i) = Q_i$ for $i = 1, 2, 3, 4$ if and only if $\beta(P_1, P_2, P_3, P_4) = \beta(Q_1, Q_2, Q_3, Q_4)$, where β denotes the cross-ratio of four points on a conic defined in Exercise 142.

Solution. (a) Let $S = \mathrm{pol}_{\mathcal{C}}(P_1) \cap \mathrm{pol}_{\mathcal{C}}(P_2)$, $T = \mathrm{pol}_{\mathcal{C}'}(Q_1) \cap \mathrm{pol}_{\mathcal{C}'}(Q_2)$. It is immediate to check that the quadruple P_1, P_2, P_3, S forms a projective frame of $\mathbb{P}^2(\mathbb{C})$; the same holds for the quadruple Q_1, Q_2, Q_3, T. Then, by the Fundamental theorem of projective transformations, it is sufficient to prove that a projectivity $f \colon \mathbb{P}^2(\mathbb{K}) \to \mathbb{P}^2(\mathbb{K})$ fulfils the required properties if and only if $f(S) = T$ and $f(P_i) = Q_i$ for $i = 1, 2, 3$ (Fig. 4.6).

If f satisfies the conditions described in the statement, then evidently $f(P_i) = Q_i$ for $i = 1, 2, 3$. Moreover, since projectivities preserve tangency conditions, we have that $f(\mathrm{pol}_{\mathcal{C}}(P_i)) = \mathrm{pol}_{\mathcal{C}'}(Q_i)$, and hence

$$f(S) = f(\mathrm{pol}_{\mathcal{C}}(P_1) \cap \mathrm{pol}_{\mathcal{C}}(P_2)) = \mathrm{pol}_{\mathcal{C}'}(Q_1) \cap \mathrm{pol}_{\mathcal{C}'}(Q_2) = T.$$

Conversely assume that $f(S) = T$ and $f(P_i) = Q_i$ for $i = 1, 2, 3$. Then, for $i = 1, 2$ we have

$$\text{pol}_{f(C)}(Q_i) = f(\text{pol}_C(P_i)) = f(L(P_i, S)) = L(Q_i, T) = \text{pol}_{C'}(Q_i),$$

so that both conics $f(C)$ and C' pass through Q_1 with tangent $\text{pol}_{C'}(Q_1)$, through Q_2 with tangent $\text{pol}_{C'}(Q_2)$ and through Q_3.

Since Q_1, Q_2, Q_3, T are in general position, there exists exactly one conic passing through Q_1 with tangent $\text{pol}_{C'}(Q_1)$, through Q_2 with tangent $\text{pol}_{C'}(Q_2)$ and through Q_3 (cf. Exercise 125). This implies that $f(C) = C'$, which proves the result.

(b) First of all assume there exists a projectivity $f: \mathbb{P}^2(\mathbb{K}) \to \mathbb{P}^2(\mathbb{K})$ such that $f(C) = C'$ and $f(P_i) = Q_i$ for $i = 1, 2, 3, 4$. If O is any point of C distinct from P_1, P_2, P_3, P_4 we have that $f(O) = O' \in C' \setminus \{Q_1, Q_2, Q_3, Q_4\}$. Furthermore, f induces a projectivity from the pencil of lines centred at O onto the pencil of lines centred at O', and this projectivity transforms $r_i = L(O, P_i)$ into $s_i = L(O', Q_i)$ for each $i = 1, 2, 3, 4$. Since the cross-ratio is invariant under projective transformations, we have $\beta(r_1, r_2, r_3, r_4) = \beta(s_1, s_2, s_3, s_4)$, that is $\beta(P_1, P_2, P_3, P_4) = \beta(Q_1, Q_2, Q_3, Q_4)$ by the definition of cross-ratio of points on a conic.

Conversely, assume that $\beta(P_1, P_2, P_3, P_4) = \beta(Q_1, Q_2, Q_3, Q_4)$. As shown in part (a), there exists a projectivity $f: \mathbb{P}^2(\mathbb{K}) \to \mathbb{P}^2(\mathbb{K})$ such that $f(C) = C'$ and $f(P_i) = Q_i$ for $i = 1, 2, 3$. We are going to show that $f(P_4) = Q_4$, which concludes the proof. As above let $O \in C \setminus \{P_1, P_2, P_3, P_4\}$, $O' = f(O) \in C'$ and, for each $i = 1, 2, 3, 4$, $r_i = L(O, P_i)$, $s_i = L(O', Q_i)$. As above, by the invariance of cross-ratio under projective transformations, we have

$$\beta(r_1, r_2, r_3, r_4) = \beta(f(r_1), f(r_2), f(r_3), f(r_4)) = \beta(s_1, s_2, s_3, L(O', f(P_4))).$$

Moreover, the hypothesis implies that $\beta(r_1, r_2, r_3, r_4) = \beta(s_1, s_2, s_3, s_4)$, so that $L(O', f(P_4)) = s_4 = L(O', Q_4)$. Hence

$$f(\{O, P_4\}) = f(C \cap L(O, P_4)) = C' \cap L(O', f(P_4)) = C' \cap L(O', Q_4) = \{O', Q_4\}.$$

As $f(O) = O'$, it follows that $f(P_4) = Q_4$, and hence the thesis.

Exercise 144. Assume that A, B, C, D are points in general position in $\mathbb{P}^2(\mathbb{K})$. Let $f: \mathbb{P}^2(\mathbb{K}) \to \mathbb{P}^2(\mathbb{K})$ be the projectivity such that $f(A) = B$, $f(B) = A$, $f(C) = D$ and $f(D) = C$. Show that $f(Q) = Q$ for every conic Q of $\mathbb{P}^2(\mathbb{K})$ passing through the points A, B, C, D.

Solution (1). The conics passing through the points A, B, C, D form a pencil $\mathcal{F}$ containing 3 degenerate conics, that is $C_1 = L(A, B) + L(C, D)$, $C_2 = L(A, D) + L(B, C)$ and $C_3 = L(A, C) + L(B, D)$ (cf. Sect. 1.9.7). The hypotheses immediately imply that each of the previous conics is invariant under f. As a consequence f acts on the pencil (which is projectively isomorphic to $\mathbb{P}^1(\mathbb{K})$) as a projectivity having 3 fixed points and hence as the identity map; therefore, each conic passing through A, B, C, D is invariant under f.

Solution (2). A different solution can be given by exploiting the notion of cross-ratio of four points on a conic (cf. Esercizio 142). As observed in Solution (1), the degenerate conics C_1, C_2, C_3 passing through A, B, C, D are evidently f-invariant, so it is sufficient to deal with the case when Q is non-degenerate. If we denote by β_Q the cross-ratio of four points on Q, using the symmetries of the usual cross-ratio of points on a line it is easy to show that $\beta_Q(A, B, C, D) = \beta_Q(B, A, D, C)$; therefore, using the result of Exercise 143 (b), there exists a projectivity $g: \mathbb{P}^2(\mathbb{K}) \to \mathbb{P}^2(\mathbb{K})$ such that $g(A) = B$, $g(B) = A$, $g(C) = D$, $g(D) = C$ and $g(Q) = Q$. Then the projectivities g and f, which coincide on A, B, C, D, must coincide on the whole $\mathbb{P}^2(\mathbb{K})$, so that $f(Q) = g(Q) = Q$, as required.

Exercise 145. Let C be a non-degenerate conic of $\mathbb{P}^2(\mathbb{C})$.

(a) Denote by $P, Q, R \in \mathbb{P}^2(\mathbb{C})$ the vertices of a self-polar triangle for C. Prove that there exists a projectivity f of $\mathbb{P}^2(\mathbb{C})$ such that $f(C) = C$, $f(P) = Q$, $f(Q) = P$ and $f(R) = R$.
(b) Given two distinct points $P, Q \notin C$ such that the line joining them is not tangent to C, prove that there exists a projectivity f of $\mathbb{P}^2(\mathbb{C})$ such that $f(C) = C$ and $f(P) = Q$.

Solution. (a) Denote by A and B the points where the line $L(P, R) = \mathrm{pol}(Q)$ meets C (cf. Fig. 4.7). Since the vertices of a self-polar triangle do not belong to the conic and its edges are not tangent to the conic, the points A and B are distinct and different both from P and from R. Similarly denote by D and E the points where the line $L(Q, R) = \mathrm{pol}(P)$ meets C; these points are distinct and different both from R and from Q.

Consider the projectivity f of $\mathbb{P}^2(\mathbb{C})$ such that $f(A) = D$, $f(D) = A$, $f(B) = E$ and $f(E) = B$ (it exists because the points A, B, D, E turn out to be in general position). Then by Exercise 144 we have that $f(C) = C$.

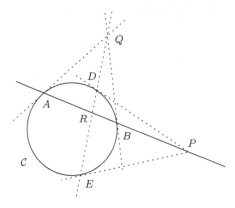

Fig. 4.7 The construction described in the solution of Exercise 145 (a)

Moreover, $f(\mathrm{pol}(Q)) = \mathrm{pol}(P)$, $f(\mathrm{pol}(P)) = \mathrm{pol}(Q)$, so that $f(Q) = P$ and $f(P) = Q$. It follows that $f(\mathrm{pol}(R)) = \mathrm{pol}(R)$ and hence $f(R) = R$.

An alternative proof of statement (a) can be given as follows. Since P, Q, R are in general position, there exists a projective frame of $\mathbb{P}^2(\mathbb{C})$ having P, Q, R as fundamental points. With respect to the homogeneous coordinates defined by this frame, the conic C is represented by the matrix $M = \begin{pmatrix} 1 & 0 & 0 \\ 0 & a & 0 \\ 0 & 0 & b \end{pmatrix}$ with $a, b \in \mathbb{C}^*$. Moreover, a projectivity $f \colon \mathbb{P}^2(\mathbb{C}) \to \mathbb{P}^2(\mathbb{C})$ satisfies the conditions $f(P) = Q$, $f(Q) = P$, $f(R) = R$ if and only if it is represented by a matrix N such that $N = \begin{pmatrix} 0 & c & 0 \\ 1 & 0 & 0 \\ 0 & 0 & d \end{pmatrix}$ with $c, d \in \mathbb{C}^*$.

Observe that $f(C) = C$ if and only if $f^{-1}(C) = C$, i.e. if and only if ${}^t N M N = \lambda M$ for some $\lambda \in \mathbb{C}^*$. An easy computation shows that the latter condition is fulfilled if and only if $a^2 = c^2$ and $bd^2 = ab$. It follows that, if $\alpha \in \mathbb{C}^*$ is a square root of a, the projectivity associated to the invertible matrix $\begin{pmatrix} 0 & a & 0 \\ 1 & 0 & 0 \\ 0 & 0 & \alpha \end{pmatrix}$ satisfies the required properties.

(b) Consider first the case when $Q \in \mathrm{pol}_C(P)$. Then by reciprocity $P \in \mathrm{pol}_C(Q)$; in addition, the lines $\mathrm{pol}_C(P)$ and $\mathrm{pol}_C(Q)$ are distinct, so that they meet at one point R. The points P, Q, R turn out to be the vertices of a self-polar triangle for C; then the thesis follows from part (a).

If instead $Q \notin \mathrm{pol}_C(P)$, let $T = \mathrm{pol}_C(P) \cap \mathrm{pol}_C(Q)$. Since $T \in \mathrm{pol}_C(P)$ and $T \notin C$ (which happens because $L(P, Q)$ is not tangent to C), the points P and T are two vertices of a self-polar triangle for C. Then, as just proved, there exists a projectivity g of $\mathbb{P}^2(\mathbb{C})$ such that $g(C) = C$ and $g(P) = T$. Similarly, since $T \in \mathrm{pol}_C(Q)$, there exists a projectivity h of $\mathbb{P}^2(\mathbb{C})$ such that $h(C) = C$ and $h(Q) = T$. Then the projectivity $f = h^{-1} \circ g$ fulfils the required properties (Fig. 4.8).

Note. Using Exercise 165, it is possible to show that the thesis of part (b) holds also when the line passing through P and Q is tangent to C.

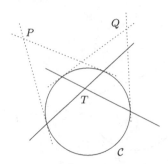

Fig. 4.8 Exercise 145: how to derive part (b) from part (a) when $Q \notin \mathrm{pol}_C(P)$

Exercise 146. Consider in $\mathbb{P}^2(\mathbb{C})$ the points $A = [1, 0, 0]$, $B = [0, 1, -1]$ and $C = [1, 2, -3]$. Find, if it exists, a non-degenerate conic $\mathcal{C}$ of $\mathbb{P}^2(\mathbb{C})$ such that

(i) A, B, C are the vertices of a self-polar triangle for $\mathcal{C}$;
(ii) $\mathcal{C}$ passes through the point $P = [1, 1, 1]$ and it is tangent at P to the line of equation $3x_0 - 4x_1 + x_2 = 0$.

Solution. The conic $\mathcal{C}$ we are looking for has equation ${}^tXMX = 0$ with

$$M = \begin{pmatrix} a & b & c \\ b & d & e \\ c & e & f \end{pmatrix}.$$

The polar of the point A with respect to $\mathcal{C}$ has equation $ax_0 + bx_1 + cx_2 = 0$; in order for condition (i) to hold, necessarily this polar must coincide with the line $L(B, C)$ which has equation $x_0 + x_1 + x_2 = 0$. Thus we find the first condition $a = b = c$, so that $M = \begin{pmatrix} a & a & a \\ a & d & e \\ a & e & f \end{pmatrix}$.

We now find further necessary conditions that a, d, e, f must fulfil if $\mathcal{C}$ satisfies the properties required in the statement of the exercise.

The polar of the point C with respect to the conic has equation $(a + 2d - 3e)x_1 + (a + 2e - 3f)x_2 = 0$. As the polar of A passes through C, by reciprocity the polar of C passes through A; in order to impose that $\mathrm{pol}_{\mathcal{C}}(C)$ is the line $L(A, B)$ it is sufficient to impose that it passes through the point B. Thus we find the condition $2d - 5e + 3f = 0$. By imposing these conditions we get, again by reciprocity, that the polar of B is the line $L(A, C)$.

The conic $\mathcal{C}$ passes through P if and only if $5a + d + 2e + f = 0$ and, if $\mathcal{C}$ is non-degenerate, $\mathrm{pol}_{\mathcal{C}}(P)$ coincides with the only tangent to $\mathcal{C}$ at P and it has equation $3ax_0 + (a + d + e)x_1 + (a + e + f)x_2 = 0$. In order for this line to coincide with the line r of equation $3x_0 - 4x_1 + x_2 = 0$, it is sufficient to impose that $\mathrm{pol}_{\mathcal{C}}(P)$ passes through another point of r distinct from P, for instance through $[1, 0, -3]$. Thus we find the additional condition $e + f = 0$.

The linear system having as equations the conditions found above has as solutions all multiples of the quadruple $a = 1$, $d = -4$, $e = -1$, $f = 1$. The conic associated to the matrix $M = \begin{pmatrix} 1 & 1 & 1 \\ 1 & -4 & -1 \\ 1 & -1 & 1 \end{pmatrix}$ is non-degenerate and satisfies all the required conditions.

Exercise 147. Let $\mathcal{C}$ and $\mathcal{D}$ be two non-degenerate conics of $\mathbb{P}^2(\mathbb{K})$ which meet at 4 distinct points.

(a) Show that no line in $\mathbb{P}^2(\mathbb{K})$ is tangent to all conics of the pencil generated by $\mathcal{C}$ and $\mathcal{D}$.
(b) Show that there exists a point $Q \in \mathbb{P}^2(\mathbb{K})$ such that $\mathrm{pol}_{\mathcal{C}}(Q) = \mathrm{pol}_{\mathcal{D}}(Q)$ and that the point Q lies neither on $\mathcal{C}$ nor on $\mathcal{D}$.

Solution. (a) Denote by A, B, C, D the four points where the conics C and D meet. Since the two conics are non-degenerate, the points A, B, C, D are in general position. The pencil $\mathcal{F}$ generated by C and D contains three degenerate conics (cf. Sect. 1.9.7), that is the conics

$$\mathcal{A}_1 = L(A, B) + L(C, D), \quad \mathcal{A}_2 = L(A, D) + L(B, C), \quad \mathcal{A}_3 = L(A, C) + L(B, D).$$

Assume by contradiction that there exists a line r tangent to all conics of $\mathcal{F}$. Since in particular r is tangent to the degenerate conic $\mathcal{A}_1$, necessarily r must pass through the point $M_1 = L(A, B) \cap L(C, D)$.

The line r is tangent also to the two degenerate conics $\mathcal{A}_2$ and $\mathcal{A}_3$. Hence r must pass also through the points $M_2 = L(A, D) \cap L(B, C)$ and $M_3 = L(A, C) \cap L(B, D)$. In particular the points M_1, M_2, M_3 are collinear, which contradicts the fact that A, B, C, D are in general position (cf. Exercise 6). Thus (a) is proved.

(b) Recall that the map $\text{pol}_C \colon \mathbb{P}^2(\mathbb{K}) \to \mathbb{P}^2(\mathbb{K})^*$ which associates to $P \in \mathbb{P}^2(\mathbb{K})$ the polar of P with respect to the non-degenerate conic C is a projective isomorphism (cf. Sect. 1.8.2). Therefore, the composite map $\text{pol}_D^{-1} \circ \text{pol}_C$ is a projectivity of $\mathbb{P}^2(\mathbb{K})$ and so it has at least a fixed point Q (cf. Sect. 1.2.5), that is a point such that $\text{pol}_C(Q) = \text{pol}_D(Q)$ as required.

Assume by contradiction that Q belongs to C. Then the line $\text{pol}_C(Q)$ is tangent to C at Q. Since $\text{pol}_C(Q) = \text{pol}_D(Q)$, the point Q belongs also to the line $\text{pol}_D(Q)$, so that Q lies also on the conic D and the line $\text{pol}_D(Q)$ is tangent also to D at the point Q which belongs to both conics. Then the line is tangent to all conics of the pencil $\mathcal{F}$, which contradicts part (a).

Exercise 148. Let P_1, P_2, P_3, P_4 be points of $\mathbb{P}^2(\mathbb{K})$ in general position. Given points O, O' in $\mathbb{P}^2(\mathbb{K})$ such that P_1, P_2, P_3, P_4, O and P_1, P_2, P_3, P_4, O' are in general position, let λ (resp. λ') denote the cross-ratio of the four lines passing through O (resp. O') and through P_1, P_2, P_3, P_4. Show that $\lambda = \lambda'$ if and only if there exists a conic passing through $P_1, P_2, P_3, P_4, O, O'$.

Solution (1). For $i = 1, 2, 3, 4$ let $t_i = L(O, P_i)$ and $s_i = L(O', P_i)$.

Assume there exists a conic C passing through the points $P_1, P_2, P_3, P_4, O, O'$; necessarily this conic is non-degenerate. Then

$$\lambda = \beta(t_1, t_2, t_3, t_4) = \beta(s_1, s_2, s_3, s_4) = \lambda'$$

as an immediate consequence of Exercise 142 (b).

Conversely suppose $\lambda = \lambda'$.

We will prove that the points $P_1, P_2, P_3, P_4, O, O'$ lie on a conic by showing analytically that the locus $\mathcal{W}$ of points Q such that P_1, P_2, P_3, P_4, Q are in general position and such that $\beta(L(Q, P_1), L(Q, P_2), L(Q, P_3), L(Q, P_4)) = \lambda$ is contained in a conic passing through P_1, P_2, P_3, P_4.

So let $Q \in \mathcal{W}$. Since the points P_i are in general position, there exists a system of homogeneous coordinates in which

$$P_1 = [1, 0, 0], \quad P_2 = [0, 1, 0], \quad P_3 = [0, 0, 1], \quad P_4 = [1, 1, 1].$$

Then the line $r = L(P_1, P_2)$ has equation $x_2 = 0$; since $Q \notin r$ we have that $Q = [y_0, y_1, y_2]$ with $y_2 \neq 0$. We can compute the cross-ratio of the four lines $L(Q, P_i)$ as the cross-ratio of their intersections with the transversal line r. We easily compute that

$$R_3 = L(Q, P_3) \cap r = [y_0, y_1, 0] \quad \text{and} \quad R_4 = L(Q, P_4) \cap r = [y_2 - y_0, y_2 - y_1, 0].$$

The hypothesis that the points P_1, P_2, P_3, P_4 are in general position implies that the points P_1, P_2, R_3, R_4 are distinct (and hence their cross-ratio is well defined and $y_2 - y_0 \neq 0$, $y_2 - y_1 \neq 0$). Therefore $\beta(L(Q, P_1), L(Q, P_2), L(Q, P_3), L(Q, P_4)) = \beta(P_1, P_2, R_3, R_4) = \beta([1, 0], [0, 1], [y_0, y_1], [y_2 - y_0, y_2 - y_1]) = \dfrac{y_2 - y_1}{y_1} \dfrac{y_0}{y_2 - y_0}$.

As $Q \in \mathcal{W}$, then $\dfrac{y_2 - y_1}{y_1} \dfrac{y_0}{y_2 - y_0} = \lambda$, that is Q belongs to the conic of equation $(x_2 - x_1)x_0 - \lambda x_1(x_2 - x_0) = 0$; one can immediately check that this conic passes through P_1, P_2, P_3, P_4.

Solution (2). Like in Solution (1) one shows that, if the points P_1, P_2, P_3, P_4, O and O' lie on a conic $\mathcal{C}$, then $\lambda = \lambda'$.

The opposite implication can be proved synthetically as follows. Let $\mathcal{C}$ be the (unique and non-degenerate) conic passing through P_1, P_2, P_3, P_4, O and let $\mathcal{C}'$ be the (unique and non-degenerate) conic passing through P_1, P_2, P_3, P_4, O' (cf. Fig. 4.9).

Let $\psi \colon \mathcal{F}_O \to \mathcal{F}_{P_1}$ denote the projective isomorphism defined in Exercise 142 (a), starting from the conic $\mathcal{C}$ and the points O and P_1. If we denote by $T_{P_1}(\mathcal{C})$ the tangent

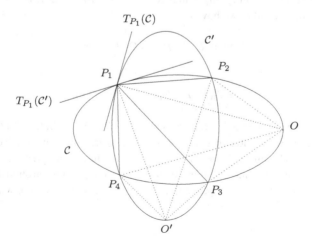

Fig. 4.9 Solution (2) of Exercise 148

to $\mathcal{C}$ at the point P_1, then

$$\psi(t_1) = T_{P_1}(\mathcal{C}) \quad \text{and} \quad \psi(t_j) = L(P_1, P_j) \text{ for } j = 2, 3, 4.$$

Since the cross-ratio is invariant under projective isomorphisms, we have that

$$\lambda = \beta(T_{P_1}(\mathcal{C}), L(P_1, P_2), L(P_1, P_3), L(P_1, P_4)).$$

Similarly, let $\psi': \mathcal{F}_{O'} \to \mathcal{F}_{P_1}$ denote the projective isomorphism defined in Exercise 142 (a), starting from the conic $\mathcal{C}'$ and the points O' and P_1. If we denote by $T_{P_1}(\mathcal{C}')$ the tangent to $\mathcal{C}'$ at the point P_1, then

$$\psi'(s_1) = T_{P_1}(\mathcal{C}') \quad \text{and} \quad \psi'(s_j) = L(P_1, P_j) \text{ for } j = 2, 3, 4$$

and hence
$$\lambda' = \beta(T_{P_1}(\mathcal{C}'), L(P_1, P_2), L(P_1, P_3), L(P_1, P_4)).$$

Since the lines $L(P_1, P_2), L(P_1, P_3), L(P_1, P_4)$ are distinct, the hypothesis $\lambda = \lambda'$ implies that $T_{P_1}(\mathcal{C}) = T_{P_1}(\mathcal{C}')$, i.e. the conics $\mathcal{C}$ and $\mathcal{C}'$ have the same tangent r at P_1. Since there exists only one conic passing through P_1, P_2, P_3, P_4 and with tangent r at P_1, necessarily $\mathcal{C} = \mathcal{C}'$.

Solution (3). Like in Solution (1) one shows that, if the points P_1, P_2, P_3, P_4, O and O' lie on a conic $\mathcal{C}$, then $\lambda = \lambda'$.

We are now going to prove the opposite implication. Evidently if $O = O'$ the result is trivially true. If $O \neq O'$, observe that the line $L(O, O')$ cannot pass through any of the P_i. Namely, suppose by contradiction that $L(O, O')$ passes through at least one of those points, for instance P_1, and set $r = L(P_2, P_3)$ and $R = r \cap L(O, O')$. The hypotheses immediately imply that $R \notin \{P_1, P_2, P_3, O, O'\}$. If we compute λ by using the transversal r, we have that

$$\lambda = \beta(R, P_2, P_3, L(O, P_4) \cap r).$$

Similarly
$$\lambda' = \beta(R, P_2, P_3, L(O', P_4) \cap r).$$

From the hypothesis $\lambda = \lambda'$ it follows that $L(O, P_4) \cap r = L(O', P_4) \cap r$, that is O, O', P_4 are collinear and hence also O, O', P_1, P_4 are collinear, which contradicts our hypotheses. Therefore, $P_1, P_2, P_3, P_4, O, O'$ are in general position.

Denote by $\mathcal{C}$ the (unique and non-degenerate) conic passing through P_1, P_2, P_3, O, O' (cf. Fig. 4.10) and let us prove that, if $\lambda = \lambda'$, then $P_4 \in \mathcal{C}$ too, which proves the thesis.

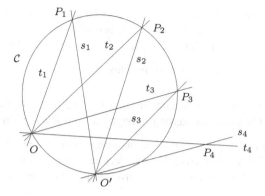

Fig. 4.10 Solution (3) of Exercise 148

Let $\psi : \mathcal{F}_O \to \mathcal{F}_{O'}$ denote the projective isomorphism defined in Exercise 142 (a), starting from the conic $\mathcal{C}$ and the points O and O'. Therefore, $\psi(t_i) = s_i$ for $i = 1, 2, 3$. Since the cross-ratio is invariant under projective isomorphisms, we have

$$\beta(t_1, t_2, t_3, t_4) = \beta(s_1, s_2, s_3, \psi(t_4)).$$

From the hypothesis $\lambda = \lambda'$, i.e. $\beta(t_1, t_2, t_3, t_4) = \beta(s_1, s_2, s_3, s_4)$, it follows that $\psi(t_4) = s_4$.

By definition, $t_4 \cap \psi(t_4)$ is a point and belongs to $\mathcal{C}$. But $P_4 \in t_4 \cap s_4 = t_4 \cap \psi(t_4)$ and hence $P_4 \in \mathcal{C}$.

Exercise 149. Consider the pencil $\mathcal{F}$ of conics of $\mathbb{P}^2(\mathbb{C})$ of equations

$$\lambda x_1(x_0 - x_1 + x_2) + \mu x_2(2x_0 - x_1) = 0, \quad [\lambda, \mu] \in \mathbb{P}^1(\mathbb{C}).$$

Let $P = [1, 3, 1]$ and let r be the line $x_0 - x_1 = 0$. Check that the map $f : \mathcal{F} \to r$ which associates to $\mathcal{C} \in \mathcal{F}$ the point $\mathrm{pol}_\mathcal{C}(P) \cap r$ is well defined and it is a projective isomorphism.

Solution. The conic $\mathcal{C}_{\lambda, \mu}$ of the pencil is represented by the matrix

$$A_{\lambda, \mu} = \begin{pmatrix} 0 & \lambda & 2\mu \\ \lambda & -2\lambda & \lambda - \mu \\ 2\mu & \lambda - \mu & 0 \end{pmatrix}.$$

Since for every $[\lambda, \mu]$ in $\mathbb{P}^1(\mathbb{C})$ the vector $A_{\lambda, \mu} \begin{pmatrix} 1 \\ 3 \\ 1 \end{pmatrix} = \begin{pmatrix} 3\lambda + 2\mu \\ -4\lambda - \mu \\ 3\lambda - \mu \end{pmatrix}$ is non-zero

(in other words P is non-singular for all conics in the pencil), then $\mathrm{pol}_{\mathcal{C}_{\lambda, \mu}}(P)$ is a line of equation ${}^tPA_{\lambda, \mu}X = 0$ (cf. Sect. 1.8.2). Moreover, $\mathrm{pol}_{\mathcal{C}_{\lambda, \mu}}(P) \neq r$ for every $[\lambda, \mu] \in \mathbb{P}^1(\mathbb{C})$, so that the map f is well defined.

Easy computations show that the lines $\mathrm{pol}_{\mathcal{C}_{\lambda,\mu}}(P)$ and r meet at the point $f(P) = [3\lambda - \mu, 3\lambda - \mu, \lambda - \mu]$. With respect to the homogeneous coordinates λ, μ on $\mathcal{F}$ and x_0, x_2 on r, the map is thus represented by $f([\lambda, \mu]) = [3\lambda - \mu, \lambda - \mu]$. As $\det \begin{pmatrix} 3 & -1 \\ 1 & -1 \end{pmatrix} \neq 0$, f is a projective isomorphism.

Exercise 150. Let $\mathcal{C}$ be a non-degenerate conic of $\mathbb{P}^2(\mathbb{C})$; assume that r is a line not tangent to $\mathcal{C}$ and s is a line not passing through the pole R of r with respect to $\overline{\mathcal{C}}$.

(a) Check that the map $f: s \to r$ which associates to every point $P \in s$ the point $\mathrm{pol}_{\mathcal{C}}(P) \cap r$ is well defined and is a projective isomorphism.
(b) If $s = r$, prove that f is a non-trivial involution and describe its fixed points.

Solution. (a) As $R \notin s$, for every $P \in s$ we have that $P \neq R$ and hence $\mathrm{pol}_{\mathcal{C}}(P) \neq \mathrm{pol}_{\mathcal{C}}(R) = r$; therefore the lines $\mathrm{pol}_{\mathcal{C}}(P)$ and r meet only at one point and thus f is well defined.

The map that associates to every point $P \in s$ the line $\mathrm{pol}_{\mathcal{C}}(P)$ is the restriction to s of the projective isomorphism $\mathrm{pol}_{\mathcal{C}}: \mathbb{P}^2(\mathbb{C}) \to \mathbb{P}^2(\mathbb{C})^*$ (cf. Sect. 1.8.2). Then this restriction is a projective isomorphism between s and a 1-dimensional subspace of $\mathbb{P}^2(\mathbb{C})^*$, that is a pencil $\mathcal{F}$ of lines of $\mathbb{P}^2(\mathbb{C})$. The fact that f is well defined ensures that the line r does not belong to the pencil. Thus the map f coincides with the composition of the projective isomorphism $\mathrm{pol}_{\mathcal{C}}|_s: s \to \mathcal{F}$ with the parametrization of $\mathcal{F}$ by means of the transversal r. Since this parametrization is an isomorphism (cf. Exercise 32 and the following Note), f is a projective isomorphism too.

(b) If $s = r$, for every $P \in r$ we have that $f(P) \in \mathrm{pol}_{\mathcal{C}}(P)$ and hence, by reciprocity, $P \in \mathrm{pol}_{\mathcal{C}}(f(P))$. As a consequence $f(f(P)) = \mathrm{pol}_{\mathcal{C}}(f(P)) \cap r = P$ and so f is an involution. Furthermore, $f(P) = P$ if and only if the line $\mathrm{pol}_{\mathcal{C}}(P)$ passes through P. This latter fact occurs if and only if P lies on the conic. Then the fixed-point set of f is $r \cap \mathcal{C}$ and hence, since r is not tangent to $\mathcal{C}$, either it is empty or it contains exactly two points. In any case f is different from the identity map.

Exercise 151. Assume that $r \subseteq \mathbb{P}^2(\mathbb{C})$ is a projective line and $f: r \to r$ is a non-trivial involution. Show that there exists a non-degenerate conic $\mathcal{C}$ not tangent to r and such that $f(P) = \mathrm{pol}_{\mathcal{C}}(P) \cap r$ for every $P \in r$.

Solution. The map f, being a non-trivial involution, has two fixed points A, B (cf. Exercise 23). Let $\mathcal{C}$ be any non-degenerate conic of $\mathbb{P}^2(\mathbb{C})$ passing through A and through B; in particular this conic is not tangent to $r = L(A, B)$. Denote by $g: r \to r$ the map which associates to every point $P \in r$ the point $\mathrm{pol}_{\mathcal{C}}(P) \cap r$; as shown in Exercise 150, g is a non-trivial involution of r which fixes A and B. Since there exists exactly one non-trivial involution of r having A and B as fixed points (cf. Exercise 24), then $f = g$ and hence the thesis.

Exercise 152. Assume that $\mathcal{C}$ is a non-degenerate conic of $\mathbb{P}^2(\mathbb{C})$ and $P \in \mathbb{P}^2(\mathbb{C})$ is a point not lying on $\mathcal{C}$. Let $f: \mathbb{P}^2(\mathbb{C}) \to \mathbb{P}^2(\mathbb{C})$ be a projectivity such that $f(s \cap \mathcal{C}) = s \cap \mathcal{C}$ for every line s passing through P. Prove that P is a fixed point of f and that the restriction of f to the polar of P with respect to $\mathcal{C}$ is the identity map.

Solution. Denote by r_1, r_2 the two tangents to C passing through P; they meet the conic at the points A and B, respectively. Any line s passing through P and not tangent to C intersects the conic at two distinct points; since by hypothesis these two points either are fixed for f or are exchanged by f, the line s joining them is invariant under f. Therefore, if s_1, s_2 are two lines passing through P and not tangent to C, we have $f(s_1) = s_1$, $f(s_2) = s_2$ and hence $f(P) = f(s_1 \cap s_2) = s_1 \cap s_2 = P$.

If we apply the hypothesis to the lines r_1, r_2, we get that also the points of tangency A and B are fixed by f; in particular the line $\mathrm{pol}_C(P) = L(A, B)$ is f-invariant. If R is any point on $\mathrm{pol}_C(P)$ distinct both from A and from B, we have $R = \mathrm{pol}_C(P) \cap L(P, R)$; since both $\mathrm{pol}_C(P)$ and $L(P, R)$ are f-invariant, then $f(R) = R$.

Exercise 153. Assume that C is a non-degenerate conic of $\mathbb{P}^2(\mathbb{K})$. Let P_1, P_2, P_3, P_4 be distinct points of C and for any $i \neq j$ let $s_{ij} = L(P_i, P_j)$. Show that the points $A = s_{12} \cap s_{34}$, $B = s_{13} \cap s_{24}$, $C = s_{14} \cap s_{23}$ are vertices of a self-polar triangle for C.

Solution (1). Since A, B and C can be interchanged in the statement, it suffices to prove that $\mathrm{pol}_C(A) = L(B, C)$.

Since C is non-degenerate, the distinct points P_1, P_2, P_3, P_4 are in general position. So let f be the only projectivity of $\mathbb{P}^2(\mathbb{C})$ such that

$$f(P_1) = P_2, \quad f(P_2) = P_1, \quad f(P_3) = P_4, \quad f(P_4) = P_3.$$

The lines $L(P_1, P_2)$ and $L(P_3, P_4)$ turn out to be f-invariant, so that A is a fixed point of f (cf. Fig. 4.11). Moreover, $f(B) = f(L(P_1, P_3) \cap L(P_2, P_4)) = L(P_2, P_4) \cap L(P_1, P_3) = B$, that is also B is fixed by f. Arguing in a similar way one shows that $f(C) = C$ too. As a consequence the line $L(B, C)$ is invariant under f.

More precisely we can see that all points of $L(B, C)$ are fixed. Namely the points $M = L(B, C) \cap L(P_1, P_2)$ and $N = L(B, C) \cap L(P_3, P_4)$, obtained as intersections of invariant lines, are fixed points for f. It can be easily seen that the points

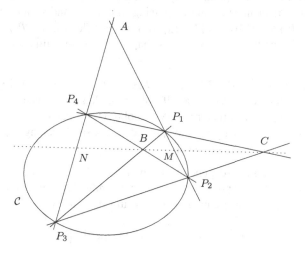

Fig. 4.11 The points A, B, C are vertices of a self-polar triangle for C

B, C, M, N are distinct, so that the restriction of f to $L(B, C)$, which is a projectivity of this line having four fixed points, is the identity map of $L(B, C)$.

Since the points P_1, P_2, P_3, P_4 are in general position and fixed for f^2, then $f^2 = \text{Id}$. Moreover, by Exercise 144, $f(\mathcal{C}) = \mathcal{C}$. Hence, by Exercise 133 (a), the fixed-point set of f is the union of a line not tangent to $\mathcal{C}$ and the pole of that line. It follows that the fixed point A is the pole of the line $L(B, C)$ consisting of fixed points, i.e. $\text{pol}_\mathcal{C}(A) = L(B, C)$, as required.

Solution (2). As observed at the beginning of Solution (1), it is sufficient to prove that $\text{pol}_\mathcal{C}(A) = L(B, C)$.

The conic $\mathcal{C}$ belongs to the pencil generated by the degenerate conics $\mathcal{D}_1 = s_{12} + s_{34}$ and $\mathcal{D}_2 = s_{13} + s_{24}$. If ${}^t X M_1 X = 0$ is an equation of $\mathcal{D}_1$ and ${}^t X M_2 X = 0$ is an equation of $\mathcal{D}_2$, then there exist $\lambda, \mu \in \mathbb{C}$ (not simultaneously zero) such that ${}^t X (\lambda M_1 + \mu M_2) X = 0$ is an equation of $\mathcal{C}$. As A is singular for $\mathcal{D}_1$, then $M_1 A = 0$; similarly, since B is singular for $\mathcal{D}_2$, then $M_2 B = 0$. As a consequence ${}^t A (\lambda M_1 + \mu M_2) B = \lambda {}^t A M_1 B + \mu {}^t A M_2 B = 0$, which proves that $B \in \text{pol}_\mathcal{C}(A)$. Arguing in a similar way for the conics $\mathcal{D}_1$ and $\mathcal{D}_3 = s_{14} + s_{23}$ (which has C as a singular point), one shows that $C \in \text{pol}_\mathcal{C}(A)$ and hence $\text{pol}_\mathcal{C}(A) = L(B, C)$.

Solution (3). Since $\mathcal{C}$ is non-degenerate, the points P_1, P_2, P_3, P_4 are in general position and hence they form a projective frame. In the induced system of homogeneous coordinates $\mathcal{C}$ is defined by an equation of the form $x_0 x_1 + a x_1 x_2 - (1 + a) x_0 x_2 = 0$, with $a \in \mathbb{K} \setminus \{0, -1\}$, and thus it is represented by the matrix
$$\begin{pmatrix} 0 & 1 & -1-a \\ 1 & 0 & a \\ -1-a & a & 0 \end{pmatrix}.$$ Moreover, $A = [1, 1, 0]$, $B = [1, 0, 1]$ and $C = [0, 1, 1]$.
Now it is immediate to check that the points A, B and C are pairwise conjugate with respect to $\mathcal{C}$, i.e. they form a self-polar triangle.

Exercise 154. Let A be a point not lying on a non-degenerate conic $\mathcal{C}$ of $\mathbb{P}^2(\mathbb{K})$. Consider two distinct lines r_1 and r_2 passing through A and secant to $\mathcal{C}$. Denote $\mathcal{C} \cap r_1 = \{P_1, P_2\}$ and $\mathcal{C} \cap r_2 = \{P_3, P_4\}$. If S_1 is the pole of $L(P_1, P_3)$ and S_2 is the pole of $L(P_2, P_4)$ with respect to $\mathcal{C}$, show that A, S_1, S_2 are collinear.

Solution (1). Let $B = L(P_1, P_3) \cap L(P_2, P_4)$. By Exercise 153 (cf. Fig. 4.11), the points A and B are two vertices of a self-polar triangle for $\mathcal{C}$; hence $A \in \text{pol}_\mathcal{C}(B)$.

Since $B \in L(P_1, P_3)$, by reciprocity $S_1 \in \text{pol}_\mathcal{C}(B)$; similarly $S_2 \in \text{pol}_\mathcal{C}(B)$. Then $\text{pol}_\mathcal{C}(B) = L(S_1, S_2)$. Having already proved that also A belongs to $\text{pol}_\mathcal{C}(B) = L(S_1, S_2)$, then A, S_1, S_2 are collinear.

Solution (2). Since $\mathcal{C}$ is non-degenerate, the points P_1, P_2, P_3, P_4 are in general position and hence they form a projective frame. In the induced system of homogeneous coordinates $\mathcal{C}$ is defined by an equation of the form $x_0 x_1 + a x_1 x_2 - (1 + a) x_0 x_2 = 0$, with $a \in \mathbb{K} \setminus \{0, -1\}$ and so it is represented by the matrix $\begin{pmatrix} 0 & 1 & -1-a \\ 1 & 0 & a \\ -1-a & a & 0 \end{pmatrix}.$

In addition, $L(P_1, P_3)$ has equation $x_1 = 0$, $L(P_2, P_4)$ has equation $x_0 - x_2 = 0$ and $A = [1, 1, 0]$. Easy computations yield that $S_1 = [a, 1 + a, 1]$ and $S_2 = [a, a - 1, -1]$; since A, S_1 and S_2 belong to the line of equation $x_0 - x_1 + x_2 = 0$, they are collinear.

Exercise 155. Given a pencil of conics $\mathcal{F}$ of $\mathbb{P}^2(\mathbb{C})$ containing at least one non-degenerate conic, show that the following conditions are equivalent:

(i) there exist $Q_1, Q_2, Q_3 \in \mathbb{P}^2(\mathbb{C})$ that form a self-polar triangle for every conic of $\mathcal{F}$;
(ii) the base-point set of $\mathcal{F}$ consists of either four distinct points or two points each counted twice (cf. Sect. 1.9.7).

Solution. Firstly observe that two points $P, Q \in \mathbb{P}^2(\mathbb{C})$ are conjugate with respect to every conic $\mathcal{C} \in \mathcal{F}$ if and only if they are conjugate with respect to two distinct conics $\mathcal{C}_1$ and $\mathcal{C}_2$ of $\mathcal{F}$. Namely, if A_i, $i = 1, 2$, is a symmetric matrix representing $\mathcal{C}_i$, the points P and Q are conjugate with respect to $\mathcal{C}_i$ if and only if ${}^t P A_i Q = 0$ for $i = 1, 2$, and hence if and only if ${}^t P(\lambda A_1 + \mu A_2)Q = 0$ for every $[\lambda, \mu] \in \mathbb{P}^1(\mathbb{C})$.

As a consequence three points of the plane are the vertices of a self-polar triangle for every conic of $\mathcal{F}$ if and only if they verify the same property for at least two conics of the pencil.

Let us now show that condition (ii) implies condition (i), by examining separately the case when $\mathcal{F}$ has four distinct base points and the case when $\mathcal{F}$ has two base points each counted twice.

If $\mathcal{F}$ has four distinct base points P_1, P_2, P_3, P_4, then by Exercise 153 the points $Q_1 = L(P_1, P_2) \cap L(P_3, P_4)$, $Q_2 = L(P_1, P_3) \cap L(P_2, P_4)$ and $Q_3 = L(P_1, P_4) \cap L(P_2, P_3)$ are the vertices of a self-polar triangle for every non-degenerate conic of $\mathcal{F}$ and so for infinitely many conics of $\mathcal{F}$. Then, because of the previous observation, Q_1, Q_2, Q_3 are the vertices of a self-polar triangle for every conic of $\mathcal{F}$.

If $\mathcal{F}$ has two base points P_1 and P_2 each counted twice, denote by $\mathcal{C}_1 \in \mathcal{F}$ the singular conic $2L(P_1, P_2)$ and by $\mathcal{C}_2 \in \mathcal{F}$ any non-degenerate conic.

Let Q_1 be the pole of the line $L(P_1, P_2)$ with respect to $\mathcal{C}_2$. As $L(P_1, P_2)$ intersects $\mathcal{C}_2$ at the two distinct points P_1 and P_2, the point Q_1 does not belong to $\mathcal{C}_2$. Then (cf. Sect. 1.8.4 or Exercise 170) there exist $Q_2, Q_3 \in \mathbb{P}^2(\mathbb{C})$ such that Q_1, Q_2, Q_3 are the vertices of a self-polar triangle for $\mathcal{C}_2$. By reciprocity $Q_2, Q_3 \in \mathrm{pol}_{\mathcal{C}_2}(Q_1) = L(P_1, P_2)$. Therefore, the points Q_2 and Q_3 are both singular for $\mathcal{C}_1$, and hence they are conjugate to any point of the plane with respect to $\mathcal{C}_1$. Then Q_1, Q_2, Q_3 are the vertices of a self-polar triangle both for $\mathcal{C}_1$ and for $\mathcal{C}_2$ and consequently for any conic $\mathcal{C} \in \mathcal{F}$.

Suppose now that Q_1, Q_2, Q_3 are points satisfying condition (i); we are going to show that condition (ii) holds. In a system of homogeneous coordinates x_0, x_1, x_2 having Q_1, Q_2, Q_3 as fundamental points, a conic $\mathcal{C} \in \mathcal{F}$ is given by an equation of the form $ax_0^2 + bx_1^2 + cx_2^2 = 0$, with $a, b, c \in \mathbb{C}$ not simultaneously zero. Observe that $\mathcal{C}$ is non-degenerate if and only if $abc \neq 0$. Since by hypothesis $\mathcal{F}$ contains a non-degenerate conic, up to scaling the coordinates we may assume that $\mathcal{F}$ contains the conic $\mathcal{C}_1$ of equation $x_0^2 + x_1^2 + x_2^2 = 0$. If $\mathcal{C}_2 \in \mathcal{F}$ is a degenerate conic, up to permuting the coordinates we may assume that $\mathcal{C}_2$ has equation $x_0^2 + dx_1^2 = 0$ for some $d \in \mathbb{C}$. It can be immediately checked that $C_1 \cap C_2$ consists of four distinct points if $d \neq 0, 1$ and consists of two points of multiplicity 2 if $d = 0$ or 1.

Note. If $\mathcal{F}$ is a pencil of conics of $\mathbb{P}^2(\mathbb{R})$, a sufficient condition for the existence of a self-polar triangle for all conics of $\mathcal{F}$ is the property that $\mathcal{F}$ contains a conic C_1 which is non-degenerate and empty. Namely, if A_1 is a symmetric matrix associated to C_1, we may assume that A_1 is positive definite. If $C_2 \in \mathcal{F}$ is a conic distinct from C_1 and represented by a symmetric matrix A_2, by the Spectral theorem there is a change of homogeneous coordinates on $\mathbb{P}^2(\mathbb{R})$ that simultaneously diagonalizes A_1 and A_2. Then the fundamental points of this projective frame are the vertices of a self-polar triangle for all conics of $\mathcal{F}$.

Note that, since C_1 is empty, the intersection set between $(C_1)_\mathbb{C}$ and $(C_2)_\mathbb{C}$ in $\mathbb{P}^2(\mathbb{C})$ consists of one or two pairs of conjugate points (cf. Sect. 1.9.7), and thus the pencil of complex conics generated by $(C_1)_\mathbb{C}$ and $(C_2)_\mathbb{C}$ fulfils condition (ii) of the statement.

Exercise 156. If C and D are two distinct non-degenerate conics of $\mathbb{P}^2(\mathbb{K})$, denote by $\mathcal{F}$ the pencil of conics generated by C and D. For every $P \in \mathbb{P}^2(\mathbb{K})$ let $f(P) = Q$ if $\mathrm{pol}_C(P) = \mathrm{pol}_D(Q)$.

(a) Show that f defines a projectivity of $\mathbb{P}^2(\mathbb{K})$.
(b) Show that P is a fixed point of f if and only if P is singular for a conic of the pencil $\mathcal{F}$.
(c) What configurations of points in $\mathbb{P}^2(\mathbb{K})$ may coincide with the fixed-point set of f?

Solution. (a) Let $^t X A X = 0$ and $^t X B X = 0$ be equations of C and D, respectively, with A, B symmetric invertible matrices of order 3. Since the line $\mathrm{pol}_C(P)$ has equation $^t P A X = 0$, then $\mathrm{pol}_C(P) = \mathrm{pol}_D(Q)$ if and only if there exists $\alpha \in \mathbb{C} \setminus \{0\}$ such that $BQ = \alpha A P$, i.e. $f(P) = Q = \alpha B^{-1} A P$. As the matrix $B^{-1} A$ is invertible, f is well defined and it is a projectivity.

(b) If P is a fixed point of f, there exists $\alpha \in \mathbb{C} \setminus \{0\}$ such that $P = \alpha B^{-1} A P$, i.e. $(B - \alpha A)P = 0$. Then P is a singular point for the conic $^t X(B - \alpha A)X = 0$ of the pencil $\mathcal{F}$.

Conversely, if P is singular for some conic of the pencil, necessarily that conic is different both from C and from D, because each of these two conics is non-degenerate and hence non-singular. Then the conic having P as a singular point has an equation of the form $^t X(B - \alpha A)X = 0$ for some $\alpha \neq 0$. So $(B - \alpha A)P = 0$, that is $P = \alpha B^{-1} A P$, and hence P is a fixed point of f.

(c) Observe first that, as shown in Exercise 44, the fixed-point set of f may coincide either with a finite set consisting of 1, 2 or 3 points, or with a line, or with the union of one line and one point not lying on that line. We are now going to prove that each of these possible situations occurs in correspondence with a suitable choice of the conics C and D, discussing separately the different cases according to the number and the nature of the base points of $\mathcal{F}$ (cf. Sect. 1.9.7).

If $\mathcal{F}$ has 4 base points, then $\mathcal{F}$ contains exactly 3 degenerate conics, and each of them has exactly one singular point. Moreover, the singular points of these conics are distinct, so that, as seen in (b), in this case f has exactly 3 fixed points.

If instead we want f to have exactly 2 fixed points, it is sufficient to choose as C a non-degenerate conic passing through three independent points P_1, P_2, P_3, to define $\mathcal{D}' = L(P_1, P_2) + L(P_1, P_3)$ and to choose as $\mathcal{D}$ any non-degenerate conic (of course distinct from C) belonging to the pencil $\mathcal{F}$ generated by C and $\mathcal{D}'$. In this case P_1, P_2, P_3 are the base points of the pencil, all conics of $\mathcal{F}$ are tangent at P_1 to the same line r and $\mathcal{F}$ contains two degenerate conics: one of them is $\mathcal{D}'$, the other one is the conic $r + L(P_2, P_3)$. Hence the fixed points of f are P_1 and $r \cap L(P_2, P_3)$.

In order to obtain only one fixed point we can argue as follows. Let C be a non-degenerate conic and A, B distinct points of C; denote by t_A the tangent to C at A. If $\mathcal{D}' = t_A + L(A, B)$, let $\mathcal{D} \neq C$ be a non-degenerate conic of the pencil $\mathcal{F}$ generated by C and by $\mathcal{D}'$. Then the only degenerate conic of $\mathcal{F}$ is $\mathcal{D}'$; so A is the only fixed point of f.

Assume now that C is a non-degenerate conic and t_A is the tangent to C at a point $A \in C$; let $\mathcal{F}$ be the pencil generated by C and $2t_A$. The only degenerate conic of $\mathcal{F}$ is $2t_A$, whose singular locus coincides with t_A. In order to obtain a line of fixed points of f it is so sufficient to choose as $\mathcal{D}$ any non-degenerate conic of $\mathcal{F}$ distinct from C.

Finally, given non-collinear points $A, B, C \in \mathbb{P}^2(\mathbb{C})$, let $\mathcal{F}$ be the pencil generated by $C' = L(A, B) + L(A, C)$ and $\mathcal{D}' = 2L(B, C)$. Then C' and $\mathcal{D}'$ are the only degenerate conics of $\mathcal{F}$; so, if C and $\mathcal{D}$ are two distinct and non-degenerate conics of $\mathcal{F}$, the fixed-point set of f coincides with $L(B, C) \cup \{A\}$.

Thus we have proved that all configurations listed at the beginning of the solution of (c) can be in fact obtained as the fixed-point set of f.

Exercise 157. Let A, B, C be points of $\mathbb{P}^2(\mathbb{K})$ in general position. Given a projectivity $f: \mathbb{P}^2(\mathbb{K}) \to \mathbb{P}^2(\mathbb{K})$ such that $f(A) = B$, $f(B) = C$ and $f(C) = A$, prove that:

(a) There is no line r in $\mathbb{P}^2(\mathbb{K})$ such that $f(P) = P$ for every $P \in r$.
(b) $f^3 = \mathrm{Id}$.
(c) There is at least one non-degenerate conic $\mathcal{Q}$ passing through A, B, C and such that $f(\mathcal{Q}) = \mathcal{Q}$.

Solution. (a) From the hypotheses it follows that

$$f(L(A, B)) = L(B, C), \quad f(L(B, C)) = L(A, C), \quad f(L(A, C)) = L(A, B). \tag{4.1}$$

If, by contradiction, there exists a line r of fixed points of f, this line cannot pass through any of the points A, B, C. If we denote by D the point of intersection between $L(A, B)$ and r, then $f(D) = D$ because $D \in r$. This gives a contradiction since $D = f(D) \in f(L(A, B)) = L(B, C)$ and $D \neq B$.

(b) Let R be a fixed point of f (recall that all projectivities of $\mathbb{P}^2(\mathbb{K})$ have a fixed point, cf. Sect. 1.2.5). As observed in the solution of (a), this point cannot lie on any of the lines $L(A, B), L(B, C), L(A, C)$. Then the points A, B, C, R are in general position; since they are fixed for f^3, then $f^3 = \mathrm{Id}$.

(c) The set $\mathcal{W}$ of all conics passing through A, B, C is a linear system of dimension 2 and hence it is projectively isomorphic to $\mathbb{P}^2(\mathbb{K})$. The map f transforms every conic of $\mathcal{W}$ into a conic of $\mathcal{W}$ and acts on the projective space $\mathcal{W}$ as a projectivity (cf. Sect. 1.9.5). Then in $\mathcal{W}$ there exists a fixed point of f, that is a conic $\mathcal{Q}$ such that $f(\mathcal{Q}) = \mathcal{Q}$. Necessarily $\mathcal{Q}$ is non-degenerate: namely, if it were degenerate, it should contain a line passing through two points among A, B, C as an irreducible component, but then by (4.1) it should contain the three lines $L(A, B)$, $L(B, C)$ and $L(A, C)$, which is not possible.

Exercise 158. Consider a non-degenerate conic $\mathcal{C}$ of $\mathbb{P}^2(\mathbb{K})$ and let A, $B \in \mathcal{C}$ be distinct points; denote by r_1 the tangent line to $\mathcal{C}$ at A, by r_2 the tangent line to $\mathcal{C}$ at B and let $R = r_1 \cap r_2$.

(a) Show that for every $X \in r_1 \setminus \{A\}$ there is exactly one line r_X passing through X, tangent to $\mathcal{C}$ and different from r_1.
(b) Let $\psi : r_1 \to r_2$ be the map defined by

    (i) $\psi(A) = R$;
    (ii) $\psi(R) = B$;
    (iii) $\psi(X) = r_X \cap r_2$ for every $X \in r_1 \setminus \{A\}$.

    Show that ψ is a projective isomorphism.

Solution. (a) Let $X \in r_1$, $X \neq A$. By reciprocity A belongs to $\mathrm{pol}_{\mathcal{C}}(X)$, hence $\mathrm{pol}_{\mathcal{C}}(X) \cap \mathcal{C} \neq \emptyset$. In addition, $\mathrm{pol}_{\mathcal{C}}(X)$ is not tangent to $\mathcal{C}$ because $X \notin \mathcal{C}$; therefore, $\mathrm{pol}_{\mathcal{C}}(X)$ intersects $\mathcal{C}$, besides in A, in a second point P_X such that the tangent line r_X to $\mathcal{C}$ at P_X passes through X. Since a point of $\mathbb{P}^2(\mathbb{K})$ lies on at most two lines tangent to a non-degenerate conic, the lines tangent to $\mathcal{C}$ and passing through X are precisely r_1 and r_X. (Note that, if $X = R$, by construction we have $P_X = B$ and $r_X = r_2$).

(b) Since by construction the points A, B and R are not collinear, there exists in $\mathbb{P}^2(\mathbb{K})$ a system of homogeneous coordinates in which $A = [1, 0, 0]$, $B = [0, 1, 0]$, $R = [0, 0, 1]$. As a consequence, r_1 has equation $x_1 = 0$ and r_2 has equation $x_0 = 0$. Imposing that the conic passes through A and B and that the polar of R is the line $L(A, B) = \{x_2 = 0\}$, we obtain that $\mathcal{C}$ has equation ${}^t X M X = 0$ with M of the form

$$M = \begin{pmatrix} 0 & a & 0 \\ a & 0 & 0 \\ 0 & 0 & b \end{pmatrix}$$

with $a, b \in \mathbb{K} \setminus \{0\}$.

Let $X = [y_0, 0, y_2] \in r_1 \setminus \{R, A\}$ (so that $y_0 \neq 0$ and $y_2 \neq 0$). The point P_X where the line r_X is tangent to $\mathcal{C}$ is the point distinct from A where the line $\mathrm{pol}_{\mathcal{C}}(X)$ intersects the conic. As $\mathrm{pol}_{\mathcal{C}}(X)$ has equation $y_0 a x_1 + y_2 b x_2 = 0$, it turns out that P_X has coordinates $[a y_0^2, -2b y_2^2, 2a y_0 y_2]$. Then the line $r_X = L(X, P_X)$ has equation

$$\det \begin{pmatrix} y_0 & ay_0^2 & x_0 \\ 0 & -2by_2^2 & x_1 \\ y_2 & 2ay_0y_2 & x_2 \end{pmatrix} = 0 \quad \text{that is} \quad 2by_2^2x_0 - ay_0^2x_1 - 2by_0y_2x_2 = 0.$$

Hence $\psi(X) = r_X \cap r_2 = [0, 2by_2, -ay_0]$.

With respect to the system of coordinates x_0, x_2 induced on r_1 and to the system of coordinates x_1, x_2 induced on r_2, we have that $\psi([y_0, y_2]) = [2by_2, -ay_0]$; so the restriction of ψ to $r_1 \setminus \{R, A\}$ coincides with the restriction to $r_1 \setminus \{R, A\}$ of the projective isomorphism $f : r_1 \to r_2$ induced by the invertible matrix $\begin{pmatrix} 0 & 2b \\ -a & 0 \end{pmatrix}$. On the other hand it can be easily checked that $f(A) = R$ and $f(R) = B$; therefore, f coincides with ψ and hence ψ is a projective isomorphism too.

Note. Observe that the statement of the previous exercise is the dual statement of Exercise 142 (a).

Exercise 159. Assume that $\mathcal{L}$ is a linear system of conics of $\mathbb{P}^2(\mathbb{C})$ of dimension 2 containing at least one non-degenerate conic. Prove that

$$B(\mathcal{L}) = \{P \in \mathbb{P}^2(\mathbb{C}) \mid P \in \mathcal{C} \ \forall \mathcal{C} \in \mathcal{L}\}$$

is a finite set containing at most 3 points.

Solution. Let $[F_1], [F_2], [F_3]$ be three projectively independent conics of the linear system $\mathcal{L}$, so that every conic of $\mathcal{L}$ has equation $aF_1 + bF_2 + cF_3 = 0$ as $[a, b, c]$ varies in $\mathbb{P}^2(\mathbb{C})$. Since $\mathcal{L}$ contains at least one non-degenerate conic, we can assume that the conic $[F_1]$ is non-degenerate.

If $P \in B(\mathcal{L})$, in particular P belongs to the supports of the conics $[F_1]$ and $[F_2]$. Since $[F_1]$ is non-degenerate, $[F_1]$ and $[F_2]$ meet in at most 4 points (cf. Sect. 1.9.7) and hence $B(\mathcal{L})$ is a finite set containing at most 4 points. On the other hand, if $B(\mathcal{L})$ contained 4 points, these points would be in general position, so $B(\mathcal{L})$ would coincide with the set of base points of the pencil of conics $\mathcal{F}$ generated by $[F_1]$ and $[F_2]$ (cf. Sect. 1.9.7). Therefore, the conic $[F_3]$, passing through the base points of $\mathcal{F}$, would belong to that pencil, and the linear system $\mathcal{L}$ would coincide with the pencil $\mathcal{F}$, contradicting the hypothesis it has dimension 2.

Exercise 160. (*Pappus-Pascal's Theorem*) Let $\mathcal{C}$ be a non-degenerate conic of $\mathbb{P}^2(\mathbb{K})$ and assume that $P_1, P_2, P_3, Q_1, Q_2, Q_3$ are distinct points of $\mathcal{C}$. Show that the points $R_1 = L(P_3, Q_2) \cap L(P_2, Q_3)$, $R_2 = L(P_1, Q_3) \cap L(P_3, Q_1)$, $R_3 = L(P_2, Q_1) \cap L(P_1, Q_2)$ are collinear.

Solution (1). Consider the points $A = L(P_1, Q_3) \cap L(P_3, Q_2)$ and $B = L(P_1, Q_2) \cap L(P_3, Q_1)$ (cf. Fig. 4.12). As seen in Exercise 142, we can compute the cross-ratio of the points P_3, P_2, P_1, Q_2 on the conic as the cross-ratio of the lines joining the four points with Q_3. If we consider the intersections of the four lines with the line $L(P_3, Q_2)$, we get that

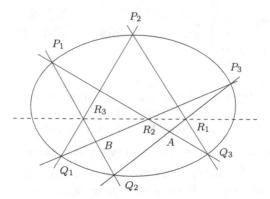

Fig. 4.12 Pappus-Pascal's Theorem

$$\beta(P_3, P_2, P_1, Q_2) = \beta(P_3, R_1, A, Q_2).$$

If instead we compute the cross-ratio of P_3, P_2, P_1, Q_2 as the cross-ratio of the lines joining the four points with Q_1 and we consider the intersections of these lines with the line $L(P_1, Q_2)$, we get that

$$\beta(P_3, P_2, P_1, Q_2) = \beta(B, R_3, P_1, Q_2).$$

Then $\beta(P_3, R_1, A, Q_2) = \beta(B, R_3, P_1, Q_2)$, so there exists a projective isomorphism $f: L(P_3, Q_2) \to L(P_1, Q_2)$ such that

$$f(P_3) = B, \quad f(R_1) = R_3, \quad f(A) = P_1, \quad f(Q_2) = Q_2.$$

As Q_2 is fixed for f, then f is the perspectivity centred at the point $L(P_3, B) \cap L(A, P_1) = R_2$ (cf. Exercise 31). Since the perspectivity centred at R_2 transforms R_1 into R_3, the points R_1, R_2, R_3 are collinear.

Solution (2). The cubics $\mathcal{D}_1 = L(P_1, Q_3) + L(P_2, Q_1) + L(P_3, Q_2)$ and $\mathcal{D}_2 = L(P_1, Q_2) + L(P_2, Q_3) + L(P_3, Q_1)$ meet at the nine points P_1, P_2, P_3, Q_1, Q_2, Q_3, R_1, R_2, R_3. The first eight points belong to the cubic $\mathcal{D} = \mathcal{C} + L(R_1, R_2)$. By Exercise 111, the cubic $\mathcal{D}$ has to pass through the point R_3 too. Since R_3 cannot lie on the irreducible conic $\mathcal{C}$, then $R_3 \in L(R_1, R_2)$, which implies the thesis.

Solution (3). The cubics $\mathcal{D}_1 = L(P_1, Q_3) + L(P_2, Q_1) + L(P_3, Q_2)$ and $\mathcal{D}_2 = L(P_1, Q_2) + L(P_2, Q_3) + L(P_3, Q_1)$ meet at the nine points P_1, P_2, P_3, Q_1, Q_2, Q_3, R_1, R_2, R_3. The first six points belong to the irreducible cubic $\mathcal{C}$; so by Exercise 106 the other three points R_1, R_2, R_3 lie on a curve of degree $3 - 2 = 1$, that is on a line.

Note. If in the statement of Pappus-Pascal's Theorem we consider, instead of a nondegenerate conic $\mathcal{C}$, a reducible conic union of two lines, we obtain Pappus' Theorem (cf. Exercise 13).

Exercise 161. Assume that C is a non-degenerate conic of $\mathbb{P}^2(\mathbb{K})$, that A, B, C are the vertices of a self-polar triangle for C and that r is a line passing through A and intersecting C at two distinct points P and Q. Prove that $L(B, P) \neq L(C, Q)$, and that $L(B, P) \cap L(C, Q) \in C$.

Solution (1). First of all we observe that $L(B, P) \neq L(C, Q)$. Namely, otherwise the points B, C would lie on r, and hence A, B, C would be collinear, in contradiction with the hypothesis that they are the vertices of a self-polar triangle for C.

In order to show that $L(B, P) \cap L(C, Q) \in C$, we first investigate the case $r = L(A, B)$ and $r = L(A, C)$. If $r = L(A, B)$ then $Q \in L(B, P)$, so $L(B, P) \cap L(C, Q) = Q \in C$. Similarly if $r = L(A, C)$ we have $L(B, P) \cap L(C, Q) = P \in C$.

So we can assume that $B \notin r$ and $C \notin r$. If $L(B, P)$ were tangent to C, then $P \in \mathrm{pol}_C(B) = L(A, C)$, and hence $C \in L(A, P) = r$, in contradiction with the assumption just made. Similarly one can show that $L(B, Q)$ is not tangent to C. Therefore there are exactly two points R, S such that $\{R\} = (L(B, P) \cap C) \setminus \{P\}$, $\{S\} = (L(B, Q) \cap C) \setminus \{Q\}$ (cf. Fig. 4.13). As $r \neq L(A, B)$, then $R \neq Q$ and $S \neq P$; in addition, the fact that $B \notin C$ easily implies that $R \neq S$. Therefore P, Q, R, S are distinct.

As proved in Exercise 153, the points $A' = L(P, Q) \cap L(R, S) = r \cap L(R, S)$, $C' = L(P, S) \cap L(Q, R)$ and $B = L(P, R) \cap L(Q, S)$ are the vertices of a self-polar triangle for C. Note that both A and A' lie on r and also on $\mathrm{pol}_C(B)$. Moreover, $\mathrm{pol}_C(B) \neq r$ because $C \in \mathrm{pol}_C(B)$, $C \notin r$. It follows that $A = A'$, and hence $C = \mathrm{pol}_C(A) \cap \mathrm{pol}_C(B) = \mathrm{pol}_C(A') \cap \mathrm{pol}_C(B) = C'$. Therefore we have $L(C, Q) = L(C', Q) = L(R, Q)$, so $L(B, R) \cap L(C, Q) = R \in C$, as required.

Solution (2). As observed at the beginning of Solution (1), the cases when either $r = L(A, B)$ or $r = L(A, C)$ can be easily discussed preliminarily. Thus assume that $B \notin r$ and $C \notin r$.

Since r is not tangent to C, it is immediate to check that the points A, B, C, P form a projective frame of $\mathbb{P}^2(\mathbb{K})$. Denote by x_0, x_1, x_2 the homogeneous coordinates induced by that frame. As A, B, C are the vertices of a self-polar triangle for C, the equation of C has the form $F(x_0, x_1, x_2) = x_0^2 + ax_1^2 + bx_2^2 = 0$ with $ab \neq 0$.

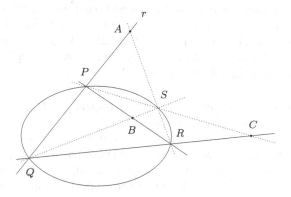

Fig. 4.13 The configuration described in Exercise 161

In addition, $P \in C$ implies that $b = -1 - a$, so that $F(x_0, x_1, x_2) = x_0^2 + ax_1^2 - (1+a)x_2^2$. The line $r = L(A, P)$ has equation $x_1 = x_2$, which easily yields $Q = [1, -1, -1]$. Therefore $L(C, Q)$ and $L(B, P)$ have equations $x_0 + x_1 = 0$ and $x_0 = x_2$, respectively. It follows that $L(B, P) \cap L(C, Q) = [1, -1, 1]$, which actually belongs to C.

Exercise 162. (*Chasles' Theorem*) Let C be a non-degenerate conic of $\mathbb{P}^2(\mathbb{K})$, and assume that P_1, P_2, P_3 are distinct points of $\mathbb{P}^2(\mathbb{K})$ such that $\mathrm{pol}_C(P_i) \neq L(P_j, P_k)$ if $\{i, j, k\} = \{1, 2, 3\}$. For $i \in \{1, 2, 3\}$ let $T_i = \mathrm{pol}_C(P_i) \cap L(P_j, P_k)$, with $\{i, j, k\} = \{1, 2, 3\}$. Show that T_1, T_2, T_3 are collinear.

Solution. If P_1, P_2, P_3 lie on a line r, then evidently $\{T_1, T_2, T_3\} \subseteq r$, which proves the thesis. Therefore we may assume that the points P_i are in general position, and we may choose homogeneous coordinates in $\mathbb{P}^2(\mathbb{K})$ so that $P_1 = [1, 0, 0]$, $P_2 = [0, 1, 0]$, $P_3 = [0, 0, 1]$.

Denote by M a symmetric matrix representing C with respect to the coordinates just chosen, and denote by $m_{i,j}$ the entry of M lying on the ith row and on the jth column. The line $L(P_2, P_3)$ has equation $x_0 = 0$, while $\mathrm{pol}_C(P_1)$ has equation $m_{1,1}x_0 + m_{1,2}x_1 + m_{1,3}x_2 = 0$ (with $m_{1,2}$, $m_{1,3}$ not simultaneously zero because $\mathrm{pol}_C(P_1) \neq L(P_2, P_3)$). Hence $T_1 = [0, m_{1,3}, -m_{1,2}]$. Similarly $T_2 = [m_{2,3}, 0, -m_{2,1}] = [m_{2,3}, 0, -m_{1,2}]$, and $T_3 = [m_{3,2}, -m_{3,1}, 0] = [m_{2,3}, -m_{1,3}, 0]$. As $(0, m_{1,3}, -m_{1,2}) - (m_{2,3}, 0, -m_{1,2}) + (m_{2,3}, -m_{1,3}, 0) = (0, 0, 0)$, the points T_1, T_2, T_3 are collinear.

Note. If the points P_1, P_2, P_3 lie on C, the line $\mathrm{pol}_C(P_i)$ is the tangent to C at P_i. If Q_1, Q_2, Q_3 are points of C such that P_1, P_2, P_3, Q_1, Q_2, Q_3 are distinct, the line $\mathrm{pol}_C(P_1)$ can be seen as the limit of the line $L(P_1, Q_2)$ when Q_2 tends to P_1. Similarly $\mathrm{pol}_C(P_2)$ (resp. $\mathrm{pol}_C(P_3)$) is the limit of the line $L(P_2, Q_3)$ (resp. $L(P_3, Q_1)$) when Q_3 tends to P_2 (resp. when Q_1 tends to P_3). Therefore the points T_1, T_2, T_3 can be obtained as limits of the points $L(P_1, Q_2) \cap L(P_2, Q_1)$, $L(P_2, Q_3) \cap L(P_3, Q_2)$, $L(P_3, Q_1) \cap L(P_1, Q_3)$, respectively. Thus the thesis of the exercise can be considered as a "limit version" of Pappus-Pascal's Theorem (cf. Exercise 160).

Exercise 163. (*Steiner construction*) Let A and B be distinct points of $\mathbb{P}^2(\mathbb{K})$ and $l = L(A, B)$; denote by $\mathcal{F}_A$ (resp. $\mathcal{F}_B$) the pencil of lines centred at A (resp. B). Let $f : \mathcal{F}_A \to \mathcal{F}_B$ be a projective isomorphism such that $f(l) \neq l$. Prove that $Q = \{r \cap f(r) \mid r \in \mathcal{F}_A\}$ is the support of a non-degenerate conic passing through A with tangent $f^{-1}(l)$ and through B with tangent $f(l)$.

Solution (1). The conics passing through A with tangent $f^{-1}(l)$ and through B with tangent $f(l)$ form a pencil $\mathcal{G}$ (cf. Exercise 125).

Let $s \in \mathcal{F}_A$ be a line different both from l and from $f^{-1}(l)$, and let $P = s \cap f(s) \in Q$. It is easy to check that A, B, P are in general position, that $P \notin f^{-1}(l)$ and that $P \notin f(l)$. Hence there exists exactly one conic C passing through P in the pencil $\mathcal{G}$ and this conic is non-degenerate. In order to prove the thesis it is enough to show that $Q = C$.

Consider the map $\psi\colon \mathcal{F}_A \to \mathcal{F}_B$ defined by $\psi(f^{-1}(l)) = l$, $\psi(l) = f(l)$ and, if $r \in \mathcal{F}_A \setminus \{l, f^{-1}(l)\}$, $\psi(r) = L(B, Q)$, where Q denotes the intersection point between r and C different from A. As shown in Exercise 142, ψ is a projective isomorphism and $\{r \cap \psi(r) \mid r \in \mathcal{F}_A\} = C$.

Moreover, by construction we have $\psi(f^{-1}(l)) = l = f(f^{-1}(l))$, $\psi(l) = f(l)$, $\psi(s) = f(s)$. Since ψ and f coincide on three distinct elements of $\mathcal{F}_A$, we deduce that $\psi = f$. Therefore $Q = \{r \cap \psi(r) \mid r \in \mathcal{F}_A\} = C$, as required.

Solution (2). If $C = f^{-1}(l) \cap f(l)$, it can be easily checked that the points A, B, C are in general position. Thus we can choose homogeneous coordinates on $\mathbb{P}^2(\mathbb{K})$ in such a way that $A = [1, 0, 0]$, $B = [0, 1, 0]$, $C = [0, 0, 1]$. These coordinates induce on the pencil $\mathcal{F}_A$ (resp. $\mathcal{F}_B$) homogeneous coordinates such that the line $ax_1 + bx_2 = 0$ (resp. $cx_0 + dx_2 = 0$) has coordinates $[a, b]$ (resp. $[c, d]$).

By construction the lines l, $f^{-1}(l)$ in $\mathcal{F}_A$ have coordinates $[0, 1]$, $[1, 0]$ respectively, while the lines l, $f(l)$ in $\mathcal{F}_B$ have coordinates $[0, 1]$, $[1, 0]$ respectively. Therefore in the chosen coordinates f can be represented by $M = \begin{pmatrix} 0 & \lambda \\ 1 & 0 \end{pmatrix}$ for some $\lambda \in \mathbb{K}^*$. Hence, if $r \in \mathcal{F}_A$ is the generic line of equation $ax_1 + bx_2 = 0$, then the line $f(r)$ has equation $\lambda b x_0 + a x_2 = 0$. It follows that $r \cap f(r) = [a^2, \lambda b^2, -\lambda ab]$, so that $Q = \{[a^2, \lambda b^2, -\lambda ab] \mid [a, b] \in \mathbb{P}^1(\mathbb{K})\}$.

Since $\lambda(a^2)(\lambda b^2) = (-\lambda ab)^2$, the set Q is contained in the support of the conic C of equation $F(x_0, x_1, x_2) = \lambda x_0 x_1 - x_2^2 = 0$. In addition, if $T = [y_0, y_1, y_2] \in C$ the following situations can occur:

(i) if $y_2 = 0$, then either $T = [1, 0, 0] = A = f^{-1}(l) \cap f(f^{-1}(l)) \in Q$ or $T = [0, 1, 0] = B = l \cap f(l) \in Q$;

(ii) if $y_2 \neq 0$, then $y_0 \neq 0$ and, if $a \neq 0$ is such that $a^2 = y_0$ and $b = -\frac{y_2}{\lambda a}$, then $T = [a^2, \lambda b^2, -\lambda ab] \in Q$.

Therefore $C \subseteq Q$, and hence $Q = C$.

In order to achieve the thesis it suffices to observe that C is non-degenerate, and that $\nabla F(1, 0, 0) = (0, \lambda, 0)$, $\nabla F(0, 1, 0) = (\lambda, 0, 0)$, so that the tangents to C at A, B are actually $f^{-1}(l)$ and $f(l)$.

Note. Exercise 43 (c) deals with the case when the line l is invariant under the isomorphism f; also in that case the set Q is a conic, but a degenerate one.

Exercise 164. Let C be a non-degenerate conic of $\mathbb{P}^2(\mathbb{C})$ and assume that f is a projectivity of $\mathbb{P}^2(\mathbb{C})$ such that $f(C) = C$. Show that there exists $P \in C$ such that $f(P) = P$.

Solution. As seen in Sect. 1.2.5, there exists a point $Q \in \mathbb{P}^2(\mathbb{C})$ such that $f(Q) = Q$. If $Q \in C$ the thesis holds, hence we can assume that $Q \notin C$. The projectivity f induces a projectivity of the pencil $\mathcal{F}_Q$ centred at Q, and also this projectivity has a fixed point, because $\mathcal{F}_Q$ is a complex projective space. Then there is a line $r \in \mathcal{F}_Q$ such that $f(r) = r$. If r is tangent to C and if P denotes the point of intersection between r and C, then $f(P) = f(C \cap r) = C \cap r = P$, and thus the thesis is proved.

Therefore suppose now that r intersects C in two distinct points A, B. Then $f(\{A, B\}) = f(C \cap r) = C \cap r = \{A, B\}$; so either $f(A) = A$ and $f(B) = B$, or $f(A) = B$ and $f(B) = A$. In both cases $f^2(A) = A$ and $f^2(B) = B$. Let $s = \mathrm{pol}_C(Q)$ and denote by C, D the points of intersection between s and C (these points are distinct because $Q \notin C$). As $f(C) = C$ and $f(Q) = Q$, we have $f(s) = s$ so that, arguing as above, $f^2(C) = C$ and $f^2(D) = D$. Moreover, the points A, B, C, D are distinct and, since they lie on a non-degenerate conic, they form a projective frame of $\mathbb{P}^2(\mathbb{C})$. Then the Fundamental theorem of projective transformations implies that $f^2 = \mathrm{Id}$. Taking into account the solution of Exercise 44, it is immediate to check that f is a projectivity of type (b); so there exists a line of fixed points of f. This line intersects C in at least one point, and so C contains at least one point which is fixed for f.

Exercise 165. Let C be a non-degenerate conic of $\mathbb{P}^2(\mathbb{K})$; assume that P is a point of C and denote $r = \mathrm{pol}_C(P)$. Let $g \colon r \to r$ be a projectivity such that $g(P) = P$. Show that there is exactly one projectivity $f \colon \mathbb{P}^2(\mathbb{K}) \to \mathbb{P}^2(\mathbb{K})$ such that $f(C) = C$, $f(r) = r$ and $f|_r = g$.

Solution. Let A_1, A_2 be distinct points of $r \setminus \{P\}$, and for $i = 1, 2$ denote by A_i' the point of intersection between C and the tangent to C different from r and passing through A_i, that is the point defined by the condition $\{A_i'\} = (\mathrm{pol}_C(A_i) \cap C) \setminus \{P\}$ (cf. Fig. 4.14). Moreover let $B_1 = g(A_1)$ and $B_2 = g(A_2)$; in a similar way, consider the points B_1', $B_2' \in C$ defined by $\{B_i'\} = (\mathrm{pol}_C(B_i) \cap C) \setminus \{P\}$.

Let us now see that a projectivity $f \colon \mathbb{P}^2(\mathbb{K}) \to \mathbb{P}^2(\mathbb{K})$ fulfils the requirements of the statement if and only if $f(C) = C$, $f(P) = P$, $f(A_1') = B_1'$ and $f(A_2') = B_2'$. At this point, the thesis is a consequence of Exercise 143 (a).

If $f \colon \mathbb{P}^2(\mathbb{K}) \to \mathbb{P}^2(\mathbb{K})$ satisfies the requirements of the statement, necessarily we have $f(C) = C$ and $f(P) = g(P) = P$. In addition $f(A_i) = g(A_i) = B_i$ for $i = 1, 2$, so that f transforms $L(A_i, A_i')$ into the tangent to $f(C) = C$ passing through $f(A_i) = B_i$ and distinct from $f(r) = r$. By construction this tangent coincides with the line $L(B_i, B_i')$, so that $f(A_i') = f(L(A_i, A_i') \cap C) = L(B_i, B_i') \cap C = B_i'$, as required.

Conversely assume that $f(C) = C$, $f(P) = P$, $f(A_1') = B_1'$ and $f(A_2') = B_2'$. Then $f(r) = f(\mathrm{pol}_C(P)) = \mathrm{pol}_{f(C)}(f(P)) = \mathrm{pol}_C(P) = r$. For $i = 1, 2$ similarly we have $f(L(A_i, A_i')) = f(\mathrm{pol}_C(A_i')) = \mathrm{pol}_C(B_i') = L(B_i, B_i')$, and hence $f(A_i) = $

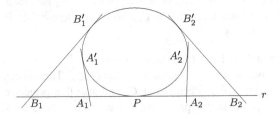

Fig. 4.14   The construction described in the solution of Exercise 165

$f(r \cap L(A_i, A_i')) = r \cap L(B_i, B_i') = B_i$. It follows that the projectivities $f|_r$ and g coincide on the three distinct points P, A_1, A_2, and therefore they are equal as a consequence of the Fundamental theorem of projective transformations.

Exercise 166. Denote by $\mathcal{F}$ a pencil of degenerate conics of $\mathbb{P}^2(\mathbb{C})$ such that there is no line which is an irreducible component of all conics in the pencil. Prove that the base-point set of $\mathcal{F}$ consists of a single point and that the pencil contains exactly two doubly degenerate conics.

Solution. First of all observe that $\mathcal{F}$ can contain at most two doubly degenerate conics. Namely, let $\mathcal{C}_1 = 2r$ and $\mathcal{C}_2 = 2s$ be two such conics in $\mathcal{F}$, where r and s are lines. In a system of homogeneous coordinates where r has equation $x_0 = 0$ and s has equation $x_1 = 0$, the generic conic $\mathcal{C}_{\lambda,\mu} \in \mathcal{F}$ is given by the equation $\lambda x_0^2 + \mu x_1^2 = 0$, $[\lambda, \mu] \in \mathbb{P}^1(\mathbb{C})$. It can be immediately checked that $\mathcal{C}_{\lambda,\mu}$ is doubly degenerate if and only if $[\lambda, \mu] \in \{[1, 0], [0, 1]\}$, i.e. if and only if $\mathcal{C}_{\lambda,\mu} \in \{\mathcal{C}_1, \mathcal{C}_2\}$.

We are now going to show that $\mathcal{F}$ has exactly one base point. As observed above, we can choose as generators of the pencil two simply degenerate conics, $\mathcal{C}_1 = r_1 + r_2$ and $\mathcal{C}_2 = s_1 + s_2$, where r_1, r_2, s_1, s_2 are distinct lines. Denote by $A = r_1 \cap r_2$ the singular point of $\mathcal{C}_1$ and by $B = s_1 \cap s_2$ the singular point of $\mathcal{C}_2$. If $A = B$, the pencil has exactly one base point, as required. Since by hypothesis $\mathcal{C}_1$ and $\mathcal{C}_2$ do not share any component, if A and B are distinct, up to exchanging $\mathcal{C}_1$ with $\mathcal{C}_2$, we may assume that A does not belong to the support of $\mathcal{C}_2$. Therefore, we can distinguish two cases depending on whether B lies on the support of $\mathcal{C}_1$ or not.

In the first case we can suppose, for instance, that B belongs to r_1. If $P_1 = r_2 \cap s_1$ and $P_2 = r_2 \cap s_2$, we can immediately check that B, P_1 and P_2 are in general position and that P_1, $P_2 \notin r_1$. All conics of $\mathcal{F}$ pass through B, P_1 and P_2 and they are tangent at B to r_1; hence, by Exercise 121, $\mathcal{F}$ coincides with the pencil of conics that satisfy these linear conditions. In other words we are in case (b) of Sect. 1.9.7 and $\mathcal{F}$ contains only two degenerate conics, in contradiction with the hypothesis.

In the second case $\mathcal{C}_1$ and $\mathcal{C}_2$ meet at four points P_1, P_2, P_3, P_4 in general position and hence $\mathcal{F}$ is the pencil of conics through P_1, P_2, P_3, P_4. Thus we are in case (a) of Sect. 1.9.7 and $\mathcal{F}$ contains only three degenerate conics, in contradiction with the hypothesis.

We have thus seen that $\mathcal{F}$ has only one base point A. Choose a system of homogeneous coordinates x_0, x_1, x_2 in $\mathbb{P}^2(\mathbb{C})$ such that $A = [1, 0, 0]$ and $[0, 1, 0] \in r_1$, $[0, 0, 1] \in r_2$ and $[1, 1, 1] \in s_1$. In this system of coordinates $\mathcal{C}_1$ is defined by $x_1 x_2 = 0$ and $\mathcal{C}_2$ is defined by $(x_1 - x_2)(x_1 + \alpha x_2)$, with $\alpha \neq 0, -1$. The generic conic $\mathcal{C}_{\lambda,\mu}$ of $\mathcal{F}$ is thus represented, as $[\lambda, \mu]$ varies in $\mathbb{P}^1(\mathbb{C})$, by the matrix

$$M_{\lambda,\mu} = \begin{pmatrix} 0 & 0 & 0 \\ 0 & 2\mu & \lambda + \mu(\alpha - 1) \\ 0 & \lambda + \mu(\alpha - 1) & -2\mu\alpha \end{pmatrix},$$ whose rank is equal to 1 if and only

if $\lambda^2 + (\alpha + 1)^2 \mu^2 + 2(\alpha - 1)\lambda\mu = 0$. This latter equality, considered as an equation in $[\lambda, \mu] \in \mathbb{P}^1(\mathbb{C})$, admits two distinct solutions, because $(\alpha - 1)^2 - (\alpha + 1)^2 = -4\alpha \neq 0$. Therefore $\mathcal{F}$ contains exactly two doubly degenerate conics, as required.

Exercise 167. Let Q be a non-degenerate quadric of $\mathbb{P}^n(\mathbb{K})$ and let H be a hyperplane. Then $Q \cap H$ is a degenerate quadric of H if and only if H is tangent to Q at a point P. In addition, in that case the quadric $Q \cap H$ has rank $n - 1$ and P is its only singular point.

Solution. The first statement of the thesis easily follows from the fact that $Q \cap H$ is a degenerate quadric of H if and only if it is singular and, by Exercise 58, this happens if and only if the hyperplane H is tangent to Q at a point P.

In that case we can choose a system of homogeneous coordinates in $\mathbb{P}^n(\mathbb{K})$ where $H = \{x_0 = 0\}$ and $P = [0, \ldots, 0, 1]$. If in these coordinates Q has equation ${}^t X A X = 0$, where A is a symmetric matrix, the quadric $Q \cap H$ is represented by the matrix $c_{0,0}(A)$ (cf. 1.1 for the definition of $c_{0,0}(A)$). Since H is tangent to Q at P, then $\mathrm{pol}(P) = H$ and hence $a_{n,1} = \ldots = a_{n,n} = 0$ and $a_{n,0} \neq 0$, i.e. the symmetric matrix A is of the form

$$A = \begin{pmatrix} a_{0,0} & \cdots & a_{n,0} \\ a_{1,0} & & 0 \\ \vdots & B & \vdots \\ a_{n-1,0} & & 0 \\ a_{n,0} & 0 \ldots 0 & 0 \end{pmatrix}$$

where B denotes a symmetric matrix of order $n - 1$. As $\det A = -a_{n,0}^2 \det B$ and $\det A \neq 0$, it follows that $\det B \neq 0$ and so $\mathrm{rk}(c_{0,0}(A)) = n - 1$. As a consequence $\mathrm{Sing}(Q \cap H)$ consists of a single point; since $(0, \ldots, 0, 1) \in \mathbb{K}^n$ belongs to the kernel of $c_{0,0}(A)$, then $\mathrm{Sing}(Q \cap H) = \{P\}$.

Exercise 168. Let Q be a quadric of rank r of $\mathbb{P}^n(\mathbb{K})$ and let $P \in \mathbb{P}^n(\mathbb{K}) \setminus Q$. Prove that, if $r = 1$, then $Q = 2 \mathrm{pol}(P)$, and that, if $r > 1$, then the quadric $Q' = Q \cap \mathrm{pol}(P)$ of $\mathrm{pol}(P)$ has rank $r - 1$.

Solution. Since $P \notin Q$, the subspace $\mathrm{pol}(P)$ is a hyperplane not passing through P. Then there are homogeneous coordinates in $\mathbb{P}^n(\mathbb{K})$ such that $P = [1, 0, \ldots, 0]$ and $\mathrm{pol}(P) = \{x_0 = 0\}$. In this system of coordinates Q is represented by a matrix of the form $A = \left(\begin{array}{c|c} a_{0,0} & 0 \\ \hline 0 & C \end{array} \right)$, where C denotes a symmetric matrix of order n and $a_{0,0} \in \mathbb{K}^*$. Since $r = \mathrm{rk}(A) = \mathrm{rk}(C) + 1$, then $C = 0$ if and only if $r = 1$, and in this case $Q = 2 \mathrm{pol}(P)$. For $r > 1$, the matrix C defines the quadric Q' in $\mathrm{pol}(P)$, which consequently has rank $r - 1$.

Exercise 169. Consider a quadric Q of $\mathbb{P}^n(\mathbb{K})$ and let $H \subset \mathbb{P}^n(\mathbb{K})$ be a hyperplane not contained in Q. Denote by Q' the quadric $Q \cap H$ of H. Prove that $\mathrm{pol}_{Q'}(P) = \mathrm{pol}_Q(P) \cap H$ for every $P \in H$.

Solution. We can choose homogeneous coordinates $x_0, \ldots, x_n$ so that $P = [1, 0, \ldots, 0]$ and $H = \{x_n = 0\}$. If $A = (a_{i,j})_{i,j=0,\ldots,n}$ is a symmetric matrix representing Q, then $\mathrm{pol}_Q(P)$ has equation $\sum_{i=0}^n a_{0,i} x_i = 0$. In the coordinates $x_0, \ldots, x_{n-1}$ induced on H, the quadric Q' is defined by the matrix $A' = c_{n,n}(A)$, obtained from A by deleting its last row and its last column (the matrix thus obtained is non-zero because $H \not\subseteq Q$). Then $\text{pol}_{Q'}(P)$ is the subspace of H defined by the equation $\sum_{i=0}^{n-1} a_{0,i} x_i = 0$, so it coincides with $H \cap \text{pol}_Q(P)$.

Exercise 170. *(Construction of a self-polar $(n + 1)$-hedron)* Let Q be a quadric of rank r of $\mathbb{P}^n(\mathbb{K})$ and assume that $P_0, \ldots, P_k \in \mathbb{P}^n(\mathbb{K}) \setminus Q$ are points such that P_i and P_j are conjugate with respect to Q for any $i \neq j$ with $0 \leq i, j \leq k$. Show that:

(a) $P_0, \ldots, P_k$ are projectively independent and $k + 1 \leq r$.
(b) The hyperplanes $\text{pol}(P_0), \ldots, \text{pol}(P_k)$ are independent and $L(P_0, \ldots, P_k) \cap \text{pol}(P_0) \cap \cdots \cap \text{pol}(P_k) = \emptyset$.
(c) There exist $P_{k+1}, \ldots, P_n \in \mathbb{P}^n(\mathbb{K})$ such that $P_0, \ldots, P_k, P_{k+1}, \ldots, P_n$ are the vertices of a self-polar $(n + 1)$-hedron for Q.

Solution. (a) Observe that, for each $i \in \{0, \ldots, k\}$, the subspace $S_i = L(P_0, \ldots, P_{i-1}, P_{i+1}, \ldots P_k)$ is contained in $\text{pol}(P_i)$. Since by hypothesis $P_i \notin Q$, then $\text{pol}(P_i)$ is a hyperplane and $P_i \notin \text{pol}(P_i)$; in particular $P_i \notin S_i$. It follows that $P_0, \ldots, P_k$ are projectively independent. Let us complete $P_0, \ldots, P_k$ to a projective frame $\{P_0, \ldots, P_k, Q_{k+1}, \ldots, Q_{n+1}\}$ and let A be a matrix defining Q in the associated system of homogeneous coordinates. Then $A = \left(\begin{array}{c|c} D & {}^t\!B \\ \hline B & C \end{array} \right)$, where D is a symmetric matrix of order $k + 1$, C is a symmetric matrix of order $n - k$ and B is a matrix $(n - k) \times (k + 1)$. As the points $P_0, \ldots, P_k$ do not belong to Q and are pairwise conjugate, one can easily check that $D = \text{diag}(d_0, \ldots, d_k)$ is a diagonal matrix whose elements d_i on the diagonal are all non-zero. Hence $r = \text{rk } A \geq \text{rk } D = k + 1$.

(b) In the system of coordinates already used in the solution of (a) the point of the dual space $\mathbb{P}^n(\mathbb{K})^*$ corresponding to $\text{pol}(P_i)$, $i = 0, \ldots, k$, is given by the i-th column of A. Having seen that the first $k + 1$ columns of A are independent, then $\text{pol}(P_0), \ldots, \text{pol}(P_k)$ are independent too. The subspace $L(P_0, \ldots, P_k)$ is defined by $x_{k+1} = \cdots = x_n = 0$ and the subspace $L(P_0, \ldots, P_k) \cap \text{pol}(P_0) \cap \cdots \cap \text{pol}(P_k)$ is defined in $L(P_0, \ldots, P_k)$ by the equations $d_0 x_0 = \cdots = d_k x_k = 0$, so it is empty.

(c) Denote by T the projective subspace $\text{pol}(P_0) \cap \cdots \cap \text{pol}(P_k)$; by point (b) T has dimension $n - k - 1$ and does not intersect $L(P_0, \ldots, P_k)$. By reciprocity, $\text{Sing}(Q)$ is contained in the polar space of any point of $\mathbb{P}^n(\mathbb{K})$, so $\text{Sing}(Q) \subseteq T$. Moreover, Grassmann's formula implies that $\dim L(L(P_0, \ldots, P_k), T) = \dim T + k + 1 = n$, and therefore $L(L(P_0, \ldots, P_k), T) = \mathbb{P}^n(\mathbb{K})$.

Let us proceed by induction on $m = r - k - 1$, which is a non-negative integer by part (a).

If $m = 0$, let us prove that, however we choose projectively independent points $P_{k+1}, \ldots, P_n \in T$, the points $P_0, \ldots, P_k, P_{k+1}, \ldots, P_n$ form a self-polar $(n + 1)$-hedron. As observed above, $L(P_0, \ldots, P_k, P_{k+1}, \ldots, P_n) = L(L(P_0, \ldots, P_k), T) = \mathbb{P}^n(\mathbb{K})$, so the points $P_0, \ldots, P_n$ are projectively independent. In addition, since $m = r - k - 1 = 0$, the projective subspaces $\text{Sing}(Q)$ and T have the same dimension, so $\text{Sing}(Q) = T$. Consider i, j such that $1 \leq i < j \leq n$. If $j \leq k$ (and hence $i \leq k$ too), the points P_i and P_j are conjugate by hypothesis, while $P_j \in T = \text{Sing}(Q)$ if $j > k$; so $\text{pol}(P_j) = \mathbb{P}^n(\mathbb{K})$ and the points P_i, P_j are conjugate.

Let now $m > 0$, and let us first prove that $T \not\subseteq Q$. Assume by contradiction that $T \subseteq Q$ and choose $P \in T$. Then $L(P, P') \subseteq T \subseteq Q$ for every $P' \in T$; as a consequence $L(P, P') \subseteq T_P(Q) = \text{pol}(P)$ and $T \subseteq \text{pol}(P)$, using the fact that P' can be arbitrarily chosen. On the other hand, by reciprocity $P_i \in \text{pol}(P)$ for each $i = 0, \ldots, k$, so $\text{pol}(P) \supseteq T \cup \{P_0, \ldots, P_k\}$ and $\text{pol}(P) \supseteq L(T, L(P_0, \ldots, P_k)) = \mathbb{P}^n(\mathbb{K})$. This implies that $P \in \text{Sing}(Q)$, so $T \subseteq \text{Sing}(Q)$ since P can be arbitrarily chosen; therefore $T = \text{Sing}(Q)$ by the observations made at the beginning of the proof of (c). Finally it follows that $n - r = \dim \text{Sing}(Q) = \dim T = n - k - 1$, contradicting the hypothesis $m = r - k - 1 > 0$.

We have thus seen that $T \setminus Q$ is non-empty, so that it contains a point P_{k+1}. Then P_{k+1} is conjugate with $P_0, \ldots, P_k$ and, by the inductive hypothesis, there exist $P_{k+2}, \ldots, P_n \in \mathbb{P}^n(\mathbb{K})$ such that $P_0, \ldots, P_n$ are the vertices of a self-polar $(n + 1)$-hedron.

Exercise 171. Let Q be a quadric of $\mathbb{P}^n(\mathbb{K})$ and assume that $P_1, P_2 \in \mathbb{P}^n(\mathbb{K}) \setminus \text{Sing}(Q)$ are distinct points. Prove that $\text{pol}(P_1) = \text{pol}(P_2)$ if and only if $L(P_1, P_2) \cap \text{Sing}(Q) \neq \emptyset$.

Solution (1). Let $v_1, v_2 \in \mathbb{K}^{n+1} \setminus \{0\}$ be vectors such that $P_i = [v_i], i = 1, 2$. Since by hypothesis $P_1, P_2 \notin \text{Sing}(Q)$, for $i = 1, 2$ $\text{pol}(P_i)$ is the hyperplane defined by ${}^t v_i A X = 0$. Therefore, if $\text{pol}(P_1) = \text{pol}(P_2)$ then there exists $\lambda \in \mathbb{K}^*$ such that $A v_1 = \lambda A v_2$. The vector $w = v_1 - \lambda v_2$ is non-zero, because v_1, v_2 represent distinct points of $\mathbb{P}^n(\mathbb{K})$ and so they are linearly independent. The point $Q = [w]$ is singular for Q, because $Aw = 0$, and it lies on the line $L(P_1, P_2)$.

Conversely, assume there exists $Q \in L(P_1, P_2) \cap \text{Sing}(Q)$ and let $w \in \mathbb{K}^{n+1} \setminus \{0\}$ be a vector such that $Q = [w]$. By hypothesis Q, P_1 and P_2 are distinct, so there exist $\lambda_1, \lambda_2 \in \mathbb{K}^*$ such that $w = \lambda_1 v_1 + \lambda_2 v_2$. If we multiply the latter equality by A on the left, we get $0 = Aw = \lambda_1 A v_1 + \lambda_2 A v_2$, that is $A v_1 = -\frac{\lambda_2}{\lambda_1} A v_2$. Therefore $\text{pol}(P_1) = \text{pol}(P_2)$.

Solution (2). Since $\text{pol}: \mathbb{P}^n(\mathbb{K}) \setminus \text{Sing}(Q) \to \mathbb{P}^n(\mathbb{K})^*$ is a (possibly degenerate) projective transformation, if $L(P_1, P_2) \cap \text{Sing}(Q) = \emptyset$ then the restriction of pol to $L(P_1, P_2)$ is a non-degenerate projective transformation, so it is injective, and hence $\text{pol}(P_1) \neq \text{pol}(P_2)$.

Conversely, if $L(P_1, P_2) \cap \text{Sing}(Q) \neq \emptyset$, then $L(P_1, P_2) \cap \text{Sing}(Q)$ is a point. As seen in Exercise 28, the image of $L(P_1, P_2) \setminus \text{Sing}(Q)$ through the map pol is a projective subspace of $\mathbb{P}^n(\mathbb{K})^*$ having dimension

$$\dim L(P_1, P_2) - \dim(L(P_1, P_2) \cap \text{Sing}(Q)) - 1 = 1 - 0 - 1 = 0,$$

i.e. it is a point. In particular it follows that $\text{pol}(P_1) = \text{pol}(P_2)$.

Exercise 172. If Q is a non-degenerate and non-empty quadric of $\mathbb{P}^3(\mathbb{K})$, show that one of the following statements holds:

(i)   for every $P \in Q$ the set $Q \cap T_P(Q)$ is the union of two distinct lines which meet at P;
(ii) for every $P \in Q$ we have $Q \cap T_P(Q) = \{P\}$.

(Recall that Q is said to be hyperbolic in case (i) and elliptic in case (ii), and that case (ii) can occur only if $\mathbb{K} = \mathbb{R}$ – cf. Sect. 1.8.4.)

Solution. If $P \in Q$ is a point of the quadric, since Q is non-degenerate, by Exercise 167 the conic $C_P = Q \cap T_P(Q)$ is singular at P and has rank 2. So the support of C_P either is the union of two distinct lines passing through P, and in this case P is called hyperbolic, or it consists only of the point P, and in this case P is called elliptic (cf. Sect. 1.8.4).

We are going to check that either all points of Q are hyperbolic (so statement (i) holds) or all of them are elliptic (and then we are in case (ii)). If Q does not contain any line, clearly case (ii) occurs. Therefore we can assume that Q contains a line r. For every point $P \in r$, the line r is contained in $T_P(Q)$, and hence also in C_P, which is consequently the union of two lines through P. If instead $P \in Q \setminus r$, there exists at least one point $R \in r \cap T_P(Q)$. As $R \neq P$ and $R \in C_P$, we can conclude that C_P is the union of two lines passing through P; so case (i) occurs.

Let us show an alternative way to prove that Q cannot simultaneously contain hyperbolic points and elliptic points. If Q contained an elliptic point P and also a hyperbolic point R, then $Q \cap \mathrm{pol}_Q(P) = \{P\}$ and $Q \cap \mathrm{pol}_Q(R) = r_1 \cup r_2$ with r_1, r_2 distinct lines meeting at R. As $R \notin \mathrm{pol}_Q(P)$, then $\mathrm{pol}_Q(P)$ should intersect $r_1 \cup r_2$ (and hence Q) in at least two points, contradicting the fact that P is elliptic.

Exercise 173. Let Q be a quadric of $\mathbb{P}^3(\mathbb{K})$ of rank 3. Prove that, for every smooth point $P \in Q$, the conic $Q \cap T_P(Q)$ is a double line.
(Recall that, in this case, if $Q \setminus \mathrm{Sing}(Q) \neq \emptyset$, the quadric Q is called parabolic – cf. Sect. 1.8.4.)

Solution. The quadric Q, having rank 3, has only one singular point R, and of course $\mathrm{pol}(R) = \mathbb{P}^3(\mathbb{K})$. Let $P \in Q$ be a smooth point. By reciprocity $R \in \mathrm{pol}(P) = T_P(Q)$. Then, by Exercise 58, the conic $Q \cap T_P(Q)$ is singular both at P and at R, and hence it coincides with $2L(P, R)$.

Exercise 174. Prove that:

(a) The quadric Q of $\mathbb{P}^3(\mathbb{C})$ of equation $x_0^2 + x_1^2 + x_2^2 + x_3^2 = 0$ is hyperbolic.
(b) The quadric Q of $\mathbb{P}^3(\mathbb{C})$ of equation $x_0^2 + x_1^2 + x_2^2 = 0$ is parabolic.
(c) The quadric Q of $\mathbb{P}^3(\mathbb{R})$ of equation $x_0^2 + x_1^2 + x_2^2 - x_3^2 = 0$ is elliptic.
(d) The quadric Q of $\mathbb{P}^3(\mathbb{R})$ of equation $x_0^2 + x_1^2 - x_2^2 - x_3^2 = 0$ is hyperbolic.
(e) The quadric Q of $\mathbb{P}^3(\mathbb{R})$ of equation $x_0^2 + x_1^2 - x_2^2 = 0$ is parabolic.

Solution. First of all it is immediate to check that the quadric Q is non-degenerate in cases (a), (c) and (d), and that it has rank 3 in cases (b) and (e). In particular, in all cases the quadric Q is irreducible. Therefore, by Exercise 173, in cases (b) and (e) it suffices to check that Q contains a smooth point; as a matter of fact, in case (b) such check is unnecessary, because Q has only one singular point and the support of a complex quadric contains infinitely many points. Moreover, statement (a) follows from the fact that all non-degenerate quadrics of $\mathbb{P}^3(\mathbb{C})$ are hyperbolic (cf. Exercise 172). Instead, in cases (c) and (d) in order to decide whether Q is either hyperbolic or elliptic, it is sufficient to choose a point $P \in Q$ and determine the support of the conic $Q \cap T_P(Q)$ (cf. Exercise 172).

(a) As recalled above, all non-degenerate quadrics of $\mathbb{P}^3(\mathbb{C})$ are hyperbolic. However, choose for instance the point $P = [1, 0, 0, i] \in Q$, and let us check that the conic $Q \cap T_P(Q)$ consists of two distinct lines which meet at P. The plane $T_P(Q)$ has equation $x_0 + ix_3 = 0$, and the conic $Q \cap T_P(Q)$ is defined in $T_P(Q)$ by the equation $x_1^2 + x_2^2 = 0$, whose support is the union of the lines $x_1 + ix_2 = 0$ and $x_1 - ix_2 = 0$.

(b) As observed before, it suffices to check that Q has at least a smooth point. Choose for instance $P = [1, 0, i, 0] \in Q$. Then $T_P(Q)$ is defined by the equation $x_0 + ix_2 = 0$, so P is non-singular. In addition, the conic $Q \cap T_P(Q)$ is defined in $T_P(Q)$ by the equation $x_1^2 = 0$, so in fact it is a double line.

(c) Choose $P = [1, 0, 0, 1] \in Q$. The plane $T_P(Q)$ has equation $x_0 - x_3 = 0$, and the conic $Q \cap T_P(Q)$ is defined in $T_P(Q)$ by the equation $x_1^2 + x_2^2 = 0$, whose support contains only the point P.

(d) Choose $P = [1, 0, 0, 1] \in Q$. The plane $T_P(Q)$ has equation $x_0 - x_3 = 0$, and the conic $Q \cap T_P(Q)$ is defined in $T_P(Q)$ by the equation $x_1^2 - x_2^2 = 0$, whose support is the union of the lines $x_1 + x_2 = 0$ and $x_1 - x_2 = 0$.

(e) Like in case (b), it is sufficient to check that Q has at least a smooth point. If $P = [1, 0, 1, 0] \in Q$, the tangent space $T_P(Q)$ is defined by $x_0 - x_2 = 0$, so that P is smooth. In additon, the conic $Q \cap T_P(Q)$ is defined in $T_P(Q)$ by the equation $x_1^2 = 0$, so it is a double line.

Exercise 175. Assume that Q is a non-degenerate and non-empty quadric of $\mathbb{P}^3(\mathbb{R})$. Prove that there exists a plane H of $\mathbb{P}^3(\mathbb{R})$ external to Q if and only if Q is an elliptic quadric.

Solution. Assume that Q is not elliptic. By Exercise 172, the quadric Q is hyperbolic and, given a point $P \in Q$, we have that $Q \cap T_P(Q) = r \cup s$, where r and s are distinct lines. As a consequence, for every plane H of $\mathbb{R}^3$, the set $Q \cap H$ contains $H \cap (r \cup s)$ and hence it is not empty. So no plane external to Q can exist.

Conversely, suppose that Q is an elliptic quadric. We can choose homogeneous coordinates in which Q is defined by one of the equations listed in Theorem 1.8.3. Then by Exercise 174 we may assume that Q is defined by the equation $x_0^2 + x_1^2 + x_2^2 - x_3^2 = 0$. It is now immediate to check that the plane H of equation $x_0 = 2x_3$ is external to Q.

Exercise 176. Consider a quadric Q of $\mathbb{P}^3(\mathbb{K})$ and assume that $r, s, t \subset Q$ are distinct lines. Show that:

(a) If $r \cap s = s \cap t = r \cap t = \emptyset$, then Q is non-degenerate.

(b) If r, s, t are coplanar, then Q is reducible.

(c) If r, s, t meet at a point P and are not coplanar, then P is singular for Q.

(d) If r and s are coplanar and $r \cap t = s \cap t = \emptyset$, then Q is reducible.

Solution. (a) Suppose by contradiction that Q is singular and consider a point $P \in$ Sing(Q). Since the lines r, s, t are pairwise skew, we may assume that $P \notin r$. As every quadric having a singular point is a cone with that point as a vertex (cf. Sect. 1.8.3), Q is a cone of vertex P. So the plane $H = L(P, r)$ is contained in the support of Q. By Exercise 57, we have $Q = H + K$, where K is a plane. Since $r \cap s = r \cap t = \emptyset$, the lines s and t are not contained in H. Then $s, t \subset K$ and therefore $s \cap t \neq \emptyset$, a contradiction.

(b) Denote by H the plane containing r, s and t. If we had $H \not\subset Q$, then the support of the conic $Q \cap H$ of H would contain three distinct lines, which is impossible. Therefore, H is contained in the support of Q. By Exercise 57 H is a component of Q, which is consequently reducible.

(c) Since the lines r, s, t are contained in Q, they are contained in $T_P(Q)$. Then $L(r, s, t) \subset T_P(Q)$, so $T_P(Q) = \mathbb{P}^3(\mathbb{K})$ because r, s, t are not coplanar. Then P is singular for Q.

(d) Denote by H the plane generated by r and s. If H is contained in the support of Q, then Q is reducible by Exercise 57. Otherwise the support of the conic $C = Q \cap H$ is $r \cup s$. Consider the point $R = H \cap t$. By construction, R belongs to C, that is to $r \cup s$, contradicting the hypothesis that $r \cap t = s \cap t = \emptyset$.

Exercise 177. (*Hyperbolic quadrics are ruled*) Let Q be a non-degenerate hyperbolic quadric of $\mathbb{P}^3(\mathbb{K})$. Show that the set of lines contained in Q is the union of two disjoint families X_1 and X_2 such that:

(i) for every point $P \in Q$ there exist a line of X_1 and a line of X_2 passing through P;

(ii) any two lines of the same family are skew;

(iii) if $r \in X_1$ and $s \in X_2$, then r and s meet at one point.

Solution (1). Assume that P is a point of Q. By the definition of hyperbolic quadric, Q intersects the plane $T_P(Q)$ in a pair of distinct lines which meet at P, therefore for any point of Q there are two distinct lines r and s passing through it; by Exercise 176 no other line contained in Q passes through P.

Given $P_0 \in Q$, denote by r_0 and s_0 two lines passing through P_0 and contained in Q. Let us define X_1 as the set of lines $r \subset Q$ such that $r \cap s_0$ is a point and X_2 as the set of lines $s \subset Q$ such that $s \cap r_0$ is a point.

Suppose by contradiction there exists $r \in X_1 \cap X_2$. By the definition of X_1 and X_2, the lines r, r_0 and s_0 are distinct and $r \cap r_0 \neq \emptyset$, $r \cap s_0 \neq \emptyset$. So either r, r_0 and s_0 pass through the point P_0 or they are coplanar, contradicting Exercise 176.

Let us now show that for any point P of Q there is a line of X_1 passing through it. If $P \in s_0$, the claim is true because there exists a line $r \neq s_0$ contained in Q and passing through P; by definition $r \in X_1$. If $P \notin s_0$, let $H = L(P, s_0)$. The conic $\mathcal{Q} \cap H$ contains s_0 and the point $P \notin s_0$, so it is the union of s_0 and a line $r \neq s_0$ such that $P \in r$. The lines s_0 and r, being coplanar, are incident, so $r \in X_1$. In the same way one shows that for any point of $\mathcal{Q}$ there is a line of X_2 passing through it. Observe now that, by Exercise 176 (b) and (c), for any point of $\mathcal{Q}$ there are exactly two lines contained in $\mathcal{Q}$ passing through that point; as a consequence $X_1 \cup X_2$ is the set of all lines contained in $\mathcal{Q}$.

We are now going to show that any two lines of the same family are disjoint (that is they are skew). If by contradiction there existed two incident lines $r, s \in X_1$, then the three distinct lines s_0, r, s either would be coplanar or would share a common point. In both cases we would get to a contradiction with Exercise 176. The same argument shows that any two distinct lines of X_2 are skew.

Finally suppose by contradiction that $r \in X_1$ and $s \in X_2$ are skew. Choose a point $R \in s$ and consider the plane $H = L(r, R)$. The conic $\mathcal{Q} \cap H$ is the union of r and a line t passing through R (hence distinct from r) and distinct from s (because r and s are skew). The line t does not belong to X_1, because it intersects $r \in X_1$ and it is different from r, and it does not belong to X_2, because it intersects $s \in X_2$ and it is different from s. We have so found a contradiction with the fact that $X_1 \cup X_2$ is the set of all lines contained in $\mathcal{Q}$.

Solution (2). Since all non-degenerate hyperbolic quadrics are projectively equivalent (cf. Theorem 1.8.3 and Exercise 174), it is sufficient to consider the quadric $\mathcal{Q}$ defined by

$$\det \begin{pmatrix} x_0 & x_1 \\ x_2 & x_3 \end{pmatrix} = x_0 x_3 - x_1 x_2 = 0. \tag{4.2}$$

Namely, as pointed out in Sect. 1.8.4, $\mathcal{Q}$ is non-degenerate and hyperbolic: an easy computation shows that the tangent plane to $\mathcal{Q}$ at the point $[1, 0, 0, 0]$ is the plane $\{x_3 = 0\}$ which intersects $\mathcal{Q}$ in the lines $r_0 = \{x_3 = x_1 = 0\}$ and $s_0 = \{x_3 = x_2 = 0\}$.

Define X_1 as the set of lines $r_{[a,b]} = \{[\lambda a, \lambda b, \mu a, \mu b] \mid [\lambda, \mu] \in \mathbb{P}^1(\mathbb{K})\}$, as $[a, b]$ varies in $\mathbb{P}^1(\mathbb{K})$. In a similar way, define X_2 as the set of lines $s_{[a,b]} = \{[\lambda a, \mu a, \lambda b, \mu b] \mid [\lambda, \mu] \in \mathbb{P}^1(\mathbb{K})\}$, as $[a, b]$ varies in $\mathbb{P}^1(\mathbb{K})$. Observe that the planes of the pencil centred at s_0 intersect $\mathcal{Q}$, besides in s_0, in a line of X_1. Similarly, the planes of the pencil centred at r_0 intersect $\mathcal{Q}$, besides in r_0, in a line of X_2.

By using parametrizations, one can readily check that the lines of X_1 and X_2 are contained in $\mathcal{Q}$, that two lines of the same family are skew and that for every $[a, b]$, $[c, d] \in \mathbb{P}^1(\mathbb{K})$ the lines $r_{[a,b]}$ and $s_{[c,d]}$ intersect at the point $[ca, cb, da, db]$. So the families X_1 and X_2 are necessarily disjoint.

We now show that for every point of $\mathcal{Q}$ there are a line of X_1 and a line of X_2 passing through it, and that $X_1 \cup X_2$ is the set of all lines contained in $\mathcal{Q}$.

Let $P \in \mathcal{Q}$ be a point of coordinates $[\alpha, \beta, \gamma, \delta]$. If, for instance, $\alpha \neq 0$, by using Eq. (4.2) one can easily check that P belongs to the lines $r_{[\alpha,\beta]}$ and $s_{[\alpha,\gamma]}$. If $\alpha = 0$, one can argue in the same way but considering another coordinate of P. Therefore for every point of $\mathcal{Q}$ there are a line of X_1 and a line of X_2 passing through it. Moreover, since every line $r \subset \mathcal{Q}$ is contained in $T_P(\mathcal{Q})$ for every $P \in r$ and $\mathcal{Q}$ is hyperbolic, then every point $P \in \mathcal{Q}$ belongs exactly to two lines contained in $\mathcal{Q}$. As a consequence, $X_1 \cup X_2$ is the set of all lines contained in $\mathcal{Q}$.

Exercise 178. Assume that r and s are two lines of $\mathbb{P}^3(\mathbb{K})$ such that $r \cap s = \emptyset$ and let $f : r \to s$ be a projective isomorphism. Prove that $X = \bigcup_{P \in r} L(P, f(P))$ is the support of a non-degenerate hyperbolic quadric of $\mathbb{P}^3(\mathbb{K})$.

Solution. If $P_0, P_1 \in r$ are distinct points, set $P_2 = f(P_0)$, $P_3 = f(P_1)$. The points $P_0, \ldots, P_3$ are in general position, so we can complete them to a projective frame $\{P_0, \ldots, P_4\}$. In the homogeneous coordinates $x_0, \ldots, x_3$ of $\mathbb{P}^3(\mathbb{K})$ induced by this frame, r has equations $x_2 = x_3 = 0$, s has equations $x_0 = x_1 = 0$ and the projective isomorphism f is given by $[y_0, y_1, 0, 0] \mapsto [0, 0, ay_0, by_1]$, with $a, b \in \mathbb{K}^*$. So X is the set of points of coordinates $[\lambda y_0, \lambda y_1, \mu a y_0, \mu b y_1]$, where $[\lambda, \mu]$, $[y_0, y_1]$ vary in $\mathbb{P}^1(\mathbb{K})$. It is easy to check that the points of X satisfy the equation

$$0 = a x_0 x_3 - b x_1 x_2 = \det \begin{pmatrix} a x_0 & b x_1 \\ x_2 & x_3 \end{pmatrix}, \tag{4.3}$$

which defines a non-degenerate quadric $\mathcal{Q}$ of $\mathbb{P}^3(\mathbb{K})$ of hyperbolic type.

Conversely, let $R \in \mathcal{Q}$ be a point of coordinates $[c_0, c_1, c_2, c_3]$. If $c_0 = c_1 = 0$, then R belongs to $s = f(r)$, hence to X. Otherwise, let $P \in r$ be the point of coordinates $[c_0, c_1, 0, 0]$. As Eq. (4.3) means precisely that $[a c_0, b c_1] = [c_2, c_3] \in \mathbb{P}^1(\mathbb{K})$, the point $f(P)$ has coordinates $[0, 0, c_2, c_3]$. Therefore R lies on the line $L(P, f(P)) \subseteq X$.

Note. Exercise 178 shows that it is possible to associate a hyperbolic quadric to every projective isomorphism between two skew lines of $\mathbb{P}^3(\mathbb{K})$.

Conversely, given a hyperbolic quadric $\mathcal{Q}$, denote by X_1, X_2 the two rulings of $\mathcal{Q}$ (cf. Exercise 177). If r, r', r'' are distinct lines of X_1 and $f : r \to r'$ is the perspectivity centred at r'', then f does not depend on r'' and $\mathcal{Q}$ coincides with the quadric associated to f. Namely, if $P \in r$, denote by s_P the only line of X_2 passing through P; for any $r'' \neq r, r'$, the line s_P intersects both r' and r'', and $f(P) = s_P \cap r'$. Thanks to this observation one can give an alternative proof of Exercise 39, which states that every projective isomorphism between skew lines of $\mathbb{P}^3(\mathbb{K})$ is a perspectivity, whose center can be chosen in infinitely many different ways.

Exercise 179. Let r_1, r_2, r_3 be pairwise skew lines of $\mathbb{P}^3(\mathbb{K})$. Let X be the set of lines of $\mathbb{P}^3(\mathbb{K})$ such that $s \in X$ if and only if $s \cap r_i \neq \emptyset$ for $i = 1, 2, 3$. Prove that $Z = \bigcup_{s \in X} s$ is the support of a non-degenerate hyperbolic quadric of $\mathbb{P}^3(\mathbb{K})$ that contains $r_1 \cup r_2 \cup r_3$.

Solution. Let $f : r_1 \to r_2$ be the perspectivity centred at r_3, and set $Y = \{L(P, f(P)) \mid P \in r_1\}$. We will first show that $Y = X$. Thanks to Exercise 178, this ensures that Z is the support of a non-degenerate quadric of $\mathbb{P}^3(\mathbb{K})$ of hyperbolic type. Then we will prove that $r_1 \cup r_2 \cup r_3 \subseteq Z$.

Let $s \in X$ be a line. For $i = 1, 2, 3$, we set $Q_i = r_i \cap s$. The plane $H = L(Q_1, r_3)$ contains $s = L(Q_1, Q_3)$. Since $Q_2 \in s$, H coincides with the plane $L(Q_2, r_3)$. Hence we have $f(Q_1) = Q_2$ and $s = L(Q_1, f(Q_1)) \in Y$. Therefore, we have $X \subseteq Y$.

Conversely, let $P \in r_1$. By definition of perspectivity we have $f(P) = L(P, r_3) \cap r_2$, so that the lines $L(P, f(P))$ and r_3, being coplanar, do intersect. Therefore, from the fact that $L(P, f(P)) \cap r_1 = P$ and $L(P, f(P)) \cap r_2 = f(P)$ we deduce that the line $L(P, f(P))$ belongs to X. Since P can be arbitrarily chosen, this implies that $Y \subseteq X$.

In order to conclude, we are left to show that $r_1 \cup r_2 \cup r_3 \subseteq Z$. Since $X = Y$, the set Z contains both r_1 and r_2, which are the domain and the image of f, respectively. Moreover, if $R \in r_3$ then it is immediate to check that the plane $L(r_2, R)$ intersects r_1 at one point P. Therefore, being coplanar with r_2, the line $L(P, R)$ intersects each of r_1, r_2, and r_3, hence it belongs to X. We have thus shown that every point of r_3 belongs to a line of X, whence to Z.

Exercise 180. Let r_1, r_2, r_3 be pairwise skew lines of $\mathbb{P}^3(\mathbb{K})$. Prove that there exists a unique quadric $\mathcal{Q}$ such that $r_i \subset \mathcal{Q}$ for $i = 1, 2, 3$, and that $\mathcal{Q}$ is non-degenerate.

Solution. Exercise 179 implies that at least one quadric $\mathcal{Q}$ containing r_1, r_2, r_3 exists. This quadric is non-degenerate of hyperbolic type and its support is equal to $\bigcup_{s \in X} s$, where X is the set of lines of $\mathbb{P}^3(\mathbb{K})$ that intersect each of r_1, r_2, r_3. Moreover, by Exercise 176 (a), no singular quadric can contain r_1, r_2, and r_3.

Suppose by contradiction that $\mathcal{Q}$ and $\mathcal{Q}'$ are distinct quadrics that contain r_1, r_2, r_3, and let A and A' be symmetric 4×4 matrices that define $\mathcal{Q}$ and $\mathcal{Q}'$, respectively. For every $[\lambda, \mu] \in \mathbb{P}^1(\mathbb{K})$ denote by $\mathcal{Q}_{\lambda, \mu}$ the quadric defined by the matrix $\lambda A + \mu A'$. Every quadric $\mathcal{Q}_{\lambda, \mu}$ contains r_1, r_2, and r_3. Moreover, if $\mathbb{K} = \mathbb{C}$ then there exists at least one homogeneous pair $[\lambda_0, \mu_0]$ such that $\mathcal{Q}_{\lambda_0, \mu_0}$ is degenerate, against the previous discussion.

If $\mathbb{K} = \mathbb{R}$, then $\mathcal{Q}$ is a real quadric containing r_1, r_2, and r_3, and the complexification $\mathcal{Q}_{\mathbb{C}}$ of $\mathcal{Q}$ is a complex quadric containing the complexifications of the lines r_1, r_2, and r_3, so that $\mathcal{Q}_{\mathbb{C}}$ is uniquely determined. This implies that also $\mathcal{Q}$ is uniquely determined.

Exercise 181. Consider the lines of $\mathbb{P}^3(\mathbb{R})$ having equations

$$r_1 = \{x_0 + x_1 = x_2 + x_3 = 0\}, \quad r_2 = \{x_0 + x_2 = x_1 - x_3 = 0\},$$
$$r_3 = \{x_0 - x_1 = x_2 - x_3 = 0\}.$$

Determine the equation of a quadric containing r_1, r_2 and r_3.

Solution (1). It is easy to check that the lines r_i are pairwise skew. Therefore, by Exercise 180 there exists a unique quadric $\mathcal{Q}$ containing r_1, r_2 and r_3, and this quadric is non-degenerate.

Let

$$F_1(x) = (x_0 + x_1)(x_0 + x_2), \quad F_2(x) = (x_2 + x_3)(x_0 + x_2),$$
$$F_3(x) = (x_0 + x_1)(x_1 - x_3), \quad F_4(x) = (x_2 + x_3)(x_1 - x_3)$$

and, for $i = 1, \ldots, 4$, let $\mathcal{Q}_i$ be the quadric defined by the equation $F_i(x) = 0$. Each $\mathcal{Q}_i$ has rank equal to 2, and contains both r_1 and r_2. Let us consider a linear combination $F = \alpha_1 F_1 + \alpha_2 F_2 + \alpha_3 F_3 + \alpha_4 F_4$ of the polynomials F_i, where $\alpha_i \in \mathbb{R}$, and let us impose that F vanishes on r_3. Since $\deg F = 2$, to this aim it is sufficient to require that $F(P_1) = F(P_2) = F(P_3) = 0$ for three distinct points $P_1, P_2, P_3 \in r_3$. Choosing the points $P_1 = [1, 1, 0, 0]$, $P_2 = [0, 0, 1, 1]$ and $P_3 = [1, 1, 1, 1]$, we get the equations $\alpha_1 + \alpha_3 = 0$, $\alpha_2 - \alpha_4 = 0$, $\alpha_1 + \alpha_2 = 0$.

A non-trivial solution of this system is given by $\alpha_1 = -1$, $\alpha_2 = \alpha_3 = \alpha_4 = 1$. With this choice we have $F(x) = -x_0^2 + x_1^2 + x_2^2 - x_3^2$, and this polynomial defines a quadric with the required properties.

Solution (2). By Exercise 179, the support of a quadric containing r_1, r_2 and r_3 is given by the set $\bigcup_{s \in X} s$, where X is the set of lines that intersect each of r_1, r_2 and r_3. Moreover, it is possible to characterize the lines in X by studying the projections $\pi_{ij} : \mathbb{P}^3(\mathbb{R}) \setminus r_i \to r_j$ on r_j centred at r_i, $i, j \in \{1, 2, 3\}$, $i \neq j$. This will allow us to obtain an analytic formulation of the condition that a point belongs to a line in X, thus getting an equation of the desired quadric.

A point $Q \in \mathbb{P}^3(\mathbb{R}) \setminus (r_1 \cup r_2 \cup r_3)$ belongs to $\bigcup_{s \in X} s$ if and only if $\pi_{21}(Q) = \pi_{31}(Q)$. Indeed, for a given line $s \in X$ let us set $Q_i = s \cap r_i$, $i = 1, 2, 3$. Then for every $Q \in s \setminus \{Q_1, Q_2, Q_3\}$ and for every $i \neq j \in \{1, 2, 3\}$ we have $\pi_{ij}(Q) = Q_j$. On the other hand, if $Q \in \mathbb{P}^3(\mathbb{R}) \setminus (r_1 \cup r_2 \cup r_3)$ and $\pi_{21}(Q) = \pi_{31}(Q) = Q_1$, denote by s the line $L(Q, Q_1)$. Of course we have $s \cap r_1 = Q_1$. Moreover, for $i = 2, 3$ we have $s \subset L(Q, r_i)$ so that s and r_i are incident, and s belongs to X.

Let us denote by $[y_0, y_1, y_2, y_3]$ the coordinates of a point $Q \notin r_2$. The plane $L(Q, r_2)$ has equation $(y_3 - y_1)(x_0 + x_2) + (y_0 + y_2)(x_1 - x_3) = 0$ and the point $L(Q, r_2) \cap r_1$ has coordinates

$$[y_0 - y_1 + y_2 + y_3, -y_0 + y_1 - y_2 - y_3, y_0 + y_1 + y_2 - y_3, -y_0 - y_1 - y_2 + y_3].$$

Hence the projection π_{21} maps the point of coordinates $[y_0, y_1, y_2, y_3]$ to the point of coordinates

$$[y_0 - y_1 + y_2 + y_3, -y_0 + y_1 - y_2 - y_3, y_0 + y_1 + y_2 - y_3, -y_0 - y_1 - y_2 + y_3].$$

A similar computation gives the following expression of π_{31} in coordinates:

$$[y_0, y_1, y_2, y_3] \mapsto [y_0 - y_1, -y_0 + y_1, y_2 - y_3, -y_2 + y_3].$$

Since r_1 has equations $x_0 + x_1 = x_2 + x_3 = 0$, the points $\pi_{21}(Q)$ and $\pi_{31}(Q)$ are uniquely determined by the first and the third coordinates. Therefore, if $Q \notin (r_2 \cup r_3)$ we have that $\pi_{21}(Q) = \pi_{31}(Q)$ if and only if

$$\mathrm{rk} \begin{pmatrix} y_0 - y_1 + y_2 + y_3 & y_0 + y_1 + y_2 - y_3 \\ y_0 - y_1 & y_2 - y_3 \end{pmatrix} \leq 1,$$

i.e. if and only if

$$0 = \det \begin{pmatrix} y_0 - y_1 + y_2 + y_3 & y_0 + y_1 + y_2 - y_3 \\ y_0 - y_1 & y_2 - y_3 \end{pmatrix} = -y_0^2 + y_1^2 + y_2^2 - y_3^2.$$

Then let Q be the quadric defined by the equation $-y_0^2 + y_1^2 + y_2^2 - y_3^2 = 0$. By construction, every line $s \in X$ has infinitely many points in common with Q, hence it is contained in Q. It follows that Q contains X, so Q contains r_1, r_2 and r_3.

Exercise 182. (*Polar of a line with respect to a quadric*) Let Q be a non-degenerate quadric of $\mathbb{P}^3(\mathbb{K})$ and let r be a projective line. Prove that:

(a) When P varies in r, the planes $\mathrm{pol}(P)$ all intersect in a line r' (called the *polar line* of r with respect to Q).
(b) The polar of r' is r.
(c) $r = r'$ if and only if r is contained in Q.
(d) If r and r' are distinct and incident, then r and r' are tangent to Q at the point $r \cap r'$.
(e) r is tangent to Q at a point P if and only if r' is tangent to Q at the same point P.
(f) If r is not tangent to Q, then there exist two distinct planes containing r and tangent to Q if and only if r' is secant to Q (i.e. it intersects Q exactly at two distinct points). In this case, the planes are tangent to Q exactly at the points of $Q \cap r'$.

Solution. (a) The projective isomorphism $\mathrm{pol} \colon \mathbb{P}^3(\mathbb{K}) \to \mathbb{P}^3(\mathbb{K})^*$ transforms the line r into a line of $\mathbb{P}^3(\mathbb{K})^*$, i.e. into a pencil of planes of $\mathbb{P}^3(\mathbb{K})$ centred at a line r'. In particular, for any given points M and N of r, the planes $\mathrm{pol}(M)$ and $\mathrm{pol}(N)$ are distinct and their intersection coincides with the line r'.

Statement (b) immediately follows from the reciprocity property of polarity. Hence the line r' is the locus of poles of the planes through the line r.

(c) Suppose $r = r'$. Then for every point $M \in r$ we have $r \subset \mathrm{pol}(M)$, hence $M \in \mathrm{pol}(M)$, so that $M \in Q$ and $r \subset Q$. Conversely, if $r \subset Q$, then for every point $M \in r$ we have $r \subset \mathrm{pol}(M) = T_M(Q)$, hence $r = r'$.

(d) Denote by P the point where the distinct lines r and r' intersect, and let $H = L(r, r')$ be the plane containing these lines. Since $P \in r$, by definition $r' \subset \mathrm{pol}(P)$. Since $P \in r'$, by (b) we have $r \subset \mathrm{pol}(P)$, hence $\mathrm{pol}(P) = H$. In particular, $P \in \mathrm{pol}(P)$; hence $P \in Q$ and $L(r, r') = H = \mathrm{pol}(P) = T_P(Q)$. It follows that r and r' are tangent to Q at P.

(e) If $r = r'$, the conclusion trivially follows. Then we may suppose that r and r' are distinct lines. If r is tangent to Q at P, then r is contained in the tangent plane $T_P(Q) = \text{pol}(P)$ so that for every point $M \in r$ we have $M \in \text{pol}(P)$. By reciprocity $P \in \bigcap_{M \in r} \text{pol}(M) = r'$, so r and r' are distinct and incident at P. Then by (d) also r' is tangent to Q at P.

The converse implication immediately follows from what we have just proved and from (b).

(f) If r' is secant and intersects Q at the distinct points M' and N', then by definition of polar line and by (b) the planes $T_{M'}(Q) = \text{pol}(M')$ and $T_{N'}(Q) = \text{pol}(N')$ pass through r.

Conversely, let us prove that if there exists at least one plane H passing through r and tangent to Q, then r' is secant. Namely, if H is tangent to Q at a point M, then $M \in r'$, because r' is the locus of poles of the planes through r. Therefore, $M \in Q \cap r'$. Moreover, by (e) r' is not tangent to Q, hence it is secant.

Exercise 183. Find the planes of $\mathbb{P}^3(\mathbb{C})$ that are tangent to the quadric Q of equation

$$F(x_0, x_1, x_2, x_3) = 2x_1^2 + x_2^2 - 2x_1x_2 - x_3^2 + 2x_0x_1 = 0$$

and that contain the line r of equations $x_2 - x_3 = 0$, $x_0 + 3x_1 - 3x_2 = 0$.

Solution (1). It is easy to check that the quadric Q is non-degenerate. The required planes are the planes of the pencil $\mathcal{F}$ centred at r that intersect Q in a degenerate conic. The planes of $\mathcal{F}$ have equation

$$\lambda(x_2 - x_3) + \mu(x_0 + 3x_1 - 3x_2) = 0$$

where $[\lambda, \mu]$ varies in $\mathbb{P}^1(\mathbb{C})$. Let us first discuss whether the plane $H = \{x_2 - x_3 = 0\}$ corresponding to the choice $[\lambda, \mu] = [1, 0]$ is tangent to the quadric or not. Using the homogeneous coordinates x_0, x_1, x_2 on H, the conic $Q \cap H$ has equation

$$G(x_0, x_1, x_2) = F(x_0, x_1, x_2, x_2) = 2x_1^2 - 2x_1x_2 + 2x_0x_1 = 0.$$

Since this conic is degenerate, H is tangent to Q.

After excluding H from the pencil $\mathcal{F}$, the remaining planes H_t of the pencil are described by the equation $t(x_2 - x_3) + x_0 + 3x_1 - 3x_2 = 0$, where t varies in $\mathbb{C}$. We endow H_t with the homogeneous coordinates x_0, x_1, x_2. Then, by substituting $x_0 = -3x_1 + (3 - t)x_2 + tx_3$ in F, we obtain that the conic $Q \cap H_t$ of the plane H_t has equation $-4x_1^2 + x_2^2 - x_3^2 + (4 - 2t)x_1x_2 + 2tx_1x_3 = 0$, so that it is represented by the matrix $A_t = \begin{pmatrix} -4 & 2-t & t \\ 2-t & 1 & 0 \\ t & 0 & -1 \end{pmatrix}$. Since $\det A_t = -4t + 8$, the conic $Q \cap H_t$ is degenerate (that is, H_t is tangent to Q) if and only if $t = 2$. Therefore, the only planes containing r and tangent to Q are given by the plane H_2 having equation $x_0 + 3x_1 - x_2 - 2x_3 = 0$, and the plane $x_2 - x_3 = 0$ previously found.

Solution (2). The points $R = [3, -1, 0, 0]$ and $S = [0, 1, 1, 1]$ belong to r, and hence $r' = \mathrm{pol}(R) \cap \mathrm{pol}(S)$. We thus get that r' is the line of equations $-x_0 + x_1 + x_2 = 0$, $x_0 + x_1 - x_3 = 0$. By solving the system given by the equations of r' and the equation of $\mathcal{Q}$, we obtain that $\mathcal{Q} \cap r'$ consists of the points $M' = [1, 0, 1, 1]$ and $N' = [1, 1, 0, 2]$. By Exercise 182 and using that $\mathcal{Q}$ is non-degenerate, we conclude that the required planes are the plane $\mathrm{pol}(M')$ and $\mathrm{pol}(N')$ having equation $x_2 - x_3 = 0$ and $x_0 + 3x_1 - x_2 - 2x_3 = 0$, respectively.

Exercise 184. Let $\mathcal{Q}$ be a non-degenerate hyperbolic quadric of $\mathbb{P}^3(\mathbb{R})$; let r be a line that is not tangent to $\mathcal{Q}$ and let r' be its polar (cf. Exercise 182). Prove that:

(a) r is secant if and only if r' is secant.
(b) There exist two planes of $\mathbb{P}^3(\mathbb{R})$ that pass through r and are tangent to $\mathcal{Q}$ if and only if r is secant.

Solution. (a) If r' is secant and intersects the quadric at two distinct points M' and N', then by Exercise 182 (f) there exist two planes H_1 and H_2 that pass through r and are tangent to $\mathcal{Q}$ at the points M' and N', respectively. Let X_1 and X_2 be the families of lines contained in the hyperbolic quadric $\mathcal{Q}$ that are described in Exercise 177. Since $\mathcal{Q}$ is hyperbolic, we have $\mathcal{Q} \cap H_1 = \mathcal{Q} \cap T_{M'}(\mathcal{Q}) = m_1 \cup m_2$, where $m_1 \in X_1$ and $m_2 \in X_2$. In the same way $\mathcal{Q} \cap H_2 = \mathcal{Q} \cap T_{N'}(\mathcal{Q}) = n_1 \cup n_2$, where $n_1 \in X_1$ and $n_2 \in X_2$. Observe that, since $H_1 \cap H_2 = r$ and r is not contained in $\mathcal{Q}$, the four lines m_1, m_2, n_1, n_2 are distinct. Then Exercise 176 implies that there is no point belonging to three of these four lines. It follows from Exercise 177 that the lines m_1 and n_2 intersect at a point of $\mathcal{Q}$, and on the other hand $m_1 \cap n_2 \in H_1 \cap H_2 = r$, so that the point $m_1 \cap n_2$ belongs to $\mathcal{Q} \cap r$. In the same way the point $m_2 \cap n_1$ belongs to $\mathcal{Q} \cap r$ and is distinct from $m_1 \cap n_2$ because of the previous considerations. Therefore, r is secant.

The converse implication immediately follows from the fact that r is the polar of r'.

(b) is an obvious consequence of (a) and of Exercise 182 (f).

Note. Let us consider the quadric $\mathcal{Q}$ and the line r defined in Exercise 183. In that case, both $\mathcal{Q}$ and r were defined by polynomials with real coefficients. Moreover, it is not difficult to check that $\mathcal{Q}$ is hyperbolic, and that r is secant because it intersects $\mathcal{Q}$ at the distinct real points $M = [3, 0, 1, 1]$ and $N = [0, 1, 1, 1]$. Coherently with what we have just proved, in that exercise we found two planes that pass through r and are tangent to $\mathcal{Q}$.

Exercise 185. Let $\mathcal{Q}$ be a non-empty non-degenerate elliptic quadric of $\mathbb{P}^3(\mathbb{R})$; let r be a line that is not tangent to $\mathcal{Q}$ and let r' be the polar of r with respect to $\mathcal{Q}$. Prove that:

(a) r is external if and only if r' is secant.
(b) There exist two planes of $\mathbb{P}^3(\mathbb{R})$ passing through r and tangent to $\mathcal{Q}$ if and only if r is external.

Solution. (a) Suppose that r' is secant and intersects the quadric at the distinct points M' and N'. Then $T_{M'}(\mathcal{Q}) \cap T_{N'}(\mathcal{Q}) = \mathrm{pol}(M') \cap \mathrm{pol}(N') = r$; moreover, $\mathcal{Q} \cap T_{M'}(\mathcal{Q}) = \{M'\}$ and $\mathcal{Q} \cap T_{N'}(\mathcal{Q}) = \{N'\}$ because the quadric is elliptic. The lines r and r' are distinct, because r is not contained in the quadric, and they are not incident, since r is not tangent (cf. Exercise 182). Therefore, if there exists a point $R \in \mathcal{Q} \cap r$, then necessarily $R \neq M'$, which contradicts the fact that $\mathcal{Q} \cap r \subset \mathcal{Q} \cap T_{M'}(\mathcal{Q}) = \{M'\}$.

Conversely, suppose that r is external. Then by Exercise 182 (e) the line r' is not tangent $\mathcal{Q}$.

The complexification $\mathcal{Q}_{\mathbb{C}}$ is a non-degenerate complex quadric, so it is hyperbolic. Let X_1 and X_2 be the two families of complex lines contained in $\mathcal{Q}_{\mathbb{C}}$ that are described in Exercise 177. Observe that the conjugation σ preserves incidence relations, and recall that lines of the same family are pairwise disjoint, while if $l_i \in X_i$, $i = 1, 2$, then l_1 and l_2 are incident. Therefore, σ either transforms each family of lines into itself or switches X_1 with X_2. In order to understand which of these possibilities occurs it is then sufficient to analyse the behaviour of σ on a single line. Let $P \in \mathcal{Q}$ and $H = T_P(\mathcal{Q}_{\mathbb{C}})$. Then $\sigma(H) = H$, hence the degenerate conic $\mathcal{C} = \mathcal{Q}_{\mathbb{C}} \cap H = l_1 \cup l_2$, where l_1, l_2 are conjugate complex lines, is left invariant by σ. Since $\mathcal{Q}$ is elliptic, the lines l_1, l_2 cannot be real, so $\sigma(l_i) \neq l_i$ for $i = 1, 2$. Therefore, σ switches the incident lines l_1, l_2, so that it switches X_1 with X_2.

Since in the complex case no line can be external to a projective quadric, let M and N be the points of intersection between $\mathcal{Q}_{\mathbb{C}}$ and $r_{\mathbb{C}}$. The complex points M and N are conjugate and $r'_{\mathbb{C}}$ is the intersection of the tangent spaces $H_1 = T_M(\mathcal{Q}_{\mathbb{C}})$ and $H_2 = T_N(\mathcal{Q}_{\mathbb{C}})$, that are also conjugate to each other. We have $\mathcal{Q}_{\mathbb{C}} \cap H_1 = \mathcal{Q}_{\mathbb{C}} \cap T_M(\mathcal{Q}_{\mathbb{C}}) = m_1 \cup m_2$ with $m_1 \in X_1$ and $m_2 \in X_2$. In the same way $\mathcal{Q}_{\mathbb{C}} \cap H_2 = \mathcal{Q}_{\mathbb{C}} \cap T_N(\mathcal{Q}_{\mathbb{C}}) = n_1 \cup n_2$ with $n_1 \in X_1$ and $n_2 \in X_2$. By the previous discussion we have $\sigma(m_1) = n_2$, hence the lines m_1 and n_2 intersect at a real point R that belongs to $r'_{\mathbb{C}} = H_1 \cap H_2$. Therefore $R \in \mathcal{Q} \cap r'$, so r' is secant. In fact, the same argument also shows that $m_2 \cap n_1$ belongs to $\mathcal{Q} \cap r'$.

(b) is an obvious consequence of (a) and of Exercise 182 (f).

Exercise 186. Let H be a plane of $\mathbb{P}^3(\mathbb{C})$, let $\mathcal{C} \subset H$ be a non-degenerate conic, let $P, P' \in \mathcal{C}$ be distinct points and $t, t' \subset H$ the tangent lines to $\mathcal{C}$ at P and at P', respectively. Let r and s be skew lines of $\mathbb{P}^3(\mathbb{C})$ such that $r \cap H = P'$ and $s \cap H = P$. Let $\pi : \mathcal{C} \to r$ be the map defined by $\pi(Q) = L(s, Q) \cap r$ if $Q \neq P$ and $\pi(P) = L(s, t) \cap r$ (Fig. 4.15).

(a) Prove that π is well defined and bijective.
(b) Set
$$m = L(r, t') \cap L(s, P'), \qquad X = \bigcup_{Q \in \mathcal{C} \setminus \{P'\}} L(Q, \pi(Q)) \cup m,$$

and prove that X is the support of a non-degenerate quadric $\mathcal{Q}$.

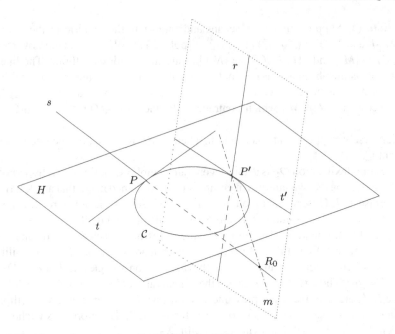

Fig. 4.15 The configuration described in Exercise 186

Solution. (a) The proof is exactly the same as in the solution of Exercise 119 (a).

(b) By Exercise 178, it is sufficient to show that there exists a projective isomorphism $\phi\colon r \to s$ such that $X = \bigcup_{R \in r} L(R, \phi(R))$.

Let us first define a projective isomorphism $\alpha\colon \mathcal{F}_s \to \mathcal{F}_r$, where $\mathcal{F}_s$ is the pencil of planes centred at s and $\mathcal{F}_r$ is the pencil of planes centred at r. Set $\alpha(L(s, t)) = L(r, P)$ and $\alpha(L(s, P')) = L(r, t')$. If $K \in \mathcal{F}_s$ is distinct both from $L(s, t)$ and from $L(s, P')$, then K intersects H in a line containing P and distinct both from t and from $L(P, P')$. This line in turn intersects $\mathcal{C}$ at P and at a point Q_K distinct both from P and from P'. We then set $\alpha(K) = L(r, Q_K)$.

Let us check that α is a projective isomorphism. If $\mathcal{F}_P$ and $\mathcal{F}_{P'}$ are the pencils of lines of H with centre at P and P', respectively, we define $\psi\colon \mathcal{F}_P \to \mathcal{F}_{P'}$ by setting $\psi(t) = L(P, P')$, $\psi(L(P, P')) = t'$, and $\psi(l) = L(P', A)$ for every $l \in \mathcal{F}_P$, $l \neq t$, $l \neq L(P, P')$, where A is the point of intersection of l with $\mathcal{C}$ that is distinct from P. Exercise 142 implies that ψ is a well-defined projective isomorphism. Moreover, if $\beta_s\colon \mathcal{F}_s \to \mathcal{F}_P$, $\beta_r\colon \mathcal{F}_r \to \mathcal{F}_{P'}$ are defined by $\beta_s(K) = K \cap H$, $\beta_r(K') = K' \cap H$ for every $K \in \mathcal{F}_s$, $K' \in \mathcal{F}_r$, then β_s and β_r are well-defined projective isomorphisms (cf. Exercise 33). Finally, it is immediate to check that $\alpha = \beta_r^{-1} \circ \psi \circ \beta_s$. Being the composition of projective isomorphisms, the map α is itself a projective isomorphism.

Let $\gamma_1\colon \mathcal{F}_s \to r$ be the projective isomorphism defined by $K \mapsto K \cap r$, and $\gamma_2\colon \mathcal{F}_r \to s$ the projective isomorphism defined by $K' \mapsto K' \cap s$ (cf. Exercise 32) and let us set $\phi = \gamma_2 \circ \alpha \circ \gamma_1^{-1}\colon r \to s$. Being the composition of projective isomorphisms, the map ϕ is a projective isomorphism.

Let us now consider the sets of lines

$$Z_1 = \{L(R, \phi(R)) \mid R \in r\}, \quad Z_2 = \{L(Q, \pi(Q)) \mid Q \in \mathcal{C} \setminus \{P'\}\} \cup \{m\}.$$

Since $X = \bigcup_{l \in Z_2} l$, in order to conclude it is sufficient to show that $Z_1 = Z_2$.

The line m belongs to Z_2 by definition. If $R_0 = L(r, t') \cap s$, then $m = L(R_0, P')$, so in order to prove that $m \in Z_1$ it is sufficient to show that $\phi(P') = R_0$. However, it readily follows from the definitions that $\phi(P') = \gamma_2(\alpha(L(s, P'))) = \gamma_2(L(r, t')) = L(r, t') \cap s = R_0$, as desired. Since $\pi(P') = P'$, the map π restricts to a bijection between $\mathcal{C} \setminus \{P'\}$ and $r \setminus \{P'\}$. For every $R \in r \setminus \{P'\}$, by construction the points $R, \pi^{-1}(R)$ and $\phi(R)$ are collinear, and we have $R \neq \pi^{-1}(R)$ and $R \neq \phi(R)$. This readily implies that $Z_1 \setminus \{m\} = Z_2 \setminus \{m\}$, so that $Z_1 = Z_2$.

Note. The construction described in Exercise 186 may be interpreted as a degeneration of the construction of Exercise 119, where the case when the point $r \cap H$ does not belong to $\mathcal{C}$ is considered. In that case, the analogue of the set X turns out to be the support of an irreducible cubic surface that, in the "limit" situation where $r \cap \mathcal{C} \neq \emptyset$, splits as the union of the plane $L(s, H \cap r)$ and the quadric $\mathcal{Q}$ described here.

Exercise 187. Let $\mathcal{Q}$ be a quadric of $\mathbb{R}^n$ having rank n and let $\overline{A} = \left(\begin{array}{c|c} c & {}^t B \\ \hline B & A \end{array} \right) \in M(n+1, \mathbb{R})$ be a matrix representing $\mathcal{Q}$. Prove that $\mathcal{Q}$ is an affine cone if and only if $\det A \neq 0$.

Solution. Since $\mathcal{Q}$ has rank n, its projective closure $\overline{\mathcal{Q}}$ has a unique singular point P of homogeneous coordinates $[z_0, \dots, z_n]$; therefore, Sect. 1.8.3 implies that $\overline{\mathcal{Q}}$ is a cone whose vertex set is given by the point $[z_0, \dots, z_n]$, so that $\mathcal{Q}$ is an affine cone if and only if $z_0 \neq 0$.

Let $Z = (z_1, \dots, z_n) \in \mathbb{R}^n$ and $\widehat{Z} = (z_0, z_1, \dots, z_n) \in \mathbb{R}^{n+1}$. The fact that P is singular for $\overline{\mathcal{Q}}$ translates into the equation $\overline{A}\widehat{Z} = 0$ or, equivalently, into the equations $AZ + z_0 B = 0$ and ${}^t B Z + z_0 c = 0$.

If $z_0 = 0$, then $Z \neq 0$ and $AZ = 0$, hence $\det A = 0$.

If $z_0 \neq 0$, since $B = -z_0^{-1} A Z$, the vector ${}^t B$ is a linear combination of the rows of A, hence every $W \in \mathbb{R}^n$ such that $AW = 0$ also satisfies ${}^t B W = 0$. Let us suppose by contradiction that $\det A = 0$ and let us denote by $Z' = (z_1', \dots, z_n') \in \mathbb{R}^n$ a vector such that $Z' \neq 0$ and $AZ' = 0$. The point $Q = [0, z_1', \dots, z_n'] \in \mathbb{P}^n(\mathbb{R})$ is singular for $\overline{\mathcal{Q}}$ and distinct from P, and this contradicts the fact that P is the unique singular point of $\overline{\mathcal{Q}}$.

Exercise 188. Let $\mathcal{Q}$ be a non-degenerate quadric of $\mathbb{R}^n$, let H be a diametral hyperplane and let $P = [(0, v)] \in \mathbb{P}^n(\mathbb{R})$ be the pole of $\overline{H}$ with respect to $\overline{\mathcal{Q}}$, where $v \in \mathbb{R}^n$.

(a) Prove that $P \in \overline{H}$ if and only if the vector v is parallel to H.

(b) Suppose $P \notin \overline{H}$ and denote by $\tau_{H,v} \colon \mathbb{R}^n \to \mathbb{R}^n$ the reflection fixing H pointwise and sending v to $-v$. Prove that $\tau_{H,v}(\mathcal{Q}) = \mathcal{Q}$.

Solution. (a) Let $v = (v_1, \ldots, v_n)$ and let $a_1 x_1 + \cdots + a_n x_n + b = 0$ be an equation of H. The hyperplane $\overline{H}$ is defined by $b x_0 + a_1 x_1 + \cdots + a_n x_n = 0$, so that $P \in \overline{H}$ if and only if $a_1 v_1 + \cdots + a_n v_n = 0$, that is, if and only if v is parallel to H.

(b) By part (a), if $P \notin \overline{H}$ then there exist affine coordinates $x_1, \ldots, x_n$ such that $v = (0, \ldots, 0, 1)$ and $H = \{x_n = 0\}$. In this system of coordinates, $\tau_{H,v}$ is described by the formula $(x_1, \ldots, x_n) \mapsto (x_1, \ldots, x_{n-1}, -x_n)$. The quadric $\mathcal{Q}$ is then represented by a matrix of the form $\begin{pmatrix} c & {}^t\!B & 0 \\ \hline B & A & 0 \\ \hline 0 & 0 & a \end{pmatrix}$, where $A \in M(n-1, \mathbb{R})$, $B \in \mathbb{R}^{n-1}$, $c \in \mathbb{R}$, and $a \in \mathbb{R}^*$. It is immediate to check that the matrix defining $\tau_{H,v}(\mathcal{Q})$ is the same as the one defining $\mathcal{Q}$, hence $\tau_{H,v}(\mathcal{Q}) = \mathcal{Q}$.

Exercise 189. (*Symmetries of quadrics of* $\mathbb{R}^n$) Let $\mathcal{Q}$ be a non-degenerate quadric of $\mathbb{R}^n$, let $H \subset \mathbb{R}^n$ be an affine hyperplane, and let $\tau_H \colon \mathbb{R}^n \to \mathbb{R}^n$ be the orthogonal reflection with respect to H. Prove that:

(a) If the projective closure $\overline{H}$ of H is tangent to $\overline{\mathcal{Q}}$, then $\tau_H(\mathcal{Q}) \neq \mathcal{Q}$.
(b) H is a principal hyperplane of $\mathcal{Q}$ if and only if $\tau_H(\mathcal{Q}) = \mathcal{Q}$.

Solution. (a) Let P be the point where $\overline{\mathcal{Q}}$ is tangent to $\overline{H}$. If P is a proper point, then up to an isometric change of coordinates on $\mathbb{R}^n$ we may suppose that P has coordinates $(0, \ldots, 0)$, and that H is defined by the equation $y_1 = 0$. In this system of coordinates, $\mathcal{Q}$ is defined by $f(y_1, \ldots, y_n) = y_1 + f_2(y_1, \ldots, y_n) = 0$, where f_2 is a homogeneous polynomial of degree 2. The orthogonal reflection τ_H is given by $(y_1, \ldots, y_n) \mapsto (-y_1, y_2, \ldots, y_n)$. Therefore, the quadric $\tau_H(\mathcal{Q})$ is defined by $-y_1 + f_2(-y_1, y_2, \ldots, y_n) = 0$. Suppose by contradiction that $\mathcal{Q} = \tau_H(\mathcal{Q})$: then there exists $\lambda \in \mathbb{R}^*$ such that $f(-y_1, y_2, \ldots, y_n) = \lambda f(y_1, \ldots, y_n)$. Since $f(-y_1, y_2, \ldots, y_n) = -y_1 + f_2(-y_1, y_2, \ldots, y_n)$, we must have $\lambda = -1$ and $f_2(-y_1, y_2, \ldots, y_n) = -f_2(y_1, y_2, \ldots, y_n)$, that is, each monomial appearing in f_2 is of type $y_1 y_j$, for some $2 \leq j \leq n$. Hence f_2 is divisible by y_1, and the same is true for f. However, this is impossible, since $\mathcal{Q}$ is non-degenerate, hence irreducible.

If P is a point at infinity, then we can suppose as above that H is the hyperplane of equation $y_1 = 0$, and that P is the point at infinity of the axis y_n, that is, $P = [0, \ldots, 0, 1]$. In this system of coordinates, $\mathcal{Q}$ is defined by $f(y_1, \ldots, y_n) = c + f_1(y_1, \ldots, y_n) + f_2(y_1, \ldots, y_n) = 0$, where $c \in \mathbb{R}$, the polynomial f_1 is either zero or homogeneous of degree 1, and $f_2 = y_1 y_n + g_2(y_1, \ldots, y_{n-1})$, where g_2 is homogeneous of degree 2 in $y_1, \ldots, y_{n-1}$. In this case $\tau_H(\mathcal{Q})$ is defined by $f(-y_1, y_2, \ldots, y_n) = c - y_1 y_n + f_1(-y_1, y_2, \ldots, y_n) + g_2(-y_1, y_2, y_{n-1}) = 0$. Then, as in the previous case, it is easy to check that if $\tau_H(\mathcal{Q}) = \mathcal{Q}$ then y_1 divides f and $\mathcal{Q}$ is reducible, against the hypothesis that $\mathcal{Q}$ is non-degenerate.

(b) Denote by $w \in \mathbb{R}^n \setminus \{0\}$ a vector orthogonal to H, and by $R = [(0, w)] \in \mathbb{P}^n(\mathbb{R})$ the point at infinity corresponding to the direction w. Of course we have $R \notin \overline{H}$; moreover, H is a principal hyperplane if and only if R is the pole of $\overline{H}$ with respect to the projective quadric $\overline{\mathcal{Q}}$.

If H is a principal hyperplane, then $\tau_H(\mathcal{Q}) = \mathcal{Q}$ by Exercise 188 (b). Conversely, suppose that $\tau_H(\mathcal{Q}) = \mathcal{Q}$ and let P be the pole of $\overline{H}$. Denote by $\widetilde{\tau_H} \colon \mathbb{P}^n(\mathbb{R}) \to \mathbb{P}^n(\mathbb{R})$ the projectivity induced by τ_H. The fixed-point set of $\widetilde{\tau_H}$ is $\overline{H} \cup \{R\}$. On the other hand, since $\widetilde{\tau_H}(\overline{\mathcal{Q}}) = \overline{\mathcal{Q}}$, we also have $\widetilde{\tau_H}(P) = P$. Since by (a) the projective hyperplane $\overline{H}$ is not tangent to $\overline{\mathcal{Q}}$, the point P does not belong to $\overline{H}$, hence $P = R$ and H is a principal hyperplane.

Exercise 190. Let $\mathcal{Q}$ be a non-degenerate quadric of $\mathbb{R}^n$, and let $\overline{A} = \begin{pmatrix} c & {}^tB \\ \hline B & A \end{pmatrix} \in$ $M(n+1, \mathbb{R})$ be a matrix representing $\mathcal{Q}$. Let $\lambda_1, \ldots, \lambda_h$ be the non-zero eigenvalues of A, where $\lambda_i \neq \lambda_j$ if $i \neq j$, let V_i be the eigenspace of A corresponding to the eigenvalue λ_i and set $d_i = \dim V_i$, $i = 1, \ldots, h$. Let also W be the set of principal hyperplanes of $\mathcal{Q}$.

(a) Prove that W is the union of h proper linear systems $W_1, \ldots, W_h$ (cf. Sect. 1.4.3) such that $\dim W_i = d_i - 1$ for every $i = 1, \ldots, h$.
(b) Prove that the dimension of the affine subspace

$$J = \bigcap_{H \in W} H$$

is equal to $n - \operatorname{rk} A$.
(c) Prove that, if $\mathcal{Q}$ has a centre, then the centre of $\mathcal{Q}$ is the intersection of n pairwise orthogonal principal hyperplanes.
(d) Suppose that $\mathcal{Q}$ is a paraboloid. Prove that $\mathcal{Q}$ has a unique axis, whose point at infinity is contained in the projective closure of $\mathcal{Q}$. Also prove that $\mathcal{Q}$ is the intersection of $n - 1$ pairwise orthogonal principal hyperplanes, and that $\mathcal{Q}$ has a unique vertex.
(e) Suppose that $\mathcal{Q}$ is a sphere. Prove that $\mathcal{Q}$ has a centre and that, if C is the centre of $\mathcal{Q}$, then every plane passing through C is principal, every line passing through C is an axis, and every point of the support of $\mathcal{Q}$ is a vertex.
(f) Let $n = 2$. Prove that $\mathcal{Q}$ admits at least one axis, that $\mathcal{Q}$ has a centre if and only if it has at least two orthogonal axes, and that $\mathcal{Q}$ is a circle if and only if it has at least three pairwise distinct axes (and in this case, $\mathcal{Q}$ has a centre and every line passing through its centre is an axis).

Solution. (a) A principal hyperplane of $\mathcal{Q}$ is the affine part of the polar hyperplane $\operatorname{pol}(P)$, where $P = [(0, v)]$ and v is an eigenvector of A corresponding to a non-zero eigenvalue (cf. Sect. 1.8.6). For every $i = 1, \ldots, h$, let then $K_i = \overline{V_i} \cap H_0$, where H_0 is the hyperplane at infinity, and let W_i be the set of hyperplanes H of $\mathbb{R}^n$ such that $\overline{H} = \operatorname{pol}(P)$ for some $P \in K_i$. This shows that $W = \bigcup_{i=1}^h W_i$. Moreover, if $v \in V_i \setminus \{0\}$ and $X = (x_1, \ldots, x_n)$ are the usual affine coordinates of $\mathbb{R}^n$, then an equation of the principal hyperplane corresponding to v is given by ${}^tvAX + {}^tBv = 0$ or, equivalently, by ${}^tvX = -\dfrac{1}{\lambda_i}{}^tvB$. It readily follows that, for every $i = 1, \ldots, h$, the linear system of projective hyperplanes $\operatorname{pol}(K_i)$ does not contain the hyperplane at infinity H_0, so that W_i is a proper linear system. Since pol: $\mathbb{P}^n(\mathbb{R}) \to \mathbb{P}^n(\mathbb{R})^*$ is an isomorphism, we finally have dim $W_i = \dim K_i = d_i - 1$ for every $i = 1, \ldots, h$.

(b) For every $i = 1, \ldots, h$, let $\mathcal{B}_i$ be an orthogonal basis of V_i, and let us set $\{v_1, \ldots, v_m\} = \mathcal{B}_1 \cup \ldots \cup \mathcal{B}_h$. By the Spectral theorem A is similar to a diagonal matrix, hence from the definition of the V_i one easily gets that $m = d_1 + \ldots + d_h =$ rk A. Moreover, the eigenspaces of A are pairwise orthogonal, hence v_i is orthogonal to v_j for every $i \neq j, i, j \in \{1, \ldots, m\}$.

We have seen in the solution of (a) that the affine subspace J is defined by the linear system

$$\begin{cases} {}^t v_1 X = c_1 \\ \quad \vdots \\ {}^t v_m X = c_m \end{cases} \tag{4.4}$$

where $c_1, \ldots, c_m$ are real numbers. Since $\{v_1, \ldots, v_m\}$ is a set of linearly independent vectors, it readily follows that J is an affine subspace of $\mathbb{R}^n$ of dimension $n - m = n -$ rk A. Also observe that, the v_i being pairwise orthogonal, the system (4.4) describes J as the intersection of m pairwise orthogonal principal hyperplanes.

(c) Since the centre C of $\mathcal{Q}$ coincides with the pole of the hyperplane at infinity (cf. Sect. 1.8.6), if H is a principal hyperplane, then by reciprocity we have $C \in H$. Therefore, $C \in \bigcap_{H \in W} H = J$. On the other hand, since $\mathcal{Q}$ has a centre, we have rk $A = n$ (cf. Sect. 1.8.5), and part (b) implies that dim $J = n - n = 0$, so that $J = \{C\}$. Moreover, the solution of (b) also shows that J is the intersection of n pairwise orthogonal principal hyperplanes.

(d) As observed in Sect. 1.8.5, since $\mathcal{Q}$ is a non-degenerate paraboloid we have rk $A = n - 1$. Therefore, by part (b) we have dim $J = 1$, and J is a line. Being the intersection of principal hyperplanes, J is an axis of $\mathcal{Q}$. Moreover, if r is any axis of $\mathcal{Q}$, by definition r is the intersection of principal hyperplanes, so that $r \supseteq J$, and $r = J$ since they have the same dimension. Hence J is the unique axis of $\mathcal{Q}$ and, as proved in the solution of (b), J is indeed the intersection of $n - 1$ pairwise orthogonal principal hyperplanes.

Let us now prove that the point at infinity of J belongs to $\mathcal{Q}$. Since rk $A = n - 1$, there exists a non-zero vector $v_0 \in \operatorname{Ker} A \subseteq \mathbb{R}^n$. Set $P_0 = [(0, v_0)] \in H_0$. With the same notation as in the solution of (b), the Spectral theorem implies that v_0 is orthogonal to v_i for every $i = 1, \ldots, m = n - 1$. From the explicit equations of J described in (b) it is now immediate to deduce that P_0 is the point at infinity of J. Moreover, since $A v_0 = 0$ we have $P_0 \in \overline{\mathcal{Q}}$, as desired.

By definition of vertex, in order to conclude it suffices to prove that $\mathcal{Q} \cap J$ consists of a single point. From $A v_0 = 0$ we deduce that $T_{P_0}(\overline{\mathcal{Q}}) = H_0$. Since of course $\overline{J}$ is not contained in H_0, we thus have $I(\overline{\mathcal{Q}}, \overline{J}, P_0) = 1$. It follows that J intersects $\mathcal{Q}$ at a unique point V, which is the unique vertex of $\mathcal{Q}$.

(e) By definition of sphere, $\mathcal{Q}$ has equation

$$(x_1 - a_1)^2 + \ldots + (x_n - a_n)^2 = \eta$$

for some $\eta \in \mathbb{R}$, so that $A = \mathrm{Id}$, ${}^t B = (-a_1 \; \ldots \; -a_n)$ and $c = a_1^2 + \ldots + a_n^2 - \eta$. In particular, A being invertible, the quadric $\mathcal{Q}$ has a centre. Since the coordinates of the centre C of $\mathcal{Q}$ satisfy the system $AX = -B$, we also have $C = (a_1, \ldots, a_n)$.

Since all vectors of $\mathbb{R}^n$ are eigenvectors for A, a hyperplane of $\mathbb{R}^n$ is principal if and only if it is equal to $\mathrm{pol}([0, v])$ for some $v \in \mathbb{R}^n \setminus \{0\}$, i.e. if and only if it is described by the equation

$$v_1 x_1 + v_2 x_2 + \ldots + v_n x_n = v_1 a_1 + v_2 a_2 + \ldots + v_n a_n$$

for some $v = (v_1, \ldots, v_n) \in \mathbb{R}^n \setminus \{0\}$. It readily follows that a hyperplane of $\mathbb{R}^n$ is principal if and only if it passes through C. Moreover, every line passing through C is the intersection of $n - 1$ hyperplanes passing through C, so the axes of $\mathcal{Q}$ are precisely the lines passing through C, and every point of the support of $\mathcal{Q}$ is a vertex of $\mathcal{Q}$.

(f) Let us first prove that for $n = 2$ the notions of axis and of principal hyperplane coincide, so that the axes of $\mathcal{Q}$ bijectively correspond to the linear subspaces spanned by an eigenvector of A corresponding to a non-zero eigenvalue (cf. Sect. 1.8.6). More precisely, the subspace spanned by an eigenvector $v \neq 0$ determines an axis orthogonal to v. Since A is symmetric and $\mathrm{rk}\, A \geq 1$, the Spectral theorem implies that $\mathcal{Q}$ admits at least one axis, and that $\mathcal{Q}$ admits two axes if and only if A is invertible, that is if and only if $\mathcal{Q}$ has a centre.

Recall that $\mathcal{Q}$ is a circle if and only if it is defined by an equation of the form $(x - x_0)^2 + (y - y_0)^2 = c$, where $x_0, y_0, c \in \mathbb{R}$. This readily implies that $\mathcal{Q}$ is a circle if and only if A is a non-zero multiple of the identity, i.e. (since $n = 2$) if and only if A admits three pairwise linearly independent eigenvectors corresponding to non-zero eigenvalues. We can thus conclude that $\mathcal{Q}$ is a circle if and only if it admits three distinct axes (and in this case, as we have seen in the solution of (e), $\mathcal{Q}$ has a centre and every line passing through the centre is an axis).

Exercise 191. Let $\mathcal{Q}$ be a non-degenerate quadric of $\mathbb{R}^n$ and let V be a vertex of $\mathcal{Q}$. Prove that there exists precisely one axis of $\mathcal{Q}$ containing V, and that this axis is orthogonal to the tangent hyperplane to $\mathcal{Q}$ at V.

Solution. Of course it is sufficient to show that, if r is an axis passing through V, then r is orthogonal to $T_V(\mathcal{Q})$. By definition, r is the intersection of $n - 1$ principal hyperplanes whose projective closures are given by $n - 1$ projective hyperplanes $\mathrm{pol}_{\overline{\mathcal{Q}}}(P_1), \ldots, \mathrm{pol}_{\overline{\mathcal{Q}}}(P_{n-1})$, where $P_i = [(0, v_i)] \in H_0$ for $i = 1, \ldots, n - 1$. More explicitly, as we proved in the solution of Exercise 190 (b), if A is a symmetric matrix representing $\mathcal{Q}$, then for every $i = 1, \ldots, n - 1$ the vector v_i is an eigenvector of A corresponding to a non-zero eigenvalue, and the principal hyperplane corresponding to P_i is described by an affine equation of the form ${}^t v_i X = c_i$ for some $c_i \in \mathbb{R}$. In particular, since $\dim r = 1$, the vectors $v_1, \ldots, v_{n-1}$ are linearly independent, so that the direction of r is spanned by the unique (up to multiplication by a non-zero scalar) vector $v_r \in \mathbb{R}^n \setminus \{0\}$ such that ${}^t v_i v_r = 0$ for every $i = 1, \ldots, n - 1$.

Since $V \in \bar{r} = \bigcap_{i=1}^{n-1} \text{pol}_{\overline{Q}}(P_i)$, by reciprocity every P_i belongs to $\text{pol}_{\overline{Q}}(V)$, i.e. to the projective hyperplane tangent to $\overline{Q}$ at V. It readily follows that every v_i belongs to the linear subspace associated to the affine hyperplane $T_V(Q)$. Being linearly independent, the vectors $v_1, \ldots, v_{n-1}$ form a basis of the linear subspace associated to $T_V(Q)$. From the fact that ${}^t v_i v_r = 0$ for every $i = 1, \ldots, n-1$ we can then deduce that r is orthogonal to $T_V(Q)$, as desired.

Exercise 192. Let $P_1, P_2, P_3 \in \mathbb{R}^2$ be non-collinear points. Prove that there exists a unique circle $\mathcal{C}$ containing P_1, P_2, P_3.

Solution. If O is the point of intersection between the axes of the segments $P_1 P_2$ and $P_2 P_3$, and ρ is the distance between O and P_1, then an easy argument in plane geometry shows that the unique circle containing P_1, P_2 and P_3 has centre in O and has radius equal to ρ.

Here we offer an alternative solution, which is based on the characterization of circles described in Sect. 1.8.7. In fact, we recall that, if $I_1 = [0, 1, i]$ and $I_2 = [0, 1, -i]$ are the cyclic points of the Euclidean plane, then a conic $\mathcal{C}$ of $\mathbb{R}^2$ is a circle if and only if the complexification $\overline{\mathcal{C}}_{\mathbb{C}}$ of the projective closure $\overline{\mathcal{C}}$ of $\mathcal{C}$ contains I_1 and I_2.

The points P_1, P_2, P_3, I_1, I_2 of $\mathbb{P}^2(\mathbb{C})$ are in general position: since P_1, P_2, P_3 are projectively independent, the line $L(I_1, I_2) = \{x_0 = 0\}$ does not contain any P_i, $i = 1, 2, 3$, and the lines $L(P_i, P_j), i, j \in \{1, 2, 3\}$, intersect the line at infinity $x_0 = 0$ of $\mathbb{P}^2(\mathbb{C})$ in real points, which of course are distinct from I_1 and I_2. Therefore, there exists a unique conic $\mathcal{Q}$ of $\mathbb{P}^2(\mathbb{C})$ such that $P_1, P_2, P_3, I_1, I_2 \in \mathcal{Q}$, and this conic is non-degenerate (cf. Sect. 1.9.6). The conjugate conic $\sigma(\mathcal{Q})$ contains the points $P_i = \sigma(P_i), i = 1, 2, 3$, and the points $I_1 = \sigma(I_2)$ and $I_2 = \sigma(I_1)$. The uniqueness of $\mathcal{Q}$ implies that $\sigma(\mathcal{Q}) = \mathcal{Q}$, hence there exists a conic $\mathcal{D}$ of $\mathbb{P}^2(\mathbb{R})$ such that $\mathcal{Q} = \mathcal{D}_{\mathbb{C}}$ (cf. Exercise 59). The affine part $\mathcal{C} = \mathcal{D} \cap \mathbb{R}^2$ of $\mathcal{D}$ is a circle passing through P_1, P_2 and P_3.

Let us now prove the uniqueness of $\mathcal{C}$. If $\mathcal{C}'$ is a circle of $\mathbb{R}^2$ containing P_1, P_2 and P_3, the complexification $\mathcal{Q}' = \overline{\mathcal{C}'}_{\mathbb{C}}$ of the projective closure of $\mathcal{C}'$ is a conic of $\mathbb{P}^2(\mathbb{C})$ containing P_1, P_2, P_3, I_1, I_2. Again by the uniqueness of $\mathcal{Q}$, we have $\mathcal{Q}' = \mathcal{Q}$, whence $\mathcal{C} = \mathcal{C}'$, that is, $\mathcal{C}$ is the unique circle of $\mathbb{R}^2$ passing through P_1, P_2 and P_3.

Exercise 193. Let $\mathcal{C}$ be a parabola of $\mathbb{R}^2$ and let P be any point of $\mathcal{C}$. Prove that there exists an affinity $\varphi \colon \mathbb{R}^2 \to \mathbb{R}^2$ such that $\varphi(P) = (0, 0)$ and $\varphi(\mathcal{C})$ has equation $x^2 - 2y = 0$.

Solution. Let $\tau_P \subseteq \mathbb{P}^2(\mathbb{R})$ be the projective line tangent to $\overline{\mathcal{C}}$ at P, let $Q \in \mathbb{P}^2(\mathbb{R})$ be the intersection point between τ_P and the line at infinity, and let $R \in \mathbb{P}^2(\mathbb{R})$ be the point at infinity of $\overline{\mathcal{C}}$. It is immediate to check that, if $S \in \mathbb{R}^2 \subseteq \mathbb{P}^2(\mathbb{R})$ is a point of $\mathcal{C}$ distinct from P, then the points P, Q, R, S form a projective frame of $\mathbb{P}^2(\mathbb{R})$. Let $f \colon \mathbb{P}^2(\mathbb{R}) \to \mathbb{P}^2(\mathbb{R})$ be the projectivity such that $f(P) = [1, 0, 0]$, $f(Q) = [0, 1, 0]$, $f(R) = [0, 0, 1]$, $f(S) = [1, 1, 1]$. Since $f(Q)$ and $f(R)$ lie on the line at infinity, f restricts to an affinity ψ of $\mathbb{R}^2$.

Let $C' = \psi(C)$, and let $g(x, y) = 0$ be an equation of C'. By construction C' passes through $O = (0, 0)$, and the tangent to C' at O is the affine part of the line passing through O and $[0, 1, 0]$, i.e. the line of equation $y = 0$. Therefore, C' is described by an equation of the form $g(x, y) = y + ax^2 + bxy + cy^2$, where at least one among a, b, c is non-zero. Since R is the unique point at infinity of $\overline{C}$, we also have that $[0, 0, 1]$ is the unique point at infinity of C', so that $b = c = 0$, while the fact that $S \in C$ implies that $(1, 1) \in C'$, so that $a = -1$. Therefore, C' has equation $x^2 - y = 0$. The required affinity is then the composition of ψ with the affinity η of $\mathbb{R}^2$ such that $\eta(x, y) = \left(x, \frac{y}{2}\right)$ for every $(x, y) \in \mathbb{R}^2$.

Note. The previous exercise readily implies that the group of the affinities of $\mathbb{R}^2$ that leave a parabola C invariant acts transitively on C. This is clearly false if we consider isometries rather than affinities, because the vertex of C is fixed by every isometry of $\mathbb{R}^2$ that leaves C invariant.

Exercise 194. Let C be a non-empty non-degenerate conic of $\mathbb{R}^2$. Prove that:

(a) If C has a centre but is not a circle, then C has two foci, each of which is distinct from the centre O of C.
(b) If C is a circle, then the center O of C is the unique focus of C.
(c) If C is a parabola, then C has a unique focus.

Solution. Let $I_1 = [0, 1, i]$ and $I_2 = [0, 1, -i]$ be the cyclic points of the Euclidean plane, as usual. Let $\mathcal{D} = C_{\mathbb{C}}$ be the complexification of C, let $\overline{\mathcal{D}}$ be the projective closure of $\mathcal{D}$, and recall that C is a circle if and only if $I_1, I_2 \in \overline{\mathcal{D}}$ (cf. Sect. 1.8.7).

(a) Since C has a centre, the line at infinity $x_0 = 0$ is not tangent to $\overline{\mathcal{D}}$ (cf. Sect. 1.8.5). Moreover, as previously observed, $\overline{\mathcal{D}}$ does not contain the cyclic points. Hence there are two distinct lines r_1 and r_2 of $\mathbb{P}^2(\mathbb{C})$ that pass through I_1 and are tangent to $\overline{\mathcal{D}}$ at the proper points $Q_1 = \overline{\mathcal{D}} \cap r_1$, $Q_2 = \overline{\mathcal{D}} \cap r_2$. Observe that if r_i, $i = 1, 2$, passes through the centre O of C, then $O \in r_i = \mathrm{pol}(Q_i)$, so that $Q_i \in \mathrm{pol}(O)$ by reciprocity, which contradicts the fact that $\mathrm{pol}(O)$ is the line at infinity. Therefore, $O \notin r_1 \cup r_2$. The conjugate lines $s_1 = \sigma(r_1)$ and $s_2 = \sigma(r_2)$ pass through I_2 and they also are tangent to $\overline{\mathcal{D}}$, since $\sigma(\overline{\mathcal{D}}) = \overline{\mathcal{D}}$. By Exercise 60, the points $F_1 = r_1 \cap s_1$ and $F_2 = r_2 \cap s_2$ are precisely the real points belonging to $r_1 \cup r_2 \cup s_1 \cup s_2$, so that $\{F_1, F_2\}$ is the set of foci of C. As previously observed, we have $O \notin r_1 \cup r_2$, so the foci F_1 and F_2 are distinct from O.

(b) If C is a circle, then I_1 and I_2 belong to $\overline{\mathcal{D}}$. Denote by r the line tangent to $\overline{\mathcal{D}}$ at I_1. Since the line at infinity $x_0 = 0$ intersects $\overline{\mathcal{D}}$ at I_1 and I_2, the line r is distinct from $\{x_0 = 0\}$. The conjugate line $s = \sigma(r)$ is tangent to $\overline{\mathcal{D}}$ at I_2, hence it is distinct from r. The lines r and s, being the polars of points at infinity, intersect in the centre O of C. By Exercise 60, O is the unique real point of $r \cup s$, so it is the unique focus of C.

(c) If C does not have a centre, then the conic $\mathcal{D}$ has precisely one point at infinity, which is real and distinct from I_1 and I_2. There are two lines that pass through I_1 and are tangent to $\overline{\mathcal{D}}$: the line at infinity $x_0 = 0$ and another line r. The lines tangent to $\overline{\mathcal{D}}$ and passing through I_2 are the lines $x_0 = 0$ and the conjugate line $s = \sigma(r)$.

By Exercise 60, the point $F = r \cap s$ is the unique real point of $r \cup s$, hence it is the unique focus of C.

Exercise 195. If C is a non-degenerate conic of $\mathbb{R}^2$, prove that:

(a) If C has a centre and is not a circle, then one of the axes of C contains both foci of C.

(b) If C is a parabola, then the focus F of C belongs to the unique axis of C.

Solution. The exercise can be readily solved analytically, by performing an isometric change of coordinates that transforms C into its canonical form and by using the formulas for computing the foci of Exercise 196. We propose here a synthetic proof.

(a) By Exercise 194, C has two foci F_1, F_2, each distinct from the centre O of C. Moreover, by Exercise 190 (f), the conic C has exactly two mutually orthogonal axes r_1 and r_2. For $i = 1, 2$ denote by $\tau_i : \mathbb{R}^2 \to \mathbb{R}^2$ the orthogonal reflection with respect to the line r_i. By Exercise 189, the isometries τ_1 and τ_2 (and hence $\tau = \tau_1 \circ \tau_2$ too) leave C invariant and therefore preserve the set $\{F_1, F_2\}$. As r_1 is orthogonal to r_2, the map τ is a central symmetry whose unique fixed point $O = r_1 \cap r_2$ is the centre of C; so τ exchanges F_1 and F_2. It follows that, for instance, τ_1 fixes F_1 and F_2 while τ_2 exchanges them. Since the fixed-point set of τ_1 is r_1, we have that $F_1, F_2 \in r_1$.

(b) Note that by Exercise 190 (f) the conic C, which does not have a centre, has a unique axis r. By Exercise 194, C has only one focus F, which is therefore a fixed point for any isometry $\tau : \mathbb{R}^2 \to \mathbb{R}^2$ such that $\tau(C) = C$. If we apply this observation to the orthogonal reflection with respect to r, we obtain that F belongs to r.

Note. If C is a non-degenerate conic with a centre and it is not a circle, the axis containing the foci of C is called *focal axis*. If C is an ellipse, the focal axis is also called *major axis*, while if C is a hyperbola it is also called *transverse axis*. In both cases, the focal axis intersects C at two vertices.

Exercise 196. In each of the following cases (a), (b), (c), (d), determine axes, vertices, foci and directrices of the non-degenerate conic C of $\mathbb{R}^2$ defined by the equation $f(x, y) = 0$:

(a) $f(x, y) = \dfrac{x^2}{a^2} + \dfrac{y^2}{b^2} - 1$, with $a > b > 0$;

(b) $f(x, y) = \dfrac{x^2}{a^2} + \dfrac{y^2}{a^2} - 1$, with $a > 0$;

(c) $f(x, y) = \dfrac{x^2}{a^2} - \dfrac{y^2}{b^2} + 1$, with $a > 0$, $b > 0$;

(d) $f(x, y) = x^2 - 2cy$, with $c > 0$.

Solution. (a) Since $f(x, y) = f(-x, -y)$, the conic C has the origin $O = (0, 0)$ of $\mathbb{R}^2$ as a centre. Moreover, C is represented by the matrix

$$\overline{A} = \begin{pmatrix} -1 & 0 & 0 \\ 0 & \dfrac{1}{a^2} & 0 \\ 0 & 0 & \dfrac{1}{b^2} \end{pmatrix}.$$

As $a \neq b$, the eigenspaces of the matrix $A = \begin{pmatrix} \frac{1}{a^2} & 0 \\ 0 & \frac{1}{b^2} \end{pmatrix}$ relative to non-zero eigenvalues are the lines generated by $(1, 0)$ and by $(0, 1)$. Since the line $\mathrm{pol}_{\overline{C}}([0, 1, 0])$ has equation $x_1 = 0$ and the line $\mathrm{pol}_{\overline{C}}([0, 0, 1])$ has equation $x_2 = 0$, the axes of C are the lines of equations $x = 0$ and $y = 0$. If we intersect these lines with C, we find the vertices of C, which have coordinates

$$(a, 0), \quad (-a, 0), \quad (0, b), \quad (0, -b).$$

Let us now compute the foci of C. If $\overline{C}_{\mathbb{C}}$ denotes the complexified conic of $\overline{C}$, we start by determining the lines of $\mathbb{C}^2$ whose projective closures are tangent to $\overline{C}_{\mathbb{C}}$ and pass through one of the cyclic points $I_1 = [0, 1, i]$, $I_2 = [0, 1, -i]$. The conic $\overline{C}_{\mathbb{C}}$ has equation $b^2 x_1^2 + a^2 x_2^2 - a^2 b^2 x_0^2 = 0$. If $\mathcal{F}_1$ is the pencil of lines of $\mathbb{P}^2(\mathbb{C})$ with centre I_1, the set of lines of $\mathcal{F}_1$ distinct from the improper line coincides with the set of lines r_λ of equation $\lambda x_0 + i x_1 - x_2 = 0$, as λ varies in $\mathbb{C}$. If we substitute the equality $x_2 = \lambda x_0 + i x_1$ into the equation of $\overline{C}_{\mathbb{C}}$, we obtain the equation

$$-(a^2 - b^2) x_1^2 + 2\lambda a^2 i x_0 x_1 + a^2 (\lambda^2 - b^2) x_0^2 = 0,$$

which has a double root if and only if $(\lambda a^2 i)^2 + a^2 (a^2 - b^2)(\lambda^2 - b^2) = 0$, that is iff $\lambda = \lambda_1 = -i\sqrt{a^2 - b^2}$ or $\lambda = \lambda_2 = i\sqrt{a^2 - b^2}$. So r_λ is tangent to $\overline{C}_{\mathbb{C}}$ if and only if either $\lambda = \lambda_1$ or $\lambda = \lambda_2$. Let now $j = 1$ or $j = 2$. By Exercise 60, the only real point of r_{λ_j} is $F_j = r_{\lambda_j} \cap \sigma(r_{\lambda_j})$. In addition, as $\sigma(\overline{C}_{\mathbb{C}}) = \overline{C}_{\mathbb{C}}$, the line $\sigma(r_{\lambda_j}) = \sigma(L(F_j, I_1)) = L(F_j, I_2)$ is tangent to $\overline{C}_{\mathbb{C}}$. Therefore the isotropic lines passing through F_j are tangent to $\overline{C}_{\mathbb{C}}$, so that F_j is a focus of C. By using the equations of r_{λ_1} and r_{λ_2} found above, it is now immediate to check that F_1 and F_2 have affine coordinates

$$F_1 = \left(\sqrt{a^2 - b^2}, 0\right), \quad F_2 = \left(-\sqrt{a^2 - b^2}, 0\right).$$

Furthermore, being the only real points that belong to lines passing through I_1 and tangent to $\overline{C}_{\mathbb{C}}$, the points F_1 and F_2 are the only foci of C.

Finally, the directrices r_1 and r_2 of C are the affine parts of the lines $\mathrm{pol}(F_1)$ and $\mathrm{pol}(F_2)$, so they have equations

$$x = \frac{a^2}{\sqrt{a^2 - b^2}}, \quad x = -\frac{a^2}{\sqrt{a^2 - b^2}},$$

respectively.

(b) In this case C is a circle. Moreover, since $f(x, y) = f(-x, -y)$, the centre of C is the origin $O = (0, 0)$ of $\mathbb{R}^2$. By Exercise 190 (f), the axes of C are precisely the lines passing through O, and all points of the support of C are vertices.

By Exercise 194 (b), the origin O is the only focus of C. Thus the polar of the focus of C, since it coincides with the polar of the centre of C, is the improper line, and consequently C has no directrix.

. (c) Proceeding as in part (a), we can easily show that C has the origin $O = (0, 0)$ as a centre, and that the axes of C are the lines of equations $x = 0$ and $y = 0$. If we intersect these lines with C, we find the vertices of C, which have coordinates

$$(0, b), \quad (0, -b).$$

In order to find the foci of C, we can proceed as in the solution of (a). Denote by $\overline{C}_\mathbb{C}$ the complexified conic of $\overline{C}$, and consider the lines r_λ of equation $\lambda x_0 + i x_1 - x_2 = 0$, with $\lambda \in \mathbb{C}$. Since $\overline{C}_\mathbb{C}$ has equation $b^2 x_1^2 - a^2 x_2^2 + a^2 b^2 x_0^2 = 0$, if we substitute the equality $x_2 = \lambda x_0 + i x_1$ into the equation of $\overline{C}_\mathbb{C}$ we obtain the equation

$$(a^2 + b^2) x_1^2 - 2\lambda a^2 i x_0 x_1 + a^2 (b^2 - \lambda^2) x_0^2 = 0,$$

which has a double root if and only if $(\lambda a^2 i)^2 + a^2 (a^2 + b^2)(\lambda^2 - b^2) = 0$, that is iff either $\lambda = \lambda_1 = \sqrt{a^2 + b^2}$ or $\lambda = \lambda_2 = -\sqrt{a^2 + b^2}$. Therefore, r_λ is tangent to $\overline{C}_\mathbb{C}$ if and only if either $\lambda = \lambda_1$ or $\lambda = \lambda_2$. Proceeding as in the solution of (a), we find that the foci of C coincide with the points $r_{\lambda_1} \cap \sigma(r_{\lambda_1})$ and $r_{\lambda_2} \cap \sigma(r_{\lambda_2})$, that is with the points

$$F_1 = \left(0, \sqrt{a^2 + b^2}\right), \quad F_2 = \left(0, -\sqrt{a^2 + b^2}\right).$$

Finally, the directrices r_1 and r_2 of C are the affine parts of the lines $\mathrm{pol}(F_1)$ and $\mathrm{pol}(F_2)$, so they have equations

$$y = \frac{b^2}{\sqrt{a^2 + b^2}}, \quad y = -\frac{b^2}{\sqrt{a^2 + b^2}},$$

respectively.

(d) The conic C is represented by the matrix

$$\overline{A} = \begin{pmatrix} 0 & 0 & -c \\ 0 & 1 & 0 \\ -c & 0 & 0 \end{pmatrix}.$$

The only eigenspace of the matrix $A = \begin{pmatrix} 1 & 0 \\ 0 & 0 \end{pmatrix}$ relative to a non-zero eigenvalue is the line generated by $(1, 0)$, so that C has a unique axis, of equation $x = 0$. If we intersect this line with C, we obtain the vertex of C, which coincides with the origin O of $\mathbb{R}^2$.

Let us now determine the foci of C (coherently with what we proved in Exercise 194 (c), we will show that actually C has precisely one focus). The complexified conic $\overline{C}_C$ of $\overline{C}$ has equation $x_1^2 - 2cx_0x_2 = 0$. As C has no centre, the improper line is tangent to C. Therefore, since $I_1 \notin \overline{C}_C$ and $\overline{C}_C$ is non-degenerate, there exists exactly one line $r \subseteq \mathbb{P}^2(\mathbb{C})$ distinct from the improper line, containing I_1 and tangent to $\overline{C}_C$. This line has equation $\lambda x_0 + ix_1 - x_2 = 0$ for some $\lambda \in \mathbb{C}$. If we substitute the equality $x_2 = \lambda x_0 + ix_1$ into the equation of $\overline{C}_C$, we obtain the equation

$$x_1^2 - 2icx_0x_1 - 2\lambda cx_0^2 = 0,$$

which, as $c \neq 0$, has a double root if and only if $\lambda = \frac{c}{2}$. Thus the line r has equation $cx_0 + 2ix_1 - 2x_2 = 0$; hence, as $r \cap \sigma(r) = [2, 0, c]$, the only focus of C is the point

$$F = \left(0, \frac{c}{2}\right).$$

Finally, the directrix of C is the affine part of the line $\mathrm{pol}(F)$, so it has equation

$$y = -\frac{c}{2}.$$

Exercise 197. (*Metric characterization of conics*) Assume that F_1, F_2 are distinct points of $\mathbb{R}^2$ and that r is a line of $\mathbb{R}^2$ such that $F_1 \notin r$; denote by d the usual Euclidean distance.

(a) Given $\kappa > \dfrac{d(F_1, F_2)}{2}$, let

$$X = \{P \in \mathbb{R}^2 \mid d(P, F_1) + d(P, F_2) = 2\kappa\}.$$

Show that X is the support of a real ellipse having F_1 and F_2 as foci. Moreover, if C is a real ellipse which is not a circle, show that the support of C coincides with the locus of points of the plane for which the sum of the distances to the foci of C is constant.

(b) Given $\kappa \in \mathbb{R}$ such that $0 < \kappa < \dfrac{d(F_1, F_2)}{2}$, let

$$X = \{P \in \mathbb{R}^2 \mid |d(P, F_1) - d(P, F_2)| = 2\kappa\}.$$

Show that X is the support of a hyperbola having F_1 and F_2 as foci. Show also that the support of every hyperbola C of $\mathbb{R}^2$ coincides with the locus of points of the plane for which the absolute value of the difference of the distances to the foci of C is constant.

(c) Let

$$X = \{P \in \mathbb{R}^2 \mid d(P, F_1) = d(P, r)\}.$$

Show that X is the support of a parabola having F_1 as focus and r as directrix. Show also that the support of every parabola C of $\mathbb{R}^2$ coincides with the locus of points of the plane that are equidistant from the focus of C and from the directrix of C.

Solution. (a) Up to performing a change of coordinates by means of an isometry of $\mathbb{R}^2$, we may assume that the points F_1 and F_2 have coordinates $(\mu, 0)$ and $(-\mu, 0)$, respectively, with $\mu > 0$. Then $\mu < \kappa$, and the point of the plane of coordinates (x, y) belongs to X if and only if

$$\sqrt{(x - \mu)^2 + y^2} + \sqrt{(x + \mu)^2 + y^2} = 2\kappa, \tag{4.5}$$

that is iff

$$\sqrt{(x - \mu)^2 + y^2} = 2\kappa - \sqrt{(x + \mu)^2 + y^2}.$$

Squaring both sides of this equation, dividing by 4 and rearranging the terms, we obtain the equation

$$\kappa^2 + \mu x = \kappa\sqrt{(x + \mu)^2 + y^2}.$$

Squaring again, after some simplifications we find the equation

$$\frac{x^2}{\kappa^2} + \frac{y^2}{\kappa^2 - \mu^2} = 1, \tag{4.6}$$

and one can check that Eq. (4.6) is actually equivalent to Eq. (4.5), so that, since $\kappa > \mu > 0$, the set X is the support of a real ellipse.

Then from Exercise 196 (a) one can deduce that the foci of the ellipse of Eq. (4.6) are the points $(\sqrt{\kappa^2 - (\kappa^2 - \mu^2)}, 0) = (\mu, 0) = F_1$ and $(-\sqrt{\kappa^2 - (\kappa^2 - \mu^2)}, 0) = (-\mu, 0) = F_2$, as desired.

Assume now that $\mathcal{C}$ is a real ellipse which is not a circle. By Theorem 1.8.8 we can choose Euclidean coordinates where $\mathcal{C}$ has equation $\frac{x^2}{a^2} + \frac{y^2}{b^2} = 1$, with $a > b > 0$. Let $F_1 = (\sqrt{a^2 - b^2}, 0)$, $F_2 = (-\sqrt{a^2 - b^2}, 0)$ be the foci of $\mathcal{C}$ (cf. Exercise 196). The computations just carried out show that the set

$$\{P \in \mathbb{R}^2 \mid d(P, F_1) + d(P, F_2) = 2a\}$$

coincides with the support of $\mathcal{C}$, which concludes the proof of (a).

(b) Up to performing a change of coordinates by means of an isometry of $\mathbb{R}^2$, we may assume that the points F_1 and F_2 have coordinates $(0, \mu)$ and $(0, -\mu)$, respectively, with $\mu > 0$. Then $\mu > \kappa$, and the point of the plane of coordinates (x, y) belongs to X if and only if

$$\left| \sqrt{x^2 + (y - \mu)^2} - \sqrt{x^2 + (y + \mu)^2} \right| = 2\kappa, \tag{4.7}$$

that is iff

$$\left(\sqrt{x^2 + (y - \mu)^2} - \sqrt{x^2 + (y + \mu)^2} \right)^2 = 4\kappa^2.$$

This latter equation is equivalent to

$$x^2 + y^2 + \mu^2 - 2\kappa^2 = \sqrt{(x^2 + (y-\mu)^2)(x^2 + (y+\mu)^2)}.$$

Squaring both sides of this equation, after several but trivial simplifications we get to the equation

$$\frac{x^2}{\mu^2 - \kappa^2} - \frac{y^2}{\kappa^2} + 1 = 0, \tag{4.8}$$

which is actually equivalent to Eq. (4.7). As $\mu > \kappa > 0$, it follows that X is the support of a hyperbola.

Then from Exercise 196 (d) one can deduce that the foci of the hyperbola of Eq. (4.8) are the points $(0, \sqrt{\kappa^2 + (\mu^2 - \kappa^2)}) = (0, \mu) = F_1$ and $(0, -\sqrt{\kappa^2 + (\mu^2 - \kappa^2)}) = (0, -\mu) = F_2$, as desired.

Assume now that C is a hyperbola of $\mathbb{R}^2$. By Theorem 1.8.8 we can choose Euclidean coordinates where C has equation $\frac{x^2}{a^2} - \frac{y^2}{b^2} + 1 = 0$, with $a > 0, b > 0$. If $F_1 = (0, \sqrt{a^2 + b^2})$, $F_2 = (0, -\sqrt{a^2 + b^2})$ are the foci of C (cf. Exercise 196), the computations just performed show that the set

$$\{P \in \mathbb{R}^2 \mid |d(P, F_1) - d(P, F_2)| = 2b\}$$

coincides with the support of C. Therefore, the support of every hyperbola coincides with the locus of points of the plane for which the absolute value of the difference of the distances to the foci of C is constant.

(c) Up to performing a change of coordinates by means of an isometry of $\mathbb{R}^2$, we may assume that there exists $\mu > 0$ such that F_1 has coordinates $(0, \mu)$ and r has equation $y = -\mu$. So the point of the plane of coordinates (x, y) belongs to X if and only if

$$\sqrt{x^2 + (y - \mu)^2} = |y + \mu|,$$

that is iff

$$x^2 + (y - \mu)^2 = (y + \mu)^2.$$

This condition is in turn equivalent to the equation

$$x^2 - 4\mu y = 0.$$

So, from Exercise 196 (d) one can deduce that X is the support of a parabola having F_1 as focus and r as directrix.

Assume now that C is a parabola of $\mathbb{R}^2$. By Theorem 1.8.8 we can choose Euclidean coordinates in $\mathbb{R}^2$ where C has equation $x^2 - 2cy = 0$ with $c > 0$. Since $F = \left(0, \frac{c}{2}\right)$ and the line r of equation $y = -\frac{c}{2}$ are the focus and the directrix of C respectively (cf. Exercise 196), the computations just performed show that the set

$$\{P \in \mathbb{R}^2 \mid d(P, F) = d(P, r)\}$$

coincides with the support of C. From this we deduce that the support of every parabola of $\mathbb{R}^2$ coincides with the locus of points of the plane that are equidistant from the focus and from the directrix.

Exercise 198. (*Eccentricity of a conic*) Assume that C is a non-degenerate conic of $\mathbb{R}^2$ which is not a circle. Denote by F a focus of C and by r the corresponding directrix.

(a) Prove that $F \notin C$ and r does not intersect C.
(b) Prove that there exists a positive real constant e, called *eccentricity* of the conic, such that

$$\frac{d(P, F)}{d(P, r)} = e \qquad \forall P \in C$$

and check that $e = 1$ if C is a parabola, $e < 1$ if C is an ellipse, $e > 1$ if C is a hyperbola.
(c) If C is not a parabola, F_1, F_2 are the foci of C and V_1, V_2 are the vertices of C that lie on the line $L(F_1, F_2)$ (cf. Exercise 195 and the Note following it), prove that

$$e = \frac{d(F_1, F_2)}{d(V_1, V_2)}.$$

Solution. (a) Denote by I_1 and I_2 the cyclic points of the Euclidean plane; let $\mathcal{D} = C_{\mathbb{C}}$ be the complexified conic of C and $\overline{\mathcal{D}}$ its projective closure. As seen in the solution of Exercise 194, the focus F is the intersection of a proper line r of $\mathbb{P}^2(\mathbb{C})$ passing through I_1 and tangent to $\overline{\mathcal{D}}$ at a proper point Q with the line $\sigma(r)$ passing through I_2 and tangent to $\overline{\mathcal{D}}$ at the point $\sigma(Q)$. Moreover, $Q \neq \sigma(Q)$ and $r \neq \sigma(r)$. So there are two distinct lines tangent to $\overline{\mathcal{D}}$ and passing through F; as a consequence $F \notin \overline{\mathcal{D}}$ and in particular $F \notin C$.

In addition, the fact that $F = \mathrm{pol}_{\overline{\mathcal{D}}}(Q) \cap \mathrm{pol}_{\overline{\mathcal{D}}}(\sigma(Q))$ implies that $\mathrm{pol}_{\overline{\mathcal{D}}}(F) = L(Q, \sigma(Q))$. Therefore, $\mathrm{pol}_{\overline{\mathcal{D}}}(F)$ intersects $\mathcal{D}$ at the non-real points Q and $\sigma(Q)$, so that the directrix relative to F does not intersect C.

If C is a parabola, part (b) follows immediately from Exercise 197 (c), taking $e = 1$.

Let us now prove parts (b) and (c) when C is an ellipse which is not a circle. By Theorem 1.8.8 we may assume that C has equation $\frac{x^2}{a^2} + \frac{y^2}{b^2} = 1$, with $a > b > 0$. By Exercise 196 (a) the foci of C are the points

$$F_1 = \left(\sqrt{a^2 - b^2}, 0\right), \quad F_2 = \left(-\sqrt{a^2 - b^2}, 0\right)$$

and the equations of the directrices r_1 and r_2 relative to such foci are

$$x = \frac{a^2}{\sqrt{a^2 - b^2}}, \qquad x = -\frac{a^2}{\sqrt{a^2 - b^2}},$$

respectively.

Let now $P = (x_0, y_0) \in C$; since the coordinates of P satisfy the equation of C, then $y_0^2 = b^2 - \frac{b^2}{a^2}x_0^2$. If for instance we take $F = F_1$ and $r = r_1$, then we have

$$d(P, F)^2 = \left(x_0 - \sqrt{a^2 - b^2}\right)^2 + y_0^2 = \left(1 - \frac{b^2}{a^2}\right)x_0^2 - 2\sqrt{a^2 - b^2}\,x_0 + a^2$$

and

$$d(P, r)^2 = \left(x_0 - \frac{a^2}{\sqrt{a^2 - b^2}}\right)^2 = x_0^2 - 2\frac{a^2 x_0}{\sqrt{a^2 - b^2}} + \frac{a^4}{a^2 - b^2}.$$

As

$$\frac{a^2 - b^2}{a^2}d(P, r)^2 = d(P, F)^2$$

it suffices to take

$$e = \frac{\sqrt{a^2 - b^2}}{a}$$

in order to have $e^2 d(P, r)^2 = d(P, F)^2$, and therefore $d(P, F) = e\,d(P, r)$ for every point $P \in C$, as desired. It turns out that $e < 1$.

A similar computation shows that, choosing e as above, then we have $\frac{d(P, F_2)}{d(P, r_2)} = e$ too.

We also observe that, again by Exercise 196, the vertices V_1, V_2 of C that lie on the line $L(F_1, F_2)$ are the points $V_1 = (a, 0), V_2 = (-a, 0)$, so that $e = \frac{d(F_1, F_2)}{d(V_1, V_2)}$.

Finally let us prove parts (b) and (c) when C is a hyperbola. In this case we may assume that C has equation $\frac{x^2}{a^2} - \frac{y^2}{b^2} + 1 = 0$, with $a > 0, b > 0$. By Exercise 196 (c) the foci of C are the points

$$F_1 = \left(0, \sqrt{a^2 + b^2}\right), \qquad F_2 = \left(0, -\sqrt{a^2 + b^2}\right),$$

and the equations of the directrices r_1 and r_2 relative to such foci are

$$y = \frac{b^2}{\sqrt{a^2 + b^2}}, \qquad y = -\frac{b^2}{\sqrt{a^2 + b^2}},$$

respectively.

Let now $P = (x_0, y_0) \in \mathcal{C}$; since the coordinates of P satisfy the equation of $\mathcal{C}$, then $x_0^2 = \frac{a^2}{b^2} y_0^2 - a^2$. If for instance we take $F = F_1$ and $r = r_1$, then we have

$$d(P, F)^2 = x_0^2 + \left(y_0 - \sqrt{a^2 + b^2}\right)^2 = \left(1 + \frac{a^2}{b^2}\right) y_0^2 - 2\sqrt{a^2 + b^2}\, y_0 + b^2$$

and

$$d(P, r)^2 = \left(y_0 - \frac{b^2}{\sqrt{a^2 + b^2}}\right)^2 = y_0^2 - 2\frac{b^2 y_0}{\sqrt{a^2 + b^2}} + \frac{b^4}{a^2 + b^2}.$$

As

$$\frac{a^2 + b^2}{b^2} d(P, r)^2 = d(P, F)^2$$

it suffices to take

$$e = \frac{\sqrt{a^2 + b^2}}{b}$$

in order to have $e^2 d(P, r)^2 = d(P, F)^2$, and therefore $d(P, F) = e\, d(P, r)$ for every point $P \in \mathcal{C}$, as desired. In this case it turns out that $e > 1$.

A similar computation shows that, choosing e as above, one gets $\dfrac{d(P, F_2)}{d(P, r_2)} = e$.

Finally, again by Exercise 196, the vertices V_1, V_2 of $\mathcal{C}$ that lie on the line $L(F_1, F_2)$ are the points $V_1 = (0, b)$, $V_2 = (0, -b)$, so that $e = \dfrac{d(F_1, F_2)}{d(V_1, V_2)}$.

Note. In Exercise 197 we showed that, given a point $F \in \mathbb{R}^2$ and a line r not passing through F, the set $X = \{P \in \mathbb{R}^2 \mid d(P, F) = d(P, r)\}$ is the support of a parabola. More in general, easy computations allow us to check that, given a positive real number e, the set $X = \{P \in \mathbb{R}^2 \mid d(P, F) = e\, d(P, r)\}$ is the support of a non-degenerate conic which is an ellipse if $e < 1$, a hyperbola if $e > 1$ and evidently a parabola if $e = 1$.

Exercise 199. Consider the conic $\mathcal{C}$ of $\mathbb{R}^2$ of equation

$$5x^2 + 5y^2 - 10x - 8y + 8xy - 4 = 0.$$

(a) Find the affine canonical form of $\mathcal{C}$.
(b) Find the metric canonical form $\mathcal{D}$ of $\mathcal{C}$ and an isometry φ such that $\varphi(\mathcal{D}) = \mathcal{C}$.
(c) Determine axes, vertices, foci and directrices of $\mathcal{C}$.

Solution. (a) Recall (cf. Sect. 1.8.5) that the pair $(\text{sign}(\overline{A}), \text{sign}(A))$ is a complete system of affine invariants, which determines the affine canonical form of the conic.

In the case we are studying the conic $\mathcal{C}$ is represented by the symmetric matrix

$$\overline{A} = \left(\begin{array}{c|c} c & {}^t B \\ \hline B & A \end{array}\right) = \begin{pmatrix} -4 & -5 & -4 \\ -5 & 5 & 4 \\ -4 & 4 & 5 \end{pmatrix}.$$

As det $A = 9$, tr$A = 10$ and det $\overline{A} = -81$, we immediately obtain that sign$(A) = (2, 0)$ and sign$(\overline{A}) = (2, 1)$. So $\mathcal{C}$ is a real ellipse and its affine canonical form has equation $x^2 + y^2 = 1$.

(b) In order to transform $\mathcal{C}$ into its metric canonical form, as a first step we diagonalize the matrix A by means of an orthogonal matrix. The eigenvalues of A are 1 and 9 and an orthonormal basis of eigenvectors of A is formed by the vectors $v_1 = \left(\frac{1}{\sqrt{2}}, -\frac{1}{\sqrt{2}}\right)$ and $v_2 = \left(\frac{1}{\sqrt{2}}, \frac{1}{\sqrt{2}}\right)$. Let ψ be the linear isometry of $\mathbb{R}^2$ that transforms the canonical basis into the basis $\{v_1, v_2\}$; if we consider ψ as an affinity and we use the notation fixed in Sect. 1.8.5, ψ is represented by the matrix

$$M_\psi = \begin{pmatrix} 1 & 0 & 0 \\ 0 & \frac{1}{\sqrt{2}} & \frac{1}{\sqrt{2}} \\ 0 & -\frac{1}{\sqrt{2}} & \frac{1}{\sqrt{2}} \end{pmatrix}.$$

Therefore the conic $\mathcal{C}_1 = \psi^{-1}(\mathcal{C})$ is represented by the matrix

$$\overline{A_1} = {}^tM_\psi \overline{A} M_\psi = \begin{pmatrix} -4 & -\frac{1}{\sqrt{2}} & -\frac{9}{\sqrt{2}} \\ -\frac{1}{\sqrt{2}} & 1 & 0 \\ -\frac{9}{\sqrt{2}} & 0 & 9 \end{pmatrix}.$$

Solving the linear system $A_1 \begin{pmatrix} x \\ y \end{pmatrix} = -B_1$, that is $\begin{pmatrix} 1 & 0 \\ 0 & 9 \end{pmatrix}\begin{pmatrix} x \\ y \end{pmatrix} = \begin{pmatrix} \frac{1}{\sqrt{2}} \\ \frac{9}{\sqrt{2}} \end{pmatrix}$, we

obtain (cf. Sect. 1.8.5) that the conic $\mathcal{C}_1$ has the point $C_1 = \left(\frac{1}{\sqrt{2}}, \frac{1}{\sqrt{2}}\right)$ as centre. Thus we can eliminate the linear part of the equation by means of the translation $\tau(X) = X + C_1$. If we set $M_\tau = \begin{pmatrix} 1 & 0 & 0 \\ \frac{1}{\sqrt{2}} & 1 & 0 \\ \frac{1}{\sqrt{2}} & 0 & 1 \end{pmatrix}$, then the conic $\mathcal{D} = \tau^{-1}(\mathcal{C}_1)$ is rep-

resented by the matrix

$$\overline{A_2} = {}^tM_\tau \overline{A_1} M_\tau = \begin{pmatrix} -9 & 0 & 0 \\ 0 & 1 & 0 \\ 0 & 0 & 9 \end{pmatrix}.$$

We have so found that the metric canonical form $\mathcal{D}$ of $\mathcal{C}$ has equation $\frac{x^2}{9} + y^2 = 1$ and that the isometry $\varphi = \psi \circ \tau$ is such that $\varphi(\mathcal{D}) = \mathcal{C}$.

(c) By Exercise 196 the conic $\mathcal{D}$ has the origin $O = (0, 0)$ as a centre, has the lines of equations $x = 0$ and $y = 0$ as axes, the points $(3, 0)$, $(-3, 0)$, $(0, 1)$ and $(0, -1)$ as vertices; moreover, its foci are the points $(2\sqrt{2}, 0)$ and $(-2\sqrt{2}, 0)$, and its directrices are the lines of equations $x = \dfrac{9}{2\sqrt{2}}$ and $x = -\dfrac{9}{2\sqrt{2}}$.

Since the isometry φ transforms the centre, the axes, the vertices, the foci and the directrices of $\mathcal{D}$ into those of $\mathcal{C}$, respectively, we readily obtain that

- the centre of $\mathcal{C}$ is the point $C = \varphi(0, 0) = \psi(\tau(0, 0)) = (1, 0)$;
- the axes of $\mathcal{C}$ are the lines $a_1 = \{x + y - 1 = 0\}$ and $a_2 = \{x - y - 1 = 0\}$;
- the vertices are the points

$$V_1 = \varphi(3, 0) = \left(1 + \frac{3}{\sqrt{2}}, -\frac{3}{\sqrt{2}}\right), \quad V_2 = \varphi(-3, 0) = \left(1 - \frac{3}{\sqrt{2}}, \frac{3}{\sqrt{2}}\right);$$

$$V_3 = \varphi(0, 1) = \left(1 + \frac{1}{\sqrt{2}}, \frac{1}{\sqrt{2}}\right), \quad V_4 = \varphi(0, -1) = \left(1 - \frac{1}{\sqrt{2}}, -\frac{1}{\sqrt{2}}\right)$$

- the foci of $\mathcal{C}$ are the points

$$F_1 = \varphi(2\sqrt{2}, 0) = (3, -2), \quad F_2 = \varphi(-2\sqrt{2}, 0) = (-1, 2);$$

- the directrices are the lines

$$d_1 = \{2x - 2y - 11 = 0\}, \quad d_2 = \{2x - 2y + 7 = 0\}.$$

In addition we observe that, after determining the foci, the centre can be also determined as the midpoint of the segment $F_1 F_2$, an axis as the line $L(F_1, F_2)$, the other axis as the line orthogonal to $L(F_1, F_2)$ and passing through the centre, the vertices as the intersections of the axes with the conic.

Note. (*Metric invariants of conics*) If the exercise had not explicitly asked to determine an isometry φ between the conic $\mathcal{C}$ and its metric canonical form $\mathcal{D}$, we could have found an equation of $\mathcal{D}$ by looking at how the matrix associated to $\mathcal{C}$ changes under an isometry.

Namely, if $\mathcal{C}$ has equation ${}^t\widetilde{X} A \widetilde{X} = 0$ and if $\varphi(X) = MX + N$ is an isometry of $\mathbb{R}^2$ represented by the matrix $\overline{M}_N = \left(\begin{array}{c|c} 1 & 0\,0 \\ \hline N & M \end{array}\right)$ with $M \in O(2)$, we know (cf. Sect. 1.8.5) that the conic $\varphi^{-1}(\mathcal{C})$ is represented by the matrix $\overline{A}' = {}^t\overline{M}_N \overline{A}\, \overline{M}_N$ and, in particular, $A' = {}^t M A M$. As M is orthogonal, then $A' = M^{-1} A M$, i.e. A and A' are similar matrices. So it is possible to determine a diagonal matrix $A' = \begin{pmatrix} \lambda_1 & 0 \\ 0 & \lambda_2 \end{pmatrix}$ similar to A by merely using the trace and the determinant of A, avoiding the explicit computation of the orthogonal matrix M. In addition, since $\det \overline{M}_N = \pm 1$, then $\det \overline{A}' = \det \overline{A}$.

Recall that, if $\mathcal{C}$ is a non-degenerate conic, by means of an isometry, it is possible to transform $\mathcal{C}$ into the conic represented either by a matrix of the form

$$\overline{A'} = \begin{pmatrix} c & 0 & 0 \\ 0 & \lambda_1 & 0 \\ 0 & 0 & \lambda_2 \end{pmatrix}$$ if C has a centre, or by a matrix of the form $\overline{A'} = \begin{pmatrix} 0 & 0 & c \\ 0 & \lambda_1 & 0 \\ c & 0 & 0 \end{pmatrix}$

if C is a parabola.

In the first case one can determine c by using that $\det \overline{A'} = \det \overline{A}$; then one can determine the metric canonical form of C (cf. Theorem 1.8.8) by choosing a suitable multiple of $\overline{A'}$ and exchanging, if necessary, the coordinates via the linear isometry $(x, y) \mapsto (y, x)$.

In the second case (when C is a parabola), we have $\det \overline{A'} = \det \overline{A} = -\lambda_1 c^2$ and

therefore $c^2 = -\dfrac{\det \overline{A}}{\lambda_1}$. In addition, the matrices $\begin{pmatrix} 0 & 0 & c \\ 0 & \lambda_1 & 0 \\ c & 0 & 0 \end{pmatrix}$ and $\begin{pmatrix} 0 & 0 & -c \\ 0 & \lambda_1 & 0 \\ -c & 0 & 0 \end{pmatrix}$

are congruent, so that, if λ_1 is positive (resp. negative), it suffices to choose as c the negative (resp. positive) square root of $-\dfrac{\det \overline{A}}{\lambda_1}$ in such a way that the metric

canonical form of C is the one represented by the matrix $\begin{pmatrix} 0 & 0 & \frac{c}{\lambda_1} \\ 0 & 1 & 0 \\ \frac{c}{\lambda_1} & 0 & 0 \end{pmatrix}$.

In the particular case examined in the exercise we have $\operatorname{tr} A = 10$ and $\det A = 9$, so that the characteristic polynomial of A is $t^2 - 10t + 9$ and the eigenvalues of A are 1 and 9. The conic C, having a centre, is metrically equivalent to the conic

represented by $\overline{A'} = \begin{pmatrix} c & 0 & 0 \\ 0 & 1 & 0 \\ 0 & 0 & 9 \end{pmatrix}$ where we can determine c by means of the condition

$\det \overline{A'} = \det \overline{A}$, that is $9c = -81$ and hence $c = -9$. Thus we find again that the metric canonical form of C has equation $\dfrac{x^2}{9} + y^2 = 1$.

We can now complete the previous observations by showing how the data required in part (c) of the exercise can be computed avoiding the use of an explicit isometry between the conic and its metric canonical form.

The centre $C = (1, 0)$ can be readily computed by solving the system $A\begin{pmatrix} x \\ y \end{pmatrix} = -B$.

The eigenspaces of the matrix A are the lines generated by the vectors $(1, -1)$ and $(1, 1)$. Since $\operatorname{pol}_{\overline{C}}([0, 1, -1]) = \{-x_0 + x_1 - x_2 = 0\}$ and $\operatorname{pol}_{\overline{C}}([0, 1, 1]) = \{-x_0 + x_1 + x_2 = 0\}$, the axes of C are the lines of equations $x - y - 1 = 0$ and $x + y - 1 = 0$.

Intersecting these lines with the conic, we obtain that the vertices of C are the points $V_1 = \left(1 + \dfrac{3}{\sqrt{2}}, -\dfrac{3}{\sqrt{2}}\right)$, $V_2 = \left(1 - \dfrac{3}{\sqrt{2}}, \dfrac{3}{\sqrt{2}}\right)$, $V_3 = \left(1 + \dfrac{1}{\sqrt{2}}, \dfrac{1}{\sqrt{2}}\right)$ and $V_4 = \left(1 - \dfrac{1}{\sqrt{2}}, -\dfrac{1}{\sqrt{2}}\right)$.

We now compute the foci of C as the points of intersection in $\mathbb{R}^2$ of the isotropic lines whose projective closures are tangent to the complexification of $\overline{C}$. (cf. Sect. 1.8.7).

The affine lines of $\mathbb{C}^2$ having the cyclic point $I_1 = [0, 1, i]$ as improper point have equations $y = ix - a$ with $a \in \mathbb{C}$; the ones whose projective closures are tangent to $\overline{C}_{\mathbb{C}}$ correspond to the values $a = 2 + 3i$ and $a = -2 - i$, so that we obtain the lines $l_1 = \{y = ix - 2 - 3i\}$ and $l_2 = \{y = ix + 2 + i\}$.

The lines of $\mathbb{C}^2$ having the cyclic point $I_2 = [0, 1, -i]$ as improper point and whose projective closures are tangent to $\overline{C}_{\mathbb{C}}$ are the conjugate lines of the lines l_1, l_2 (cf. Exercise 194), that is the lines $l_3 = \sigma(l_1) = \{y = -ix - 2 + 3i\}$ and $l_4 = \sigma(l_2) = \{y = -ix + 2 - i\}$. Thus we find as foci the points

$$F_1 = l_1 \cap l_3 = (3, -2), \quad F_2 = l_2 \cap l_4 = (-1, 2).$$

The directrices d_1 and d_2 are the affine parts of the lines $\mathrm{pol}_{\overline{c}}(F_1)$ and $\mathrm{pol}_{\overline{c}}(F_2)$; so they are the lines of equations $2x - 2y - 11 = 0$ and $2x - 2y + 7 = 0$, respectively.

Exercise 200. Find the affine canonical form and the metric canonical form of the conic C of $\mathbb{R}^2$ of equation $x^2 + y^2 + 6xy + 2x + 6y - 2 = 0$.

Solution. The conic C is represented by the symmetric matrix $\begin{pmatrix} -2 & 1 & 3 \\ 1 & 1 & 3 \\ 3 & 3 & 1 \end{pmatrix}$. This matrix, which has a positive determinant but is not positive definite, has signature $(1, 2)$. Since we adopted the convention to represent the conic by means of a matrix $\overline{A}$ such that $i_+(\overline{A}) \geq i_-(\overline{A})$ and $i_+(A) \geq i_-(A)$, we multiply the equation of the conic by -1 so that it is represented by the matrix

$$\overline{A} = \begin{pmatrix} 2 & -1 & -3 \\ -1 & -1 & -3 \\ -3 & -3 & -1 \end{pmatrix}$$

for which $\mathrm{sign}(\overline{A}) = (2, 1)$. As $\det A = -8$, it follows at once that $\mathrm{sign}(A) = (1, 1)$. Therefore the conic C is a hyperbola and its affine canonical form has equation $x^2 - y^2 + 1 = 0$.

In order to determine the metric canonical form, we can use the considerations made in the Note following Exercise 199. As $\mathrm{tr} A = -2$, the characteristic polynomial of A is $t^2 + 2t - 8$, so that the eigenvalues of A are 2 and -4. Therefore the conic is metrically equivalent to the hyperbola represented by the matrix $\overline{A'} = \begin{pmatrix} c & 0 & 0 \\ 0 & 2 & 0 \\ 0 & 0 & -4 \end{pmatrix}$ where c is determined by the condition $\det \overline{A'} = \det \overline{A}$, that is $-8c = -24$ and hence $c = 3$. So the metric canonical form of C turns out to have equation $\frac{2}{3}x^2 - \frac{4}{3}y^2 + 1 = 0$.

Exercise 201. Consider the conic C_α of $\mathbb{R}^2$ of equation

$$x^2 + \alpha y^2 + 2(1 - \alpha)x - 2\alpha y + \alpha + 1 = 0$$

with $\alpha \in \mathbb{R}$.

(a) Determine the affine type of C_α as α varies in $\mathbb{R}$.
(b) Find the values of $\alpha \in \mathbb{R}$ such that C_α is metrically equivalent to the conic D of equation $3x^2 + y^2 - 6x - 2y + 1 = 0$.

Solution. (a) The conic C_α is represented by the matrix

$$\overline{A_\alpha} = \begin{pmatrix} 1+\alpha & 1-\alpha & -\alpha \\ 1-\alpha & 1 & 0 \\ -\alpha & 0 & \alpha \end{pmatrix}.$$

In order to determine the affine type of C_α it is sufficient to compute $\text{sign}(\overline{A_\alpha})$ and $\text{sign}(A_\alpha)$. Since

$$\text{tr} A_\alpha = 1 + \alpha, \quad \det A_\alpha = \alpha, \quad \det \overline{A_\alpha} = \alpha^2(2 - \alpha),$$

we obtain:

- if $\alpha < 0$, after multiplying the equation of C_α by -1 we have that $\text{sign}(-A_\alpha) = (1, 1)$ and $\text{sign}(-\overline{A_\alpha}) = (2, 1)$, so C_α is a hyperbola;
- if $\alpha = 0$, the matrix $\overline{A_0}$ has rank 1, so that C_0 is doubly degenerate (in fact its equation turns out to be $(x + 1)^2 = 0$);
- if $0 < \alpha < 2$, then $\text{sign}(A_\alpha) = (2, 0)$ and $\text{sign}(\overline{A_\alpha}) = (3, 0)$, so C_α is an imaginary ellipse;
- if $\alpha = 2$, then $\text{sign}(A_2) = (2, 0)$ and $\text{sign}(\overline{A_2}) = (2, 0)$, so C_2 is a simply degenerate conic whose real support consists of a single point;
- if $\alpha > 2$, we have $\text{sign}(A_\alpha) = (2, 0)$ and $\text{sign}(\overline{A_\alpha}) = (2, 1)$, and hence C_α is a real ellipse.

(b) By the metric classification Theorem 1.8.8, the conics C_α and D are metrically equivalent if and only if they have the same metric canonical forms. Thus one could solve the exercise by finding and comparing the canonical forms of C_α and D.

Though, in the Note following Exercise 199 we saw that it is possible to determine the metric canonical form of a non-degenerate conic provided we know $\text{tr} A$, $\det A$ and $\det \overline{A}$. Adapting those considerations to the problem of deciding whether two conics C and D of equations ${}^t X \overline{A} X = 0$ and ${}^t X \overline{A'} X = 0$, respectively, are metrically equivalent, we observe that, if there exists a isometry $\varphi(X) = MX + N$ that transforms D into C, then there is a real number $\rho \neq 0$ such that ${}^t \overline{M}_N \overline{A} \, \overline{M}_N = \rho \overline{A'}$. Since M is orthogonal, the matrix A is similar to the matrix $\rho A'$ and hence $\text{tr} A = \rho \text{tr} A'$ and $\det A = \rho^2 \det A'$. In addition, as $\det \overline{M}_N = \pm 1$, then $\det \overline{A} = \rho^3 \det \overline{A'}$. If for instance $\text{tr} A \neq 0$, then the numbers $\dfrac{\det A}{(\text{tr} A)^2}$ and $\dfrac{\det \overline{A}}{(\text{tr} A)^3}$ turn out to be invariant under isometries or *metric invariants*.

On the other hand, if there exists $\rho \neq 0$ such that $\text{tr} A = \rho \text{tr} A'$, $\det A = \rho^2 \det A'$ and $\det \overline{A} = \rho^3 \det \overline{A'}$, up to dividing by ρ the equation of C we may assume that the triple $(\text{tr} A, \det A, \det \overline{A})$ coincides with the triple $(\text{tr} A', \det A', \det \overline{A'})$. As recalled above, if C and D are non-degenerate, then we can conclude that they have the same metric canonical forms and hence by transitivity they are metrically equivalent.

Coming back to the particular case of the exercise, the conic D is represented by the matrix

$$\overline{A'} = \begin{pmatrix} 1 & -3 & -1 \\ -3 & 3 & 0 \\ -1 & 0 & 1 \end{pmatrix}$$

which has $\operatorname{tr} A' = 4$, $\det A' = 3$ and $\det \overline{A'} = -9$. It follows at once that D is non-degenerate and, more precisely, that it is a real ellipse.

So let us see for which values of α there exists $\rho \neq 0$ such that $\operatorname{tr} A_\alpha = \rho \operatorname{tr} A'$, $\det A_\alpha = \rho^2 \det A'$ and $\det \overline{A_\alpha} = \rho^3 \det \overline{A'}$, i.e.

$$1 + \alpha = 4\rho, \quad \alpha = 3\rho^2, \quad \alpha^2(2 - \alpha) = -9\rho^3.$$

With easy calculations we find that the only values that satisfy the three equations are $\rho = 1$ and $\alpha = 3$. Therefore, thanks to the previous considerations, the only conic of the family which is metrically equivalent to D is the conic C_3 of equation $x^2 + 3y^2 - 4x - 6y + 4 = 0$.

Exercise 202. Verify that the conic C obtained by intersecting the plane H of $\mathbb{R}^3$ of equation $x + y + z = 0$ with the quadric Q of $\mathbb{R}^3$ of equation

$$f(x, y, z) = xy - 2xz + yz + 2x - y + 2z - 1 = 0$$

is a hyperbola and find its centre and its asymptotes.

Solution. The points of H are of the form $(x, y, -x - y)$ and the map $\varphi \colon H \to L = \{z = 0\}$ defined by $\varphi(x, y, -x - y) = (x, y, 0)$ is an affine isomorphism that allows us to use on H the system of affine coordinates (x, y) of L. In this system of coordinates the conic $C = Q \cap H$ has equation

$$g(x, y) = f(x, y, -x - y) = 2x^2 - y^2 + 2xy - 3y - 1 = 0.$$

Equivalently we may regard $g(x, y) = 0$ as the equation of the conic $C' = \varphi(C)$ of the plane $z = 0$.

Since φ is an affine isomorphism, in order to determine the affine type of C it suffices to determine the type of the conic C' represented by the symmetric matrix

$$\overline{A'} = \left(\frac{c' \; | \; {}^t B'}{B' \; | \; A'} \right) = \begin{pmatrix} -2 & 0 & -3 \\ 0 & 4 & 2 \\ -3 & 2 & -2 \end{pmatrix}.$$

As $\det A' = -12$ and $\det \overline{A'} \neq 0$, we realize immediately that C' is a hyperbola and so C is a hyperbola too.

Since C' is a conic with centre and the notion of centre is invariant under affinities, also C is a conic with centre and, more precisely, the affine isomorphism φ transforms the centre of C into the centre of C'.

Solving the linear system $A' \begin{pmatrix} x \\ y \end{pmatrix} = \begin{pmatrix} 0 \\ 3 \end{pmatrix}$ we find that the centre of C' is the point $\left(\frac{1}{2}, -1\right)$; therefore the centre of C is the point $\varphi^{-1}\left(\frac{1}{2}, -1\right) = \left(\frac{1}{2}, -1, \frac{1}{2}\right)$.

Also the notion of asymptote is invariant under affinities, so that we may compute the asymptotes of C as the images of the asymptotes of C' through the affine isomorphism φ^{-1}. Now, the points at infinity of C' are $R = [0, \sqrt{3} - 1, 2]$ and $S = [0, -\sqrt{3} - 1, 2]$. If we compute $\mathrm{pol}_{\overline{C'}}(R)$ and $\mathrm{pol}_{\overline{C'}}(S)$, it turns out that the asymptotes of C' are the lines of equations

$$4\sqrt{3}x + (2\sqrt{3} - 6)y - 6 = 0 \quad \text{and} \quad 4\sqrt{3}x + (2\sqrt{3} + 6)y + 6 = 0$$

(which meet at the centre $\left(\frac{1}{2}, -1\right)$). So the asymptotes of C are the lines of $\mathbb{R}^3$ of equations

$$\begin{cases} 4\sqrt{3}x + (2\sqrt{3} - 6)y - 6 = 0 \\ x + y + z = 0 \end{cases} \qquad \begin{cases} 4\sqrt{3}x + (2\sqrt{3} + 6)y + 6 = 0 \\ x + y + z = 0 \end{cases}.$$

Exercise 203. For $i = 1, \ldots, 5$ determine the affine type of the quadric Q_i of $\mathbb{R}^3$ of equation $f_i(x, y, z) = 0$, where:

(a) $f_1(x, y, z) = x^2 + 3y^2 + z^2 + 2yz - 2x - 4y + 2$;
(b) $f_2(x, y, z) = 2x^2 - y^2 - 2z^2 + xy - 3xz + 3yz - x - 4y + 7z - 3$;
(c) $f_3(x, y, z) = x^2 + y^2 - 2xy - 4x - 4y - 2z + 4$;
(d) $f_4(x, y, z) = 2x^2 + y^2 - 2z^2 + 2xz - 10x - 4y + 10z - 6$;
(e) $f_5(x, y, z) = 4x^2 + y^2 + z^2 - 4xy + 4xz - 2yz - 4x + 2y - 2z + 1$.

Solution. (a) Using the convention adopted in Sect. 1.8.5, the quadric Q_1 is represented by the matrix

$$\overline{A_1} = \begin{pmatrix} 2 & -1 & -2 & 0 \\ -1 & 1 & 0 & 0 \\ -2 & 0 & 3 & 1 \\ 0 & 0 & 1 & 1 \end{pmatrix}.$$

We have $\det A_1 = 2$, $\det \overline{A_1} = -2$, $\mathrm{sign}(A_1) = (3, 0)$ and $\mathrm{sign}(\overline{A_1}) = (3, 1)$. Thus from Table 1.2 (cf. Sect. 1.8.8) we deduce that Q_1 is a real ellipsoid.

(b) The matrix

$$\overline{A_2} = \begin{pmatrix} -6 & -1 & -4 & 7 \\ -1 & 4 & 1 & -3 \\ -4 & 1 & -2 & 3 \\ 7 & -3 & 3 & -4 \end{pmatrix},$$

which represents Q_2, has rank 2. The line of equations $x = 0, z = 0$ intersects Q_2 precisely at the points $M = (0, -1, 0)$ and $N = (0, -3, 0)$, and in particular it is not contained in the quadric. This fact ensures that the support of Q_2 is the union of two distinct planes. Then these planes are the tangent planes to Q_2 at M and N, respectively. Thus we find that the quadric decomposes into the planes $T_M(Q_2) = \{x + y - 2z + 1 = 0\}$ and $T_N(Q_2) = \{2x - y + z - 3 = 0\}$. Alternately we can compute the line of the singular points $\mathrm{Sing}(Q_2)$ and then determine the irreducible components of the quadric as the planes $L(\mathrm{Sing}(Q_2), M)$ and $L(\mathrm{Sing}(Q_2), N)$.

(c) Let $\overline{A_3}$ be the symmetric matrix that represents Q_3. Clearly we can determine the affine type of Q_3 by observing that $\mathrm{rk}\ \overline{A_3} = 3$ and $\det A_3 = 0$, so that Q_3 is a cylinder: namely, being degenerate, Q_3 is either a cone or a cylinder, but it cannot be a cone because of Exercise 187. More precisely, since the conic $\overline{Q_3} \cap H_0$ has equation $x_1^2 + x_2^2 - 2x_1x_2 = (x_1 - x_2)^2 = 0$ and so it is a double line, we deduce that Q_3 is a parabolic cylinder.

We can arrive at the same conclusion, for instance, by observing that the derivative with respect to z of the polynomial f_3 never vanishes and consequently no point of Q_3 is singular. In particular the point $R = (0, 0, 2)$ is a smooth point of the quadric. If we compute $\mathrm{pol}_{\overline{Q_3}}([1, 0, 0, 2])$, we obtain that the tangent plane $T_R(Q_3)$ has equation $2x + 2y + z - 2 = 0$.

Let us now check the nature of the conic $Q_3 \cap T_R(Q_3)$ proceeding as in Exercise 202. The image of $Q_3 \cap T_R(Q_3)$ on the plane $z = 0$ through the affine isomorphism $(x, y, -2x - 2y + 2) \mapsto (x, y, 0)$ is the conic of equation

$$g_3(x, y) = f_3(x, y, -2x - 2y + 2) = (x - y)^2 = 0.$$

This conic, and hence $Q_3 \cap T_R(Q_3)$ too, is a double line, so that R is a parabolic point. Then (cf. Sect. 1.8.4) Q_3 is necessarily a quadric of rank 3, that is either a cone or a cylinder. Having already observed that no point of the affine quadric is singular, clearly Q_3 is a cylinder. Since its conic at infinity is a double line, the quadric is a parabolic cylinder.

(d) The symmetric matrix $\overline{A_4}$ that represents Q_4 has determinant zero while $\det A_4 \neq 0$. Then Exercise 187 ensures that Q_4 is a cone (in fact one can easily compute that $\mathrm{Sing}(\overline{Q_4}) = \{[1, 1, 2, 3]\}$ so that Q_4 is a cone with vertex $V = (1, 2, 3)$). In order to decide whether it is a real or an imaginary cone, it is enough to see whether the support of the quadric contains at least one real point in addition to V. For instance the line of equations $x = z = 0$ intersects Q_4 at the points $(0, y, 0)$ where y is a solution of the equation $y^2 - 4y - 6 = 0$. Since this latter equation has two distinct real roots, we realize that Q_4 is a real cone.

We could clearly get to the same result by computing that $\mathrm{sign}(A_4) = (2, 1)$ and $\mathrm{sign}(\overline{A_4}) = (2, 1)$ and making use of Table 1.2 of Sect. 1.8.8.

(e) The symmetric matrix $\overline{A_5}$ that represents Q_5 has rank 1, so that the quadric is a double plane. Since this plane coincides with the singular locus of the quadric, in order to determine it we can compute $\mathrm{Sing}(Q_5)$, which turns out to be the plane of equation $2x - y + z - 1 = 0$.

Exercise 204. Consider the quadrics Q_1 and Q_2 of $\mathbb{R}^3$ of equations

$$x^2 + 3y^2 + 3z^2 + 2yz - 2x = 0 \quad \text{and} \quad x^2 + y^2 + 2z^2 - 2xy + 3z - x - y + 1 = 0,$$

respectively. Verify that these quadrics are non-degenerate and determine their affine types and their principal planes.

Solution. The quadric Q_1 is represented by the matrix

$$\overline{A_1} = \begin{pmatrix} 0 & -1 & 0 & 0 \\ -1 & 1 & 0 & 0 \\ 0 & 0 & 3 & 1 \\ 0 & 0 & 1 & 3 \end{pmatrix}.$$

As A_1 is positive definite and $\det \overline{A_1} < 0$, it follows that the quadric is a real ellipsoid.

After realizing that A_1 is positive definite, one can also reach the same conclusion arguing as follows. The improper conic of Q_1, being represented by the positive definite matrix A_1, has no real points. So Q_1 may only be either a real ellipsoid or an imaginary ellipsoid or an imaginary cone. But Q_1 can be neither an imaginary ellipsoid nor an imaginary cone because the origin is a smooth point of the support.

In order to compute the principal planes of Q_1 we observe that A_1 has eigenvalues 1, 2, 4 and that the eigenspaces relative to them are the lines generated by the vectors $(1, 0, 0)$, $(0, 1, -1)$ and $(0, 1, 1)$, respectively. If we compute the polars of the points $[0, 1, 0, 0]$, $[0, 0, 1, -1]$ and $[0, 0, 1, 1]$ with respect to $\overline{Q_1}$, we find that the quadric has the planes of equations $x = 1$, $y - z = 0$ and $y + z = 0$ as principal planes.

Examining the matrix

$$\overline{A_2} = \begin{pmatrix} 2 & -1 & -1 & 3 \\ -1 & 2 & -2 & 0 \\ -1 & -2 & 2 & 0 \\ 3 & 0 & 0 & 4 \end{pmatrix}$$

one easily obtains that Q_2 is an elliptic paraboloid ($\det A_2 = 0$ and $\det \overline{A_2} < 0$, cf. Sect. 1.8.8).

The matrix A_2 has, in addition to the eigenvalue 0, the eigenvalue 4 with multiplicity 2 and the eigenvectors relative to 4 are all the non-zero vectors of the form $(a, -a, b)$. If we compute the polar of $[0, a, -a, b]$ with respect to $\overline{Q_2}$, it turns out that the principal planes for Q_2 are precisely the planes of equations $4ax - 4ay + 4bz + 3b = 0$ as the parameters a, b vary in $\mathbb{R}^2 \setminus \{(0, 0)\}$.

We can complete the study of the quadric Q_2 by observing that the axis of the paraboloid is the intersection of any two distinct principal planes, for instance those of equations $x - y = 0$ and $4z + 3 = 0$; by intersecting the axis with Q_2 we find that the vertex of the paraboloid is the point $\left(-\frac{1}{16}, -\frac{1}{16}, -\frac{3}{4}\right)$.

Note. The method based on the investigation of the conic at infinity used to identify the affine type of the quadric $\mathcal{Q}_1$ can also be used to determine the affine type of any non-degenerate quadric $\mathcal{Q}$. Namely, if we denote the conic at infinity by $\mathcal{Q}_\infty = \overline{\mathcal{Q}} \cap H_0$, we have:

- if $\mathcal{Q}_\infty$ is irreducible with an empty real support, then $\mathcal{Q}$ is either a real or an imaginary ellipsoid (in order to decide whether it is real, it is enough to show the existence of a real point in the support);
- if $\mathcal{Q}_\infty$ is irreducible with a non-empty real support, then $\mathcal{Q}$ is either an hyperbolic or an elliptic hyperboloid (we can decide which is the case for instance by studying the nature of a point of the quadric);
- if $\mathcal{Q}_\infty$ is a pair of distinct real lines, then $\mathcal{Q}$ is a hyperbolic paraboloid;
- if $\mathcal{Q}_\infty$ is a pair of complex conjugate lines, then $\mathcal{Q}$ is an elliptic paraboloid.

Exercise 205. Denote by $\mathcal{Q}$ the quadric of $\mathbb{R}^3$ of equation

$$f(x, y, z) = 2y^2 - x^2 + 2yz + 2z^2 - 3 = 0.$$

Find all the lines of $\mathbb{R}^3$ contained in $\mathcal{Q}$ and passing through the point $P = (1, \sqrt{2}, 0)$ of $\mathcal{Q}$.

Solution. The quadric $\mathcal{Q}$ is represented by the symmetric matrix

$$\overline{A} = \begin{pmatrix} -3 & 0 & 0 & 0 \\ 0 & -1 & 0 & 0 \\ 0 & 0 & 2 & 1 \\ 0 & 0 & 1 & 2 \end{pmatrix}.$$

As $\det \overline{A} = 9$, the quadric is non-degenerate. More precisely we observe that $\det A = -3$, $\mathrm{sign}(A) = (2, 1)$ and $\mathrm{sign}(\overline{A}) = (2, 2)$. From Table 1.2 of Sect. 1.8.8 we deduce that $\mathcal{Q}$ is a hyperbolic hyperboloid.

Then, since P is a hyperbolic point, there are two distinct real lines contained in $\mathcal{Q}$ and passing through it; we can find them by identifying the components of the degenerate conic $\mathcal{Q} \cap T_P(\mathcal{Q})$.

Easy computations yield that $\nabla f = (-2x, 4y + 2z, 2y + 4z)$ and hence $\nabla f(P) = (-2, 4\sqrt{2}, 2\sqrt{2})$. So $T_P(\mathcal{Q})$ has equation $-2(x - 1) + 4\sqrt{2}(y - \sqrt{2}) + 2\sqrt{2} z = 0$, that is $x - 2\sqrt{2} y - \sqrt{2} z + 3 = 0$.

The image of the degenerate conic $\mathcal{C} = \mathcal{Q} \cap T_P(\mathcal{Q})$ on the plane $x = 0$ through the projection p such that

$$(2\sqrt{2} y + \sqrt{2} z - 3, y, z) \mapsto (0, y, z)$$

is the degenerate conic $\mathcal{C}'$ of equation

$$g(y, z) = f(2\sqrt{2} y + \sqrt{2} z - 3, y, z) = -6y^2 - 6yz + 12\sqrt{2} y + 6\sqrt{2} z - 12 = 0$$

represented, for instance, by the matrix

$$\overline{A} = \begin{pmatrix} 4 & -2\sqrt{2} & -\sqrt{2} \\ -2\sqrt{2} & 2 & 1 \\ -\sqrt{2} & 1 & 0 \end{pmatrix}.$$

Since we know that C' is the union of two distinct lines that meet at $p(1, \sqrt{2}, 0) = (\sqrt{2}, 0)$, we easily find that the components of C' are the lines of the plane $x = 0$ of equations $y - \sqrt{2} = 0$ and $y + z - \sqrt{2} = 0$. Therefore the irreducible components of C are the lines of $\mathbb{R}^3$ of equations

$$\begin{cases} y - \sqrt{2} = 0 \\ x - 2\sqrt{2}\,y - \sqrt{2}z + 3 = 0 \end{cases} \qquad \begin{cases} y + z - \sqrt{2} = 0 \\ x - 2\sqrt{2}\,y - \sqrt{2}z + 3 = 0 \end{cases}$$

which are the lines we looked for.

Exercise 206. Denote by d the Euclidean distance of $\mathbb{R}^3$. Given two skew lines $r, s \subseteq \mathbb{R}^3$, consider the set

$$X = \{P \in \mathbb{R}^3 \mid d(P, r) = d(P, s)\}.$$

Moreover, let $l \subseteq \mathbb{R}^3$ be the only line that intersects both r and s and whose direction is orthogonal both to the direction of r and to the direction of s, and set $Q_1 = r \cap l$, $Q_2 = s \cap l$.

(a) Show that X is the support of a hyperbolic paraboloid Q.
(b) Show that l is the axis of Q, and that the vertex of Q coincides with the midpoint of the segment having Q_1 and Q_2 as endpoints.
(c) Show that Q is metrically equivalent to the quadric Q' of equation $x^2 - y^2 - 2z = 0$ if and only if the directions of r and s are orthogonal and $d(Q_1, Q_2) = 1$.

Solution. (a) It is easy to check that, up to changing coordinates by means of isometries of $\mathbb{R}^3$, we may assume that r has equations $x = y = 0$. By performing a rotation around r and a translation parallel to z (transformations that leave r invariant, it is possible to transform s into a line of equations $x - a = z - by = 0$ for some $a > 0$, $b \in \mathbb{R}$; so henceforth we will assume that r and s are defined by the equations just mentioned. Observe that, in the coordinates we have chosen, the line l has equations $y = z = 0$ and intersects r (resp. s) at the point $Q_1 = (0, 0, 0)$ (resp. $Q_2 = (a, 0, 0)$).
If $P \in \mathbb{R}^3$ has coordinates $(\hat{x}, \hat{y}, \hat{z})$, clearly one has

$$d(P, r)^2 = \hat{x}^2 + \hat{y}^2. \tag{4.9}$$

In addition, since the line s can be parametrized through $t \mapsto (a, t, bt)$, $t \in \mathbb{R}$, the squared distance between P and s coincides with the minimum of $(\hat{x} - a)^2 + (\hat{y} - t)^2 + (\hat{z} - bt)^2$ where we let t vary in $\mathbb{R}$. Then an easy computation shows that

$$d(P, s)^2 = (\hat{x} - a)^2 + \hat{y}^2 + \hat{z}^2 - \frac{(\hat{y} + b\hat{z})^2}{1 + b^2} \qquad (4.10)$$

Thus, comparing (4.9) and (4.10) we get that X coincides with the support of the quadric $\mathcal{Q}$ of equation

$$y^2 - z^2 + 2byz + 2a(1 + b^2)x - a^2(1 + b^2) = 0.$$

Let us now check that $\mathcal{Q}$ is a hyperbolic paraboloid. As $\mathcal{Q}$ is represented by the symmetric matrix

$$\overline{A} = \begin{pmatrix} -a^2(1 + b^2) & a(1 + b^2) & 0 & 0 \\ a(1 + b^2) & 0 & 0 & 0 \\ 0 & 0 & 1 & b \\ 0 & 0 & b & -1 \end{pmatrix},$$

the result follows from the fact that $\det A = 0$ and $\det \overline{A} = a^2(1 + b^2)^3 > 0$ (cf. Sect. 1.8.8).

(b) Recall that a hyperplane $H \subseteq \mathbb{R}^3$ is principal if and only if $\overline{H} = \text{pol}_{\overline{\mathcal{Q}}}(P)$ for some $P = [0, p_1, \ldots, p_n]$ such that the vector $(p_1, \ldots, p_n)$ is a eigenvector for A relative to a non-zero eigenvalue (cf. Sect. 1.8.6). It is easy to check that A has the eigenvalues $\lambda_1 = 0$, $\lambda_2 = \sqrt{1 + b^2}$, $\lambda_3 = -\sqrt{1 + b^2}$. In addition, if $b = 0$ the eigenspaces relative to $\lambda_2 = 1$ and $\lambda_3 = -1$ are generated by $(0, 1, 0)$ and $(0, 0, 1)$, respectively, so that $\mathcal{Q}$ has two principal hyperplanes H_1, H_2, which have equations $y = 0$ and $z = 0$, respectively. If instead $b \neq 0$, the eigenspaces relative to λ_2 and λ_3 are generated by $(0, b, \sqrt{1 + b^2} - 1)$ and $(0, b, -\sqrt{1 + b^2} - 1)$, respectively, so that the two principal hyperplanes H_1, H_2 of $\mathcal{Q}$ have equations

$$by + \left(\sqrt{1 + b^2} - 1\right) z = 0, \qquad by - \left(\sqrt{1 + b^2} + 1\right) z = 0,$$

respectively. Since in both cases $H_1 \cap H_2 = l$, the axis of $\mathcal{Q}$ coincides with l, as desired. Furthermore, considering the system formed by the equations of l and the equation of $\mathcal{Q}$, we immediately obtain that the vertex $V = \mathcal{Q} \cap l$ of $\mathcal{Q}$ has coordinates $\left(\frac{a}{2}, 0, 0\right)$, and so it coincides with the midpoint of the segment having Q_1, Q_2 as endpoints.

(c) Observe at first that $d(Q_1, Q_2) = a$, and that the directions of r and s are orthogonal if and only if $b = 0$. Suppose now that $\mathcal{Q}$ is metrically equivalent to the quadric $\mathcal{Q}'$, which is represented by the matrix

$$\overline{B} = \begin{pmatrix} 0 & 0 & 0 & -1 \\ 0 & 1 & 0 & 0 \\ 0 & 0 & -1 & 0 \\ -1 & 0 & 0 & 0 \end{pmatrix}.$$

Arguing as in the Note following Exercise 199, the fact that Q and Q' are metrically equivalent implies that $\det \overline{A} = \det \overline{B}$, and that the matrices A and B are similar. In particular, the eigenvalues of A must coincide with those of B, which are $0, 1$ and -1. As we have seen in the solution of part (b), this implies that $b = 0$, i.e. the directions of r and of s are orthogonal. As $\det \overline{B} = 1$ and $\det \overline{A} = a^2(1+b^2)^3$, when $b = 0$ the condition $\det \overline{A} = \det \overline{B}$ and the fact that $a > 0$ imply that $a = 1$. Therefore $d(Q_1, Q_2) = a = 1$.

Conversely, if $b = 0$ and $a = 1$ the quadric Q has equation $y^2 - z^2 + 2x - 1 = 0$. If $\varphi: \mathbb{R}^3 \to \mathbb{R}^3$ is the isometry defined by $\varphi(x, y, z) = \left(-z + \frac{1}{2}, x, y\right)$, then $\varphi(Q') = Q$, so that Q and Q' are metrically equivalent.

Index

© Springer International Publishing Switzerland 2016
E. Fortuna et al., *Projective Geometry*, UNITEXT - La Matematica per il 3+2 104,
DOI 10.1007/978-3-319-42824-6

Printed in the United States
By Bookmasters